Multi-body Dynamics:
Monitoring and Simulation Techniques – II

International Organizing Committee

Professor M Abe	Kangawa University of Technology, Japan
Dr R Aini	Rye Machinery Limited, UK
Dr M V Blundell	Coventry University, UK
Dr M Ebrahimi	University of Bradford, UK
Professor H Hamidzadeh	South Dakota State University, USA
Professor R Han	Illinois State University, USA

Sponsored and Organized by

UNIVERSITY OF
BRADFORD

INSTITUTE OF MEASUREMENT & CONTROL

Co-sponsored by

Institution of Electrical Engineers
Institute of Physics
Institution of Nuclear Engineers
Institute of Energy

Multi-body Dynamics: Monitoring and Simulation Techniques – II

Edited by

Dr Homer Rahnejat,

Dr Morteza Ebrahimi,

and

Professor Robert Whalley

Professional Engineering Publishing

Professional Engineering Publishing Limited,
London and Bury St Edmunds, UK

First Published 2000

ISBN 1 86058 258 3

A CIP catalogue record for this book is available from the British Library.

Printed and bound in Great Britain by Antony Rowe Limited, Chippenham, Wiltshire.

Related Titles of Interest

Title	Editor/Author	ISBN
IMechE Engineers' Data Book – Second Edition	C Matthews	1 86058 248 6
Advances in Vehicle Design	John Fenton	1 86058 181 1
Multi-Body Dynamics – Vehicles, Machines, and Mechanisms	H Rahnejat	1 86058 122 6
Multi-Body Dynamics: Monitoring and Simulation Techniques	H Rahnejat and R Whalley	1 86058 064 5
Multi-Body Dynamics – New Techniques and Applications	IMechE Conference	1 86058 152 8
Vibrations in Rotating Machinery	IMechE Conference	1 86058 009 2
Active Control of Vibration	C R Burrows and P S Keogh	0 85298 916 4
Vehicle Handling Dynamics	J R Ellis	0 85298 885 0

For the full range of titles published by Professional Engineering Publishing contact:

Sales Department
Professional Engineering Publishing Limited
Northgate Avenue
Bury St Edmunds
Suffolk
IP32 6BW
UK

Tel: +44 (0)1284 724384
Fax: +44 (0)1284 718692

Contents

Engine Dynamics

Powertrain Systems

Vibration Monitoring and Modelling

Preface

This volume contains refereed papers accepted for the Second International Symposium on Multi-body Dynamics: Monitoring and Simulation Techniques (MBD–MST 2000), which was held at the University of Bradford, UK on 27th–28th June 2000. The Symposium is the second event in this triennial series, jointly organized by the University of Bradford and the Institute of Measurement and Control (InstMC). As for the first Symposium, this event was co-sponsored and endorsed by a number of other professional institutions (IEE, Inst. Phys., INuclE, Inst. Energy) and industrial concerns (Mechanical Dynamics, Ford, Dunlop, AVL List, FHWA/NHTSA, RYE, and SIRIM), indicating the importance attached to this important field of science. This second international Symposium also coincided with the tri-centenary of the birth of Daniel Bernoulli (1700–1782), prompting the inclusion of a tribute to him as a forward to this volume of the Proceedings of the Symposium.

The aim and the scope of the Symposium was set by the Organizing Committee in consultation with the Editorial Board of the *Proceedings of the Institution of Mechanical Engineers, Journal of Multi-body Dynamics*. The scope of the Symposium, therefore, reflected a broad area in the field of dynamics and in-line with the current developments in both academia and industry.

The call for papers stimulated a vigorous response worldwide, and culminated in the inclusion of the contributions in this published volume by Professional Engineering Publishing, the publishers to the Institution of Mechanical Engineers, UK.

The Symposium enjoyed high quality papers, among them keynote and invited contributions by eminent researchers in the field. We are greatly indebted to our colleagues Prof. Werner Schiehlen (Editor-in-Chief, Journal of Multi-body Systems, Kluwer Press) and Prof. Peter Lugner (Editor-in-Chief, Vehicle System Dynamics, Swets and Zeitlinger) for their significant contributions. We are also grateful to our other keynote speakers; Prof. Olof Friberg and Prof. Azim Eskandarian (Editorial Board members of the *Proceedings of the Institution of Mechanical Engineers, Part K: Journal of Multi-body Dynamics*) for their valuable contributions. Other Editorial Board members have also provided high quality papers for this volume of work; Prof. Hamid Hamidzadeh, Prof. Brian Stone, Dr. Jan Welm Biermann, Dr. Patrick Kelly, Mr. Mike Menday, and Mr. Suresh Gupta. These contributions and other high quality papers, selected from a large number of submissions make this volume a unique and authoritative text.

We are also very grateful to the Institute of Measurement and Control for sponsoring the event and its continued support. Our appreciation and thanks go to all the other co-sponsoring institutions and Symposium endorsers.

The efforts of the Short Course Unit, particularly Ms Sue O'Brien of the University of Bradford is acknowledged. We would also like to express our gratitude to the editorial staff at Professional Engineering Publishing, particularly Ms Lynsey Partridge for all the effort expended in the presentation and publication of this volume of proceedings.

Homer Rahnejat,
Morteza Ebrahimi,
and Bob Whalley
MBD–MST 2000

Foreword – A tribute to Daniel Bernoulli (1700–1782) Mathematical-physical renaissance in mechanics of motion

H RAHNEJAT
Department of Mechanical and Medical Engineering, University of Bradford, UK

This second triennial symposium in multi-body dynamics, sponsored by the Institute of Measurement and Control and co-sponsored by a number of other learned and professional institutions and industrial concerns, happens to mark the new millennium. This coincidentally marks the birth of Daniel Bernoulli, born on 9th February 1700. During his life time (1700-1782), dynamics as a discipline in science witnessed the greatest advances in its history, much of which are attributed to him, his immediate family and his co-workers and collaborators Leonhard Euler and Jean D' Alembert. It is, therefore, appropriate to remember their contributions which span rigid and elastic body dynamics, fluid flow and hydrodynamics.

Daniel Bernoulli was born in Groningen, Holland to a family of distinguished mathematicians. His father Johann (Jean) Bernoulli held the chair of mathematics in Groningen at the time. When Daniel was five years of age, the family returned to their native city of Basle for Johann to ascend to the chair of mathematics there upon the death of his brother Jakob Bernoulli who held the same post until 1705. In the same year Daniel's younger brother Johann II was born. With their older brother Nikolas II, all three studied mathematics, although their father did not envisage a future in mathematics and was initially vehemently against such an outcome.

Daniel was sent to Basle University at the age of 13 to study philosophy and logic. He obtained his baccalaureate in 1715, followed by his master's degree in 1716. During his studies he became progressively more interested in the use of calculus, which he primarily learned from his older brother Nikolaus II. Although his father also coached him in mathematics, he insisted that Daniel should study medicine as he had by now failed to show any interest in his father's wishes of becoming a merchant. At first Daniel intended to resist his father's wishes and continue in mathematics. However, Johann Bernoulli was a very strict father; dogmatic in his views and quite arrogant in his protestations. For instance, as a mathematician of great repute he had decided to attribute all the advances in the calculus of variations (forming the main basis of his own research) to Gottfried Liebniz (1), denying any original contributions by Isaac Newton (2). Faced with his father's unrelenting demands, Daniel succumbed and went back to university to study medicine, initially to Heidelberg in 1718, followed by Strasburg in 1719 and finally to Basle to obtain his doctorate in 1721.

Whilst undertaking his doctoral work in Basle, young Daniel came across the writings of William Harvey, the English physician, on the subject of heat and blood motions in animals. He found solace in discovering a link between mathematics and medicine, a realisation which was to dominate much of his research thereon, and resulted in his greatest contribution in the basic rules of fluid flow. This subject had eluded such great men as Newton and his own father Johann Bernoulli. Ironically, it was the latter's tutoring of Daniel on the *Law of Vis Viva Conservation* (i.e. the law of conservation of energy) that finally led to the Bernoulli's Principles in fluid flow.

In the meantime Johann Bernoulli enounced the principle of virtual work in 1717, based on Liebniz's notations for infinitesimal changes. He had already extended the theory of infinitesimal changes to problems involving limits that could be expressed by the ratio of zero to zero, and whilst in Groningen he had sent the same to L' Hospital to be included in his book on the analysis of the infinitely small (3). This theory was subsequently extended by the latter and is now referred to as L' Hospital's rule. Since 1697 Johann had been working on the principle that a virtual displacement is an infinitesimal change in the co-ordinate that may be considered irrespective of time and must remain compatible with system constraints. This work had already been recognised as his major contribution by his election to the fellowship of the Royal Society of London in December 1712. Paradoxically in the same year, the English mathematician, Brook Taylor was also elected to the fellowship of the Royal Society and acted as the adjudicator in the dispute between Isaac Newton and Gottfried Liebniz over the originality claims regarding the theory of calculus of variations. Taylor found in favour of Newton, in the face of great objections by Johann who also set about objecting to the originality of Taylor's unpublished work on centrality of small oscillations. As a result, Taylor's masterpiece on calculus of finite differences (4) remained largely unrecognised for more than half a century, until Lagrange who had become the undisputed mathematician of his time proclaimed it as a cornerstone of differential calculus in 1772. Johann Bernoulli's important enounciations of 1717 were in fact published at the end of his illustrious career in 1742 in 4 volumes (5).

In time Johann's son, Daniel and his co-worker Leonhard Euler used his 1717 virtual work principles to formulate vibrating strings, and Jean D' Alembert (6,7) and Joseph Louis Lagrange (8,9) extended the case of static equilibrium to rigid body dynamics. At the same time Euler used D' Alembert's reasoning, that the sum of all forces acting on a particle/body induces acceleration, to develop his free body diagram formulation method employing Newton's laws of motion. This method of formulation has come to be known as the Newton-Euler method (10).

Interestingly, many of the above mentioned developments closely followed the trials and tribulations of Daniel Bernoulli's career. His two unsuccessful attempts to obtain chairs in anatomy and botany at home in Basle resulted in his departure for Venice with the ultimate aim of further studies in medicine in Padua. A severe illness put paid to his plans and he remained in Venice, mostly studying the physics of flowing water and differential calculus, and published his *Mathematical Exercise* in 1724 (11). Whilst in Venice he became exposed more closely to the works of Galileo on simple harmonic motion (12) and Leonardo Da Vinci on tidal motion (for both subjects he received prizes and also 8 others from the Paris Academy of Science later on in his career).

The duality of his interests in fluid motion and calculus of variation applied to the principle of virtual work took strong roots in Venice and was set to dominate his future career. He must have noted the parallels between Galileo's observations for the relationships between the frequency of vibration of a stretched string with its length, tension and density and his uncle Jakob's last propositions in 1705 that the curvature of a vibrating beam is proportional to its bending moment. After all such observations would have agreed well with the principles of virtual work and conservation of energy that he was taught by his father. Galileo's measurements of the frequency of oscillation of a swinging chandelier with his own pulse rate (12) would have particularly appealed to Daniel more than others.

Although his *Mathematical Exercises* (11) contained some erroneous assumptions (concerning the flow of water through a hole), his time in Venice proved to be well spent, particularly in the design of an hour-glass that could be used at sea with a constant sand trickle rate, even when the rolling ship was subject to heavy seas. The design won him a prize from the Paris Academy and because of his *Mathematical Exercises,* he received an invitation from Empress Catherine I for a chair in mathematics at St Petersburg University. Daniel was reported to be somewhat reluctant to take up this position. However, the Empress was so keen to effect the appointment that an additional academic position was offered to Daniel's brother, Nikolaus II. The brothers moved to St Petersburg late in 1725 to take up their appointments in the university. Nikolaus died eight months later and the home-sick Daniel yearned to return to Basle, only to be thwarted by his father who despatched his brightest assistant, Leonhard Euler to work with his son in St Petersburg from 1727. Empress Catherine was all too pleased to give Euler an assistantship. He later ascended to the chair of mathematics after Daniel Bernoulli left St Petersburg in 1733. The two men; Daniel Bernoulli and Leonhard Euler, worked very closely and maintained a good liaison with their contemporary worker, Jean D' Alembert. By 1728, they dominated the mechanics of motion of elastic bodies such as strings and slender beams.

Daniel Bernoulli possessed an imaginative mind, as evident from a wide ranging scientific interest. This meant that he would often not follow a piece of work to its ultimate conclusion. Fortunately his collaboration with Euler ensured that many of his ideas were rigorously formulated and investigated by the latter, who was indeed the acknowledged mathematician of his time. In 1727 with Euler as his assistant, Daniel set out to investigate the relationship between the speed at which blood flows and its pressure. He carried out a number of experiments, measuring the height of fluid column in a small open-ended straw, puncturing the wall of a pipe. Using the principle of conservation of energy, he showed that kinetic energy of a flowing fluid converts to pressure and vice-versa, a fact that was exploited by many physicians, inserting point-ended glass directly into patients' blood vessels to measure blood pressure. This practice was finally abandoned 170 years later in the mid 1890s. With this discovery he reserved for himself a well-deserved position in the history of science, although the famous Bernoulli equation was in fact derived later by Euler.

The period of collaboration between Daniel Bernoulli and Leonhard Euler spanned 1727-1733 and represents the most productive time of Bernoulli's illustrious career. His *magnum opus; Hydrodynamica* (13) was almost completed by 1733. In fact he submitted the original manuscript to the publishers in St Petersburg, before leaving for a number of destinations with his younger brother Johann II who had been staying with him towards the end of his career in St Petersburg. He made some changes and small additions in the period 1734-1736, before the treatise was finally published in 1738.

Daniel Bernoulli was a particularly observant scientist; and many practical applications of his hydrodynamic and hydrostatic theories are contained in his *Hydrodynamica,* including some quite futuristic applications at that time such as swirl flow behind a ship propeller. He kept his liaison with Euler for many years, indeed jointly winning a prize from the Paris Academy of Science for mathematical treatment of tidal motions in 1740. During their collaboration in St Petersburg, Bernoulli suggested to Euler further work on the theory of vibrating strings and slender beams. He proposed the use of partial derivatives to derive equations of motion for lateral oscillations of stretched strings. He observed the analytic agreement shown by Taylor in 1715 with the experimental observations of Galileo (12) and Mersenne (14). Most significantly, Bernoulli argued that vibrating strings have several harmonics, with their contributions summing up to the total displacement of the string at any given time. This principle has come to be termed superposition. Euler, on the other hand, doubted that the vibrating shape of a flexible element could be approximated by a series of harmonic responses. The validity of Bernoulli's harmonic superposition principle was finally settled in 1822 by Fourier (15), although Lagrange had already discretised vibrating strings to a finite series of equally spaced masses and had established in 1759 that such strings would yield a number of independent frequencies, equal to the number of identically considered point masses.

The problem of vibrating strings and thin beams occupied the attention of all the eminent mathematicians of the time, with thin beam theory published by Euler in 1744, and by Bernoulli in 1751. In the meantime D' Alembert presented his wave equation in 1750, contained in 8 volumes of work on physics of sound (15). These works resulted in the formulation of Lagrange's equation for constrained systems in 1788 (9) and torsional vibration of elastic members by Coulomb in 1784.

Thus, the period 1705-1790, spanning the life of Daniel Bernoulli (1700-1782) witnessed some of the greatest discoveries in the formulation of mechanics of motion for rigid, elastic and fluid media. It is interesting to note that Daniel Bernoulli contributed significantly to all of these discoveries, for which he received much recognition, including the fellowship of the Royal Society of London in 1750 like his father and his grandfather before him, making the Bernoullis the only family in history to have achieved such an accolade.

(1)- G. Leibniz, *"Nova methodus pro maximis et minimis itemque tangentibus, quae nec fractas, nec irrationales quantitates moratur, et singulare pro illis calculi genus"*, Acta Eruditorum, Hanover, 1684.

(2)- I. Newton, *Philosophiae Naturalis Principia Mathematica*, Royal Society, London, 1687.

(3)- L' Hospital, *Analyse des infiniment petits*, Paris, 1696.

(4)- B. Taylor, *"Methodus incrementorum directa et inversa"*, Royal Society, London, 1715.

(5)- J. Bernoulli, *Opera Johannis Bernoulli*, Basle, 1742.

(6)- J. Le R. D' Alembert, *Traité de dynamique*, Paris, 1743.

(7)- J. Le R. D' Alembert, *"Recherches sur les cordes vibrante"*, L'Académie Royal des Sciences , Paris, 1747.

(8)- J. L. Lagrange, *"Libration"*, L'Académie Royal des Sciences , Paris, 1764.

(9)- J. L. Lagrange, Mécanique Analytique, L'Académie Royal des Sciences, Paris, 1788.

(10)- L. Euler, *"Nova methods motum corporum rigidarum determinandi"*, Novi Commentarii Academiae Scientiarum Petropolitanae, 20, 1776.

(11)- D. Bernoulli, *"Exercitationes quaedam mathematicae"*, Venice, 1724.

(12)- G. Galileo, *Discourses concerning two new sciences*, Leiden, 1638.

(13)- D. Bernoulli, *Hydrodynamica*, St Petersburg, 1738.

(14)- M. Mersenne, *Harmonicorum Libri*, Paris, 1636.

(15)- J. B. Fourier, *Théorie analytique de la chaleur*, Paris, 1822.

(16)- J. Le R. D' Alembert, *Opuscules mathématiques*, Paris, 1761-1780 (in 8 Volumes)

Contact, Impact, and Flexible Multi-body Dynamics

Contact problems in multi-body dynamics

W SCHIEHLEN and **B HU**
Institute B of Mechanics, University of Stuttgart, Germany

ABSTRACT: Contact problems of mechanical systems are multiscale problems including multibody dynamics and wave propagation phenomena as well. The broad variety of applications is supplemented by a large variety of methods. The impact problems of rigid body systems result in unilateral contacts in rigid body systems. Continuous contact force models for impact analysis in multibody systems extend the Hertzian approach of elasticity to viscoelastic nonlinear constitutive laws. In this paper the different approaches are compared with respect to machine dynamics, and multirate integration methods are proposed to enhance the numerical efficiency.

1 INTRODUCTION

Multibody dynamics, finite element analysis and continuous system modeling may be applied to contact problems. Machine dynamics offers a great variety of contact problems ranging from impacts to gliding and rolling. Typical examples are cam follower and cam roller devices as well as all kinds of gears. There are numerous papers in the literature devoted to one or the other machine design with contact. Three recent papers will be mentioned with quite different techniques.

A finger–follower cam system is theoretically and experimentally analyzed by Hsu and Pisano [1]. In this paper the local deformation at the contact surface is assumed sufficiently small so that the linear Hertzian contact compliance model holds at each contact point, and only compressive forces are allowed between contact surfaces. Further, dry friction and viscous damping is considered in the gliding areas. A general approach to the determination of planar and spatial cam profiles is presented by Tsay and Wei [2]. Using the kinematical equations of cam profiles, the analytical expressions of contact forces angles and principal curvature are obtained which are important in the design process of cam follower mechanisms. The finite element method is applied by Liu [3] for the synthesis and steady–state analysis of high–speed elastic cam–actuated linkages. In this case, the unknown contact force exerted on the roller follower causes no difficulty in the synthesis due to the elastic components of the linkage.

The broad variety of application is supplemented by a large variety of methods. The impact problem of rigid body system is treated by Wittenburg [4] and other authors, e.g. Zhang [5]. Unilateral contacts in rigid body system have been investigated by Pfeiffer and Glocker [6] in the very general setting of a linear complementary problem. This approach proved to be very efficient from a computational point of view.

The sticking motion of an impact oscillator has been checked with respect to bifurcations by Toulemonde and Gontier [7] where the impacts are approximately governed by the restitution rule. The aspects of contact transition control of nonlinear mechanical systems subjected to unilateral constraints were treated by Pagilla and Tomizuka [8]. The impact model used is an extension of Newton's model with infinitely large values of the contact force during an infinitesimally small period of time.

Continuous contact force models for impact analysis in multibody systems are presented by Lankarani and Nikravesh [9]. The Hertzian approach of elasticity was extended to a viscoelastic nonlinear constitutive law resulting in hysteresis damping. Such a Kelvin–Voigt viscoelastic model was also used by Ravn [10] for the continuous analysis of planar multibody systems with joint clearances. The methods proposed have been tested experimentally for a double pendulum and a slider crank mechanism with clearance.

Wave propagation methods for impact analysis are also found in the literature. The state of the art is summarized in the textbook of Goldsmith [11]. In particular, longitudinal waves in rods excited by impacts are well–known. Just recently, these results have been reviewed and extended by means of computer algebra methods, see Hu, Eberhard and Schiehlen [12].

In this paper the different approaches are compared with respect to machine dynamics, and multirate integration methods are proposed to enhance the numerical efficiency as shown in Schiehlen [13], too.

2 CONTACT AS MULTISCALE PROBLEM

For the transition from the free motion of a multibody system to the motion with unilateral constraints, different models may be used resulting in different time scales, accuracies and efficiencies. Figure 1 presents a survey of the models and the related time scales. The free flight of the interconnected bodies is interrupted by collision. The detection of collision is a nontrivial geometrical problem which is thoroughly discussed in robotics, too. However, it is out of the scope of this paper. During collision, the bodies may be considered elastic or rigid resulting in an elastocontact or a stereocontact, respectively. The elastocontact includes dynamic deformation and wave propagation within the body and is related to the name of De Saint-Venant (1797 – 1886). It turns out that there is a finite duration of contact. After dissipation of energy by one or more contacts, a steady contact representing a unilateral constraint may occur. The stereocontact comprises two models. One model introduced by Hertz (1887 – 1894) considers the elastostatic and/or viscoelastic deformation and the rigid body inertia. Thus, there exists also a finite duration of contact and a transition to a steady contact featuring unilateral constraints. The second model does not make use of any constitutive law of the bodies involved and the loss of energy is characterized by a coefficient of restitution. This second model is due to Newton (1643 – 1727) and Poisson (1781 – 1840) with slightly different definitions. The hypothesis of an infinitely small duration of contact and an infinitely large contact force is used, resulting in the classical impact approach.

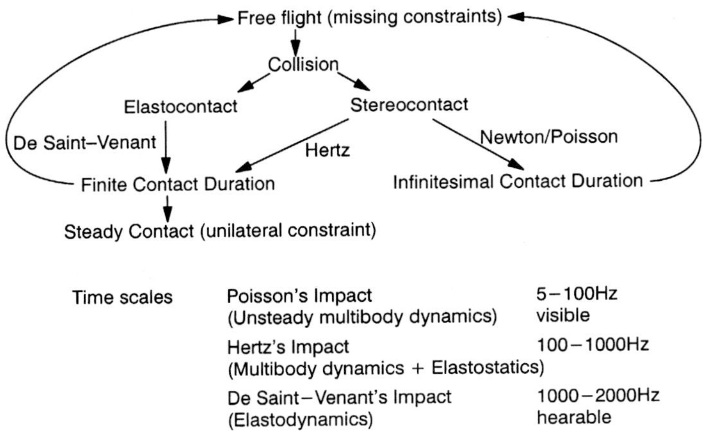

Figure 1: **Contact as a multirate problem**

Most important for the comparison of methods are the time scales which may occur, and the related mechanical models. Poisson's impact typically leads to frequencies of $5 - 100Hz$; such motions may be still visible. The method of analysis is unsteady multibody dynamics. Hertz's impact shows frequencies of $100 - 1000Hz$; the methods of mechanical analysis include multibody dynamics and elastostatics. De Saint-Venant's impacts are characterized by high frequencies with more than 1000 Hz, and they are easily audible.

During the numerical simulation of mechanical systems, different time scales of the models affect the efficiency. Therefore, the models will be discussed in more detail.

2.1 Free flight

During free flight, the multibody system is clearly divided into two parts. The equations of motion of both parts could be written independently, but they can be also combined to one set as follows:

$$\boldsymbol{M}(\boldsymbol{x})\ddot{\boldsymbol{x}} + \boldsymbol{k}(\boldsymbol{x}, \dot{\boldsymbol{x}}) = \boldsymbol{q}(\boldsymbol{x}) \tag{1}$$

where $\boldsymbol{x}(t)$ is the global position vector of both parts featuring f degrees of freedom, $\boldsymbol{M}(\boldsymbol{x})$ the inertial matrix , $\boldsymbol{k}(\boldsymbol{x}, \dot{\boldsymbol{x}})$ the vector of Coriolis and gyroscopic forces and $\boldsymbol{q}(\boldsymbol{x})$ the vector of the applied forces. For more details see Schiehlen [14].

2.2 Collision

The free flight of two parts of a mechanical system, or of one part and the environment, respectively, is stopped by collision, Figure 2. Therefore, the collision detection is fundamental in contact problems. The complete theory of collision detection in a plane was presented by Pfeiffer and Glocker [6]. The conditions read as

$$\boldsymbol{n}_1^T(s_1)\boldsymbol{t}_2(s_2) = 0\,, \quad \boldsymbol{r}^T(s_1, s_2)\boldsymbol{t}_1(s_2) = 0 \tag{2}$$

where s_1, s_2 are contour parameters, $\boldsymbol{n}_1, \boldsymbol{t}_1, \boldsymbol{t}_2$ represent the normal and tangential vectors depending on the contour parameters and $\boldsymbol{r}$ is the distance vector between the colliding bodies. Another approach was presented by Eberhard and Jiang [15].

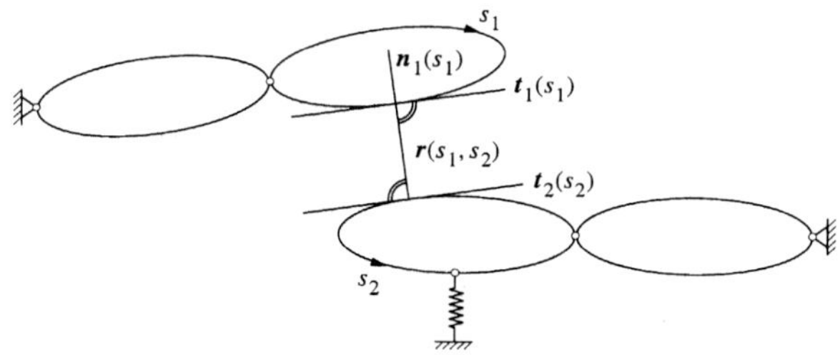

Figure 2: **Collision**

2.3 Newton's/Poisson's impact

The hypothesis of an infinitely large contact force leads to the force impulse

$$\triangle \boldsymbol{p} = \lim_{\varepsilon \to 0} \int_{t_1-\varepsilon}^{t_1+\varepsilon} \boldsymbol{f}(\tau)\, d\tau \tag{3}$$

which has to be included in the equations of motion; the result is an unsteady system

$$\boldsymbol{M}(\boldsymbol{x})\ddot{\boldsymbol{x}} + \boldsymbol{k}(\boldsymbol{x}, \dot{\boldsymbol{x}}) = \boldsymbol{q}(\boldsymbol{x})\,, \quad 0 < t < t_1\,,\; t_1 < t < \infty\,. \tag{4}$$

At the instant of the impact $t = t_1$, the initial conditions of the second time period have to be changed using the coefficient of restitution. However, the number f of degrees of freedom remains unchanged, Figure 3.

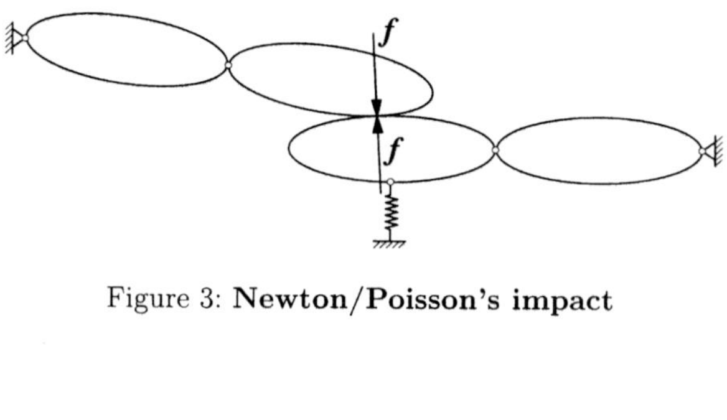

Figure 3: **Newton/Poisson's impact**

2.4 Hertz's impact

The contact force is given by the constitutive law of the colliding bodies under the hypothesis of local deformation, Figure 4. The contact force results in an occasionally stiff system represented by the equations of motion

$$\boldsymbol{M}(\boldsymbol{x})\ddot{\boldsymbol{x}} + \boldsymbol{k}(\boldsymbol{x}, \dot{\boldsymbol{x}}) = \boldsymbol{q}(\boldsymbol{x}) + \boldsymbol{Q}\boldsymbol{f} \;, \tag{5}$$

$$\boldsymbol{f} = \boldsymbol{f}_c(\boldsymbol{x}) + \boldsymbol{f}_d(\dot{\boldsymbol{x}}) \qquad \text{for} \quad t_1 < t < t_1 + t_c \tag{6}$$

where $\boldsymbol{Q}(\boldsymbol{x})$ is the distribution matrix of the contact force vector $\boldsymbol{f}$ which is active during the contact period t_c only. The number f of degrees of freedom remains unchanged, too.

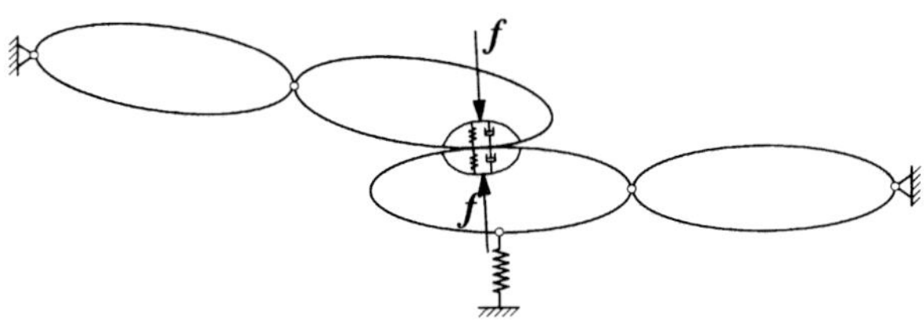

Figure 4: **Hertz's impact**

2.5 De Saint-Venant's impact

The contact force is replaced by the contact stresses and the colliding bodies are modeled as elastic bodies, Figure 5. The elasticity is considered only during the period of contact for $t_1 < t < t_1 + t_c$

$$\begin{bmatrix} \boldsymbol{M}_{xx} & \boldsymbol{M}_{ux} \\ \boldsymbol{M}_{ux} & \boldsymbol{M}_{uu} \end{bmatrix} \begin{bmatrix} \ddot{\boldsymbol{x}} \\ \ddot{\boldsymbol{u}} \end{bmatrix} + \boldsymbol{h} = \boldsymbol{q} + \begin{bmatrix} \boldsymbol{0} & \boldsymbol{0} \\ \boldsymbol{0} & \boldsymbol{K}_{uu} \end{bmatrix} \begin{bmatrix} \boldsymbol{x} \\ \boldsymbol{u} \end{bmatrix} \tag{7}$$

Here, $\boldsymbol{u}$ is the vector of the additional position coordinates of the flexible structure, $\boldsymbol{M}_{uu}$ and $\boldsymbol{K}_{uu}$ are the corresponding inertia and stiffness matrices of the structure and $\boldsymbol{M}_{ux}$ represents the coupling between rigid and elastic body motion. Obviously, the total number of degrees of freedom is larger than f.

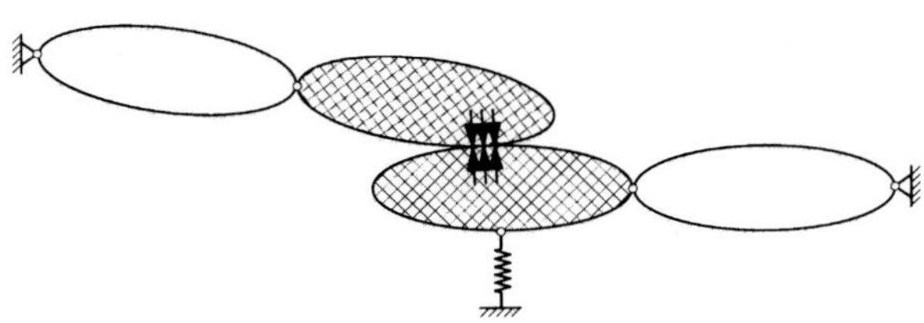

Figure 5: **De Saint-Venant's impact**

2.6 Unilateral constraints

The contact force is now replaced by a reaction force following from the constraint equations

$$\boldsymbol{\phi}(\boldsymbol{x}) = \mathbf{0} \quad \text{or} \quad \boldsymbol{x} = \boldsymbol{x}(\boldsymbol{y}) \tag{8}$$

in implicit or explicit form, respectively. Then, using d'Alembert's principle, we reduce the equations of motion by premultiplication by the Jacobian $\boldsymbol{Q} = \partial\boldsymbol{x}/\partial\boldsymbol{y}$

$$\boldsymbol{Q}^T\boldsymbol{M}(\boldsymbol{x})\boldsymbol{Q}\ddot{\boldsymbol{y}} + \boldsymbol{Q}^T\boldsymbol{k}(\boldsymbol{x},\dot{\boldsymbol{x}}) = \boldsymbol{Q}^T\boldsymbol{q}(\boldsymbol{x}) + \boldsymbol{Q}^T\boldsymbol{Q}(\boldsymbol{x})\boldsymbol{f}_r \tag{9}$$

to

$$\boldsymbol{M}(\boldsymbol{y})\ddot{\boldsymbol{y}} + \boldsymbol{k}(\boldsymbol{y},\dot{\boldsymbol{y}}) = \boldsymbol{q}(\boldsymbol{y}) \tag{10}$$

where the orthogonality condition $\boldsymbol{Q}^T(\boldsymbol{x})\boldsymbol{Q}(\boldsymbol{x})$ of the reaction force $\boldsymbol{f}_r$ is used. Then, compared to the impacts according to Sections 2.1, 2.3 and 2.4 the number of degrees of freedom is smaller than f as well as the dimensions of the matrix and the vectors in (10).

Two problems will be addressed in this paper which are related to the time scales of the impact models: the multirate integration of occasionally stiff model of Hertz's impact; the benefit of a more De Saint-Venant's model of impact.

3 SIMULATION AT DIFFERENT TIME SCALES

In problems with very different time scales, as in Hertz's impact, the integration step size is fully controlled by the fast modules. Most often, the modules with the expensive equations would not need such small time steps. To circumvent this effect, a block modular structure can be exploited by applying different discretizations to groups of blocks. This is also referred to as multirate integration (MR), see Gear [16]. Then, one must distinguish between external and internal couplings. Internal couplings are those for which all blocks involved are treated by the same integrator. The corresponding solution is exact. External couplings must be reconstructed by interpolations $\boldsymbol{\theta}$ as shown in Figure 6 for one MR time step h_{mr}. The first system with the states $\boldsymbol{x}_1$ is integrated from t_n to $t_{n+1} = t_n + h_{mr}$, judging the external inputs $\boldsymbol{u}_1$ by the interpolation $\hat{\boldsymbol{u}}_1$. After the integration step, the output equation of the first system is computed, and the second system can be integrated using interpolations $\hat{\boldsymbol{y}}^1$ of the first system's outputs $\boldsymbol{y}_1$ as input. As no MR step size ratio is given but a fixed MR step size, the integrators may be subject to any kind of internal error control.

To supply the simulation functionality, as introduced above by MR-integration, the simulator *NEWMOS* has been set up, see Rükgauer and Schiehlen [17]. The simulator *NEWMOS* is designed as a runtime system that allows for the iterative assembly of the problem to be solved. The basic elements to assemble are dynamic **systems**, **joints** and **integrators** as in common approaches, too. In addition, **buffers** and **filters** are supported to allow a MR simulation. All these elements are referred to as **services**. At startup, no services are known by the application. Rather, these can be loaded at run time from a module library. After systems are loaded, their inputs and outputs can be assigned appropriately, and node points of mechanical systems can be linked by joints. Furthermore, dynamical systems, an integrator, a buffer and a filter can be combined into a **set**. Multiple sets then can be simulated concurrently in an MR simulation.

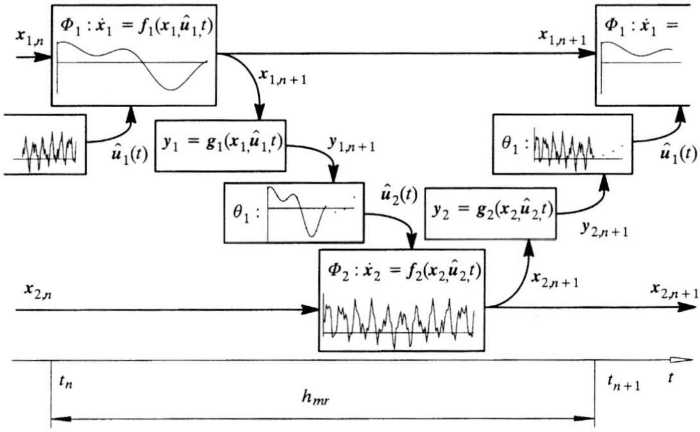

Figure 6: **Multirate integration step**

4 TRANSITION FROM ELASTODYNAMICS TO ELASTOSTATICS

Wave propagation due to longitudinal impacts on rods can be used to assess the model errors due to the Hertzian assumption of elastostatic deformation of an inertially rigid body. In principle, propagating waves remaining in the rod after separation of the impacting bodies mean energy loss, or a coefficient of restitution less than one. The problem is depicted in Figure 7.

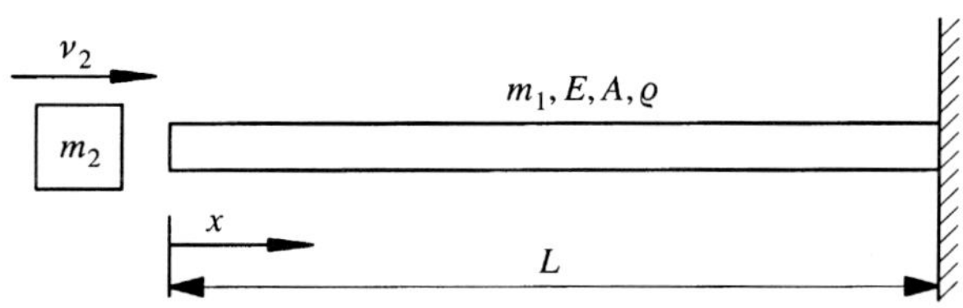

Figure 7: **Longitudinal impact of a mass on a rod**

The governing partial differential equation for the longitudinal wave reads as

$$\frac{\partial^2 u(x,t)}{\partial^2 t} = c^2 \frac{\partial u(x,t)}{\partial^2 x} \tag{11}$$

where $c = \sqrt{E/\rho}$ is the wave propagation velocity determined by Young's modulus E and density ρ. The general solution is presented in Hu, Eberhard and Schiehlen [12]. The results are related to the period $T = 2L/c$ and mass ratio $\alpha = m_1/m_2$ where L is the length of the rod and m_1 and m_2 are the mass of the rod and impacting mass, respectively. The impact time t_c and the coefficient of restitution e are shown in Figure 8 and 9.
It is clear that for small impacting masses the coefficient of restitution is smaller than one. After the duration of the impact given by time t_c the rod is still in motion, and then energy is lost for the reflection of the impacting mass m_2. As a result, it can be stated that small impacting bodies may be subject to dynamic energy loss due to the internal vibrations of a struck body.

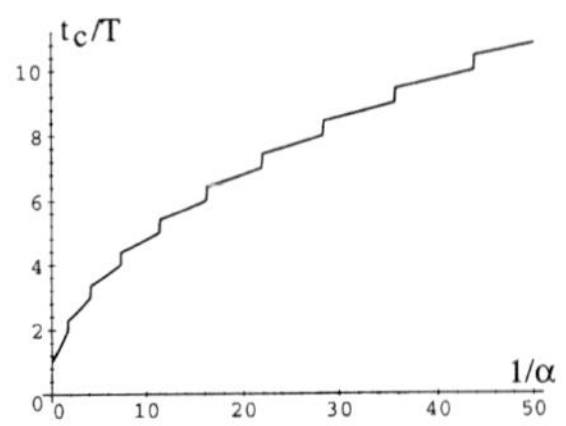

Figure 8: **Duration of impact**

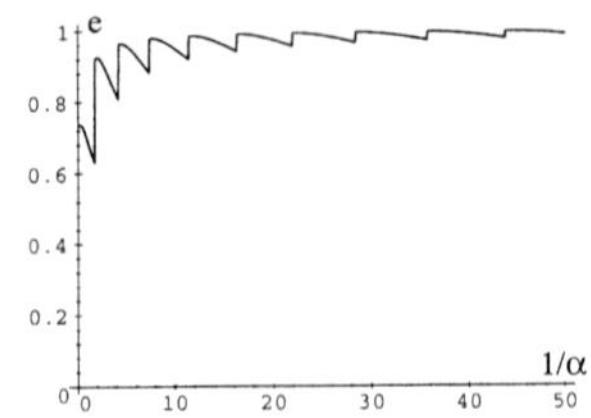

Figure 9: **Coefficient of restitution**

5 HOPPING WHEEL

A wheel suspended at a double pendulum is a mechanical system showing free flight, Hertzian contact and unilateral constraints, Figure 10. The system has $f = 3$ degrees of freedom characterized by $\alpha_1, \alpha_2, \alpha_3$ and they will be reduced to one degree of freedom by two unilateral constraints which are also used to identity the collision. The equations of motion generated by *NEWEUL* are shown in Figure 11.
The constraints conditions are

$$\Phi 1(\boldsymbol{x}) = LL - L1\cos\alpha_1 - L2\sin\alpha_2 - R = 0 \quad (12)$$
$$\Phi 2(\dot{\boldsymbol{x}}) = L1\dot{\alpha}_1\cos\alpha_1 - L2\dot{\alpha}_2\sin\alpha_2 - R\dot{\alpha}_3 = 0 \quad (13)$$

in the normal and tangential direction. Further, the impact forces affect the mechanical system by the viscoelastic force $F1$ and the slip force $F2$ as follows

$$\boldsymbol{Q}1(\boldsymbol{x})F1 = \begin{bmatrix} L1\cos\alpha_1 \\ L2\sin\alpha_2 \\ R \end{bmatrix} F1\,, \quad \boldsymbol{Q}2(\boldsymbol{x})F2 = \begin{bmatrix} L1\sin\alpha_1 \\ -L2\cos\alpha_2 \\ 0 \end{bmatrix} F2\,. \quad (14)$$

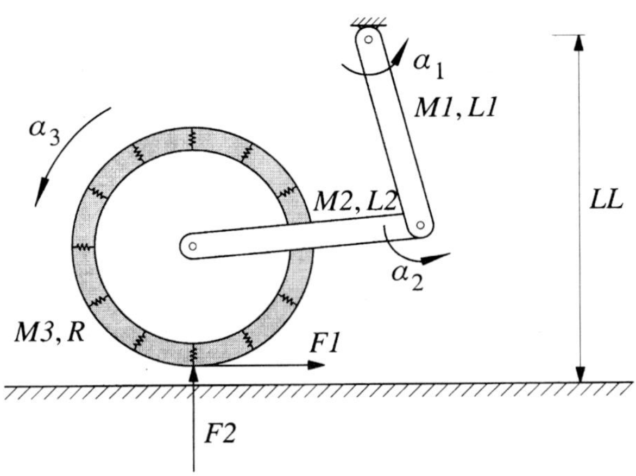

Figure 10: **Hopping wheel**

```
C> Massenmatrix
   M(1,1)=I1+0.25*M1*L1**2+M2*L1**2+M3*L1**2
C
   M(2,1)=0.5*M2*L2*L1*SIN(AL2)*COS(AL1)-0.5*M2*L2*L1*SIN(AL1)*
  & COS(AL2)+M3*L2*L1*SIN(AL2)*COS(AL1)-M3*L2*L1*SIN(AL1)*
  & COS(AL2)
   M(2,2)=I2+0.25*M2*L2**2+M3*L2**2
C
   M(3,1)=0.
   M(3,2)=0.
   M(3,3)=I3
C
C>
C> Verallgemeinerte Kreisel-, Zentrifugal- und Corioliskraefte
   K(1)=0.5*M2*L2*L1*AL2P**2*COS(AL2)*COS(AL1)+0.5*M2*L2*
  & L1*AL2P**2*SIN(AL2)*SIN(AL1)+M3*L2*L1*AL2P**2*COS(AL2)*
  & COS(AL1)+M3*L2*L1*AL2P**2*SIN(AL2)*SIN(AL1)
   K(2)=-0.5*M2*L2*L1*AL1P**2*SIN(AL2)*SIN(AL1)-0.5*M2*L2*
  & L1*AL1P**2*COS(AL2)*COS(AL1)-M3*L2*L1*AL1P**2*SIN(AL2)*
  & SIN(AL1)-M3*L2*L1*AL1P**2*COS(AL2)*COS(AL1)
   K(3)=0.
C>
C> Verallgemeinerte eingepraegte Kraefte
   QE(1)=-0.5*GE*M1*L1*SIN(AL1)-GE*M2*L1*SIN(AL1)+F1*L1*
  & COS(AL1)-GE*M3*L1*SIN(AL1)+F2*L1*SIN(AL1)
   QE(2)=0.5*GE*M2*L2*COS(AL2)+F1*L2*SIN(AL2)+GE*M3*L2*COS(AL2)-
  & F2*L2*COS(AL2)
   QE(3)=F1*R
```

Figure 11: **NEWEUL equations of motion**

The unilateral constraints (12) and (13) result in the following Jacobians

$$\boldsymbol{Q}1 = \begin{bmatrix} 1 \\ \frac{L1 \sin\alpha_1}{L2\cos\alpha_2} \end{bmatrix}, \quad \boldsymbol{Q} = \begin{bmatrix} 1 \\ \frac{L1 \sin\alpha_1}{L2\cos\alpha_2} \\ -\frac{L1}{R}(\cos\alpha_1 + \sin\alpha_1 \tan\alpha_2) \end{bmatrix}. \tag{15}$$

Hence, a two–rate problem is given:

- Free flight according to Section 2.1

$$F1 = F2 = 0\,. \tag{16}$$

- Hertz's impact according to Section 2.4

$$F1 = F1(\boldsymbol{x}, \dot{\boldsymbol{x}})\,, \qquad F2 = F2(\boldsymbol{x}, \dot{\boldsymbol{x}})\,. \tag{17}$$

  The slip force $F1$ is related to Kalker's theory, now widely used in vehicle dynamics, see e.g. Popp and Schiehlen [18], while the force $F2$ follows from the constitutive law of rubber.

- Unilateral constraints according to Section 2.6

$$\boldsymbol{Q}^T[\boldsymbol{Q}1(\boldsymbol{x}) + \boldsymbol{Q}2(\boldsymbol{x})] = 0\,. \tag{18}$$

Simulation results are available on three different time scales: the free flight phase during the beginning of the motion which is repeated four times, the Hertzian contact with the dying impacts and unbiased rigid body motion of the pendulum body, and the long term influence of the slip force on the motion of the unilateral constraint system. In this paper only the medium time and long time scales are shown, Figures 12 and 13. It turns out that for mechanical dynamic problems the transition to the unilateral constraint motion is the most complex.

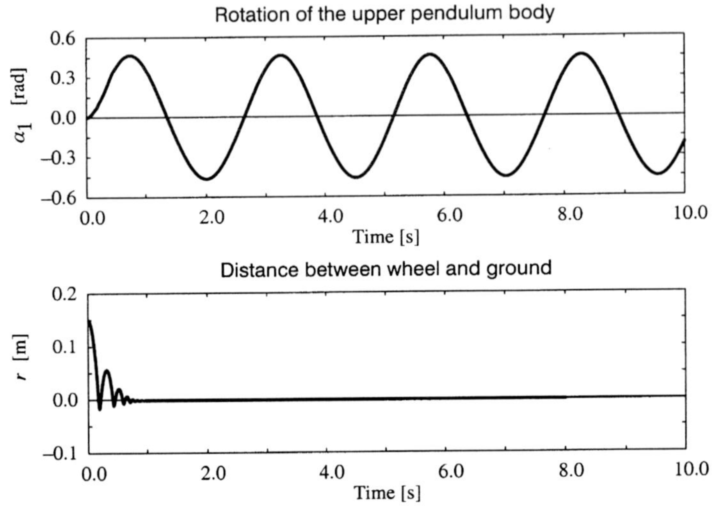

Figure 12: **Simulation of Hertz's contact: medium time scale**

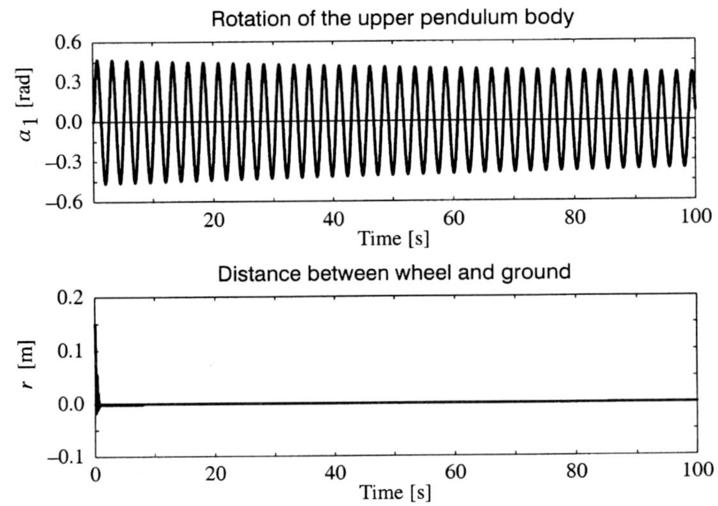

Figure 13: **Simulation of Hertz's contact: long time scale**

6 CONCLUSIONS

Contact problems may be modeled with quite different accuracy and depth as shown in the literature. The more accurate models result in different time scales for the contact. The numerical solution of the more accurate equations of motion is facilitated by multirate–multimethod integration codes. The hopping wheel with Hertzian contact represents a two–rate problem. The simulations show clearly the phenomena expected. Contact mechanics with friction is a challenging research topic for the future, in particular the micromechanics of the colliding bodies characterized by their constitutive laws.

REFERENCES

[1] Wensyang Hsu and A.P. Pisano. Modeling of a finger–follower cam system with verification in contact forces. *J. Mech. Design*, 118:132–137, 1996.

[2] Der Min Tsay and Hsien Min Wei. A general approach to the determination of planar and spatial cam profiles. *J. Mech. Design*, 118:259–265, 1996.

[3] H.-T.J. Liu. Synthesis and steady-state analysis of high-speed elastic cam-actuated linkages with fluctuated speed by a finite element method. *J. Mech. Design*, 119:395–402, 1997.

[4] J. Wittenburg. *Dynamics of Systems of Rigid Bodies.* Teubrer, Stuttgart, 1977.

[5] Dingguo Zhang. The equations of external impacted dynamics between multi-rigid body systems. *Appl. Math. Mech. (Engl. Edition)*, 18:593–598, 1997.

[6] F. Pfeiffer and C. Glocker. *Multibody Dynamics with Unilateral Contacts.* Wiley, New York, 1996.

[7] C. Toulemonde and C. Gontier. Sticking motions of impact oscillator. *Eur. J. Mech. A/Solids*, 17:339–366, 1998.

[8] P.R. Pagilla and M. Tomizuka. Contact transition control of nonlinear mechanical systems subject to a unilateral constraint. *J. Dyn. Sys. Meas. Control*, 119:749–759, 1997.

[9] H.M. Lankarani and P.E. Nikravesh. Continuous contact force models for impact analysis in multibody systems. *Nonlinear Dynamics*, 5:193–207, 1994.

[10] P. Ravn. A continuous analysis method for planar multibody systems with joints clearances. *Multibody System Dynamics*, 2:1–24, 1998.

[11] W. Goldsmith. *Impact: The Theory and Physical Behaviour of Colliding Solids.* Edward Arnold, London, 1960.

[12] B. Hu, P. Eberhard, and W. Schiehlen. Solving wave propagation problems symbolically using computer algebra. In V.I. Babitsky, editor, *Dynamics of Vibro-Impact Systems: Proceedings EUROMECH 386*, pages 231–240. Springer, Berlin, 1999.

[13] W. Schiehlen. Unilateral contacts in machine dynamics. In F. Pfeiffer, editor, *IUTAM Symp. Unilateral Multibody Dynamics, Series: Solid Mechanics and its Applications, Vol. 72*, pages 287–298. Kluwer, Dordrecht, 1999.

[14] W. Schiehlen. Multibody system dynamics: Roots and perspectives. *Multibody System Dynamics*, 1:149–188, 1997.

[15] P. Eberhard and S. Jiang. Collision detection for contact problems in mechanics with a boundary search algorithm. *Mathematical Modeling of Systems*, 3:265–281, 1997.

[16] C.W. Gear. *Multirate Methods for Ordinary Differential Equations.* Department of Computer Science, Report UIUCDCS-F-74-880, Urbana-Champaign, 1974.

[17] A. Rükgauer and W. Schiehlen. Simulation of modular mechatronic systems with application to vehicle dynamics. *Acta Mechanical*, 175:183–195, 1997.

[18] K. Popp and W. Schiehlen. *Fahrzeugdynamik.* Teubner, Stuttgart, 1993.

Multi-body impact with friction

W J STRONGE
Department of Engineering, University of Cambridge, UK

ABSTRACT

A mechanism composed of rigid bodies joined together by ideal nondissipative pinned joints has a configuation that can be described in terms of generalized coordinates q_i and time t; the system has a kinetic energy $T(\dot{q}_i, q_i, t)$. The generalized momentum of the system is defined as a vector, $\partial T / \partial \dot{q}_i$. If the system is subject to a set of independently varying forces, these give another vector — the differential of generalized impulse $\mathrm{d}\,\Pi_t$. If the applied forces act impulsively (i.e. there is a negligibly small period of force application), then the differentials of generalized momentum and generalized impulse are equal,

$$\mathrm{d}\left(\partial T / \partial \dot{q}_i\right) = \mathrm{d}\Pi_i$$

When applied to impact between systems of hard bodies where there is friction and slip that changes direction during contact, this differential relation is required. If the direction of slip is constant however, it is more convenient to use an integrated form of this generalized impulse-momentum relation. In either case, at the point of external impact the terminal impulse is obtained from the energetic coefficient of restitution.

INTRODUCTION

Impact on a mechanical system of linked rigid bodies induces reaction forces at the joints or connections between the bodies. If the compliances at all joints are small in comparison with the compliance at the point of external impact, the joint reactions are generated by kinematic constraints and the multi-body impact problem falls within the realm of analytical mechanics. For analysing multi-body dynamics, methods based on generalized coordinates and Lagrangian mechanics have the advantage of incorporating the effects of constraint reactions without explicitly including these reactions as dependent variables [Drazetic et al., 1996].

Methods for analysing impulse response of systems of rigid bodies that are joined by frictionless pinned joints (workless constraints) have been presented by Synge and Griffith [1959], Wittenberg [1977], etc. These methods equate changes in a generalized momentum to a generalized impulse, where generalized momentum and impulse are obtained from a principle of virtual work. When these methods are applied to a collision, where two systems come together at a point of external impact with a normal component of relative velocity, the analytical method can be applied directly only if the ratio of normal to tangential components of force at the impact point remains constant throughout the indefinitely small period of contact. Where this ratio of components of force remains constant, at the impact point the terminal impulse on each of the colliding bodies can be obtained from the normal impulse for compression (Souchet [1993] or Zhao [1999]). The ratio of forces is constant however only if: (i) friction is negligible or (ii) the incident tangential relative velocity is negligible or (iii) motion is constrained to be planar and the incident tangential relative velocity is sufficiently large so that sliding does not vanish before separation (Batlle [1996]). This paper obtains equations of motion from the differentials of generalized momentum and impulse — this formulation is required whenever the ratio of components of contact force is not constant.

GENERALIZED IMPULSE AND EQUATIONS OF MOTION

Equations of motion are developed first for a set of particles. These are applicable as well for a set of rigid bodies where the configuration and properties of a body are obtained by defining fixed distances between the particles which comprise the body; i.e. there are a set of constraint equations (holonomic and/or nonholonomic) which define both the relative positions of particles within each rigid body and kinematic relationships between bodies.

Let S be a set of N particles with the jth particle located at a position vector $\mathbf{r}_j$, $j = 1,\cdots,N$; each particle is subject to an external force $\mathbf{F}_j$. These forces give differential impulses $\mathrm{d}\mathbf{p}_j$ that act in the very brief period of impact, $\mathrm{d}\mathbf{p}_j = \mathbf{F}_j\,\mathrm{d}t$. The velocity $\mathbf{V}_j$ of the jth particle is the rate-of-change of the position vector and this is a function of the impulse $\mathbf{p}_j$; i.e. $\mathbf{V}_j(\mathbf{p}_j) \equiv \mathrm{d}\mathbf{r}_j(\mathbf{p}_j)/\mathrm{d}t$. Suppose the particle velocities are subject to $3N - n$ holonomic constraint equations; e.g. a fixed distance separates the jth and kth particles. Then the particle velocities can be expressed in terms of generalized coordinates q_i, generalized speeds $\dot{q}_i$, and time t; i.e. $\mathbf{V}_j = \mathbf{V}_j(q_i, \dot{q}_i, t)$. These expressions for particle velocities are consistent with $3N - n$ holonomic constraints. This holonomic system has n degrees of freedom (m degrees of freedom if nonholonomic).

Virtual displacements $\delta\mathbf{r}_j$ are any displacement field that is compatible with the displacement constraints. Similarly virtual velocities $\delta\mathbf{V}_j$ are compatible with the velocity constraints of the system. The virtual velocity of the jth particle can be expressed in terms of generalized speeds $\dot{q}_i$ as

$$\delta\mathbf{V}_j = \sum_{i=1}^{n} (\partial\mathbf{V}_j / \partial\dot{q}_i)\,\delta\dot{q}_i \tag{1}$$

During impact on a system of rigid bodies the virtual differential of work $\delta(\mathrm{d}W)$ done by external forces $\mathrm{d}\mathbf{p}_j / \mathrm{d}t$ is used to define a *differential of generalized impulse* $\mathrm{d}\Pi_i$,

$$\delta(\mathrm{d}W) = \sum_{j=1}^{N}\sum_{i=1}^{n} \mathrm{d}\mathbf{p}_j \cdot (\partial\mathbf{V}_j / \partial\dot{q}_i)\,\delta\dot{q}_i = \sum_{i=1}^{n} (\mathrm{d}\Pi_i)\,\delta\dot{q}_i \tag{2}$$

where $\mathrm{d}\Pi_i \equiv \sum_{j=1}^{N} \mathrm{d}\mathbf{p}_j \cdot (\partial\mathbf{V}_j / \partial\dot{q}_i)$. Generalized active forces $F_i = \sum_{j=1}^{N} (\mathrm{d}\mathbf{p}_j / \mathrm{d}t) \cdot (\partial\mathbf{V}_j / \partial\dot{q}_i)$ are the only forces that contribute to the differential of generalized impulse; other forces, which do not contribute, include any equal but opposite forces of interaction at rigid (i.e. non-compliant) constraints and external body forces or pressures which remain constant during impact.

A system of N particles connected by $3N-n$ holonomic velocity constraints has a kinetic energy T which is a scalar that varies with the applied impulse. This kinetic energy can be expressed in either a global coordinate system or as a function of the generalized speeds,

$$T = \frac{1}{2}\sum_{j=1}^{N} M_j(\mathbf{V}_j \cdot \mathbf{V}_j) = \frac{1}{2}\sum_{i=1}^{n}\sum_{s=1}^{n} m_{is}\dot{q}_i\dot{q}_s \tag{3}$$

where M_j is the mass of the jth particle. The inertia matrix m_{is} for the constrained system with generalized speeds $\dot{q}_i$ can be obtained from the expression for kinetic energy of the constrained motion.

For a system subject to velocity constraints, the equations of motion in terms of generalized speeds $\dot{q}_i$ are obtained directly from the kinetic energy T and the differential of generalized impulse $\mathrm{d}\Pi_i$.

Theorem — *If an impulsively loaded system can be represented by* n *generalized coordinates* q_i *and* $n-m$ *nonholonomic velocity constraints* $\sum_{i=1}^{n} a_{si}\dot{q}_i + b_s = 0,\ s = 1, \dots,$ $n-m$ *and this system is subject to a differential of generalized impulse* $\mathrm{d}\Pi_i$, *then the equations of motion in terms of generalized speeds* $\dot{q}_i$ *are obtained as,*

$$\mathrm{d}(\partial T / \partial \dot{q}_i) = \mathrm{d}\Pi_i$$

The term $\mathrm{d}(\partial T / \partial \dot{q}_i)$ is termed a *differential of generalized momentum*; it can be expressed in terms of generalized speeds as $\mathrm{d}(\partial T / \partial \dot{q}_i) = m_{is}\mathrm{d}\dot{q}_s$ where m_{is} are inertia coefficients of the system for the impact point.

This formulation of equations of motion uses two distinct scales for the effect of displacements. Infinitesimal displacements generate interaction forces at compliant constraints so it is necessary that they be included in order to represent the interaction forces which prevent overlap or interference. These infinitesimal displacements are assumed to be sufficiently small however, so that they have no affect on inertia or the kinetic energy T of the system. Thus the equations of motion (4) do not have terms arising from changes during contact of the impact configuration.

IMPACT PROCESS

Impact initiates when two colliding bodies B and B′ first come into contact at C, an initial point of contact. Each body has a point of contact, C or C′, and at incidence these points have velocities $\mathbf{V}_C(0)$ and $\mathbf{V}_{C'}(0)$, respectively. Between the contact points there is a relative velocity $\mathbf{v}(t)$ defined as $\mathbf{v}(t) \equiv \mathbf{V}_C(t) - \mathbf{V}_{C'}(t)$. If at least one of the bodies is smooth in a neighbourhood of C, there is a common tangent plane (c.t.p.) and perpendicular to this plane there is a normal direction $\mathbf{n}_3$. At incidence the bodies come together with a negative relative velocity at C; i.e. $\mathbf{n}_3 \cdot \mathbf{v}(0) < 0$.

Analyses of 'rigid' body impact can follow the process of velocity change at the contact point by introducing an infinitesimal deformable particle between the bodies at C — this particle simulates a small deforming region. The deforming region is assumed to have negligible mass because it is small in comparison with the size of either body; consequently the contact forces on either side of the particle will be equal but opposite. Noting that the normal contact force F_3 is always compressive we recognize that the normal component of impulse $p_3(t)$ is a monotonously increasing function of time; thus the normal impulse $p \equiv p_3$ can replace time as an independent variable.

TERMINAL IMPULSE

The key to calculating changes in velocity during impact is to find a means of evaluating the terminal impulse p_f at separation. The theory of rigid body impact will be more useful if the terminal impulse can be based on physical considerations. Here we relate the terminal impulse to the *energetic coefficient of restitution*; this coefficient represents dissipation of (kinetic) energy due to inelastic deformation in the region surrounding the contact point.

Compression and restitution phases of collision

After the colliding bodies first touch, the contact force $F(t)$ rises as the deformable particle is compressed. Let δ be the indentation or compression of the deformable particle. (The particle represents the compliance of the small part of the total mass near C which has significant deformation.) If compliance is rate-independent, the maximum indentation and maximum force occur simultaneously when the normal component of relative velocity vanishes. Figure 1b illustrates the normal contact force as a function of indentation δ while Fig. 1a shows this force as a function of time. The latter figure shows the separation of the contact period into an initial phase of approach or *compression* and a subsequent phase of *restitution*. During compression, *kinetic energy of relative motion is transformed into internal energy of deformation*— the normal contact force does work that reduces the initial normal relative velocity of the colliding bodies while simultaneously, an equal but opposite contact force does work that increases the internal energy of the deformable particle. The compression phase terminates and restitution begins when the normal relative velocity at the contact point vanishes. Elastic strain energy stored during compression generates the force that drives the bodies apart during the subsequent phase of restitution — the work done by this force restores part of the initial kinetic energy of relative motion. When contact terminates there is finally some residual compression δ_f of the deformable particle since the compliance during restitution is smaller than that during compression.

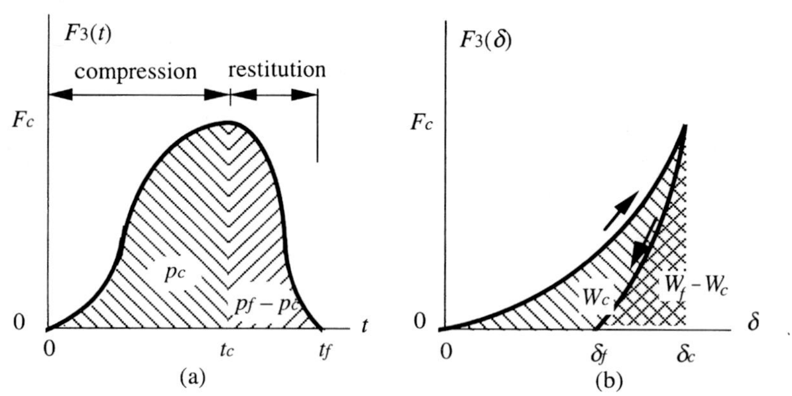

Fig. 1 Variation of contact force during impact

At any time t after incidence, the normal component of contact force F has an impulse p which equals the area under the curve of force shown in Fig. 1a. Let the instant when indentation changes from compression to restitution be t_c. The colliding bodies have a relative velocity between contact points that vanishes at the end of compression, $v(t_c)=0$; i.e. compression terminates when the contact points have the same speed V_c in the normal direction. The reaction impulse $p_c=\int_0^{t_c} F(t)dt$ which brings the two bodies to a common speed is termed the *normal impulse for compression*; this characteristic impulse is useful for analyzing collision processes.

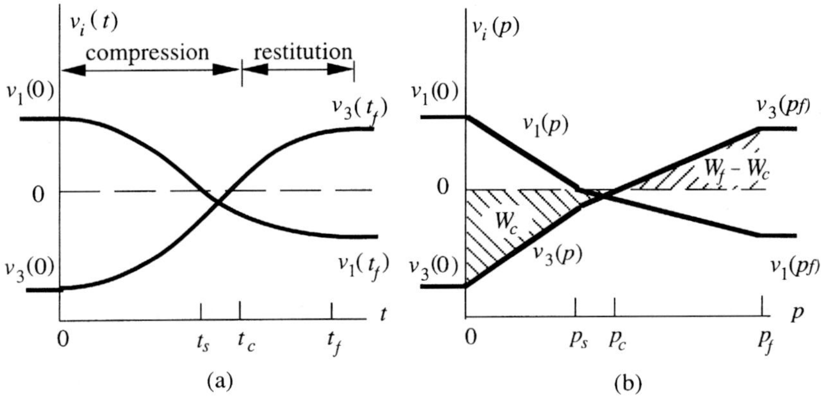

Fig. 2 Var iation of components of relative velocity at contact point during impact with small initial speed of sliding.

Coefficient of restitution and kinetic energy absorbed in collision
Dissipation of energy during collision results in smaller compliance during unloading (restitution) than was present during loading (compression); i.e. the force-deflection curve given in Fig. 1b exhibits hysteresis. The kinetic energy of relative motion that is transformed to internal energy of deformation during loading equals the area under the loading curve in Fig. 1b; this area is denoted by $W_c \equiv W_3(p_c)$. On the other hand, the area under the unloading curve equals the elastic strain energy released from the deforming region during restitution; in Fig. 1b this is denoted by $W_f - W_c \equiv W_3(p_f) - W_3(p_c)$. In the restitution phase the contact force generated by elastic unloading increases the kinetic energy of relative motion. These transformations of energy are due to work done by the contact force. This work done by the reaction force can easily be calculated for the separate phases of compression and restitution if changes in relative velocity are obtained as a function of normal impulse as illustrated in Fig. 2b; after initiation of contact the work done during these separate phases is proportional to the area between the horizontal axis and the line describing the normal relative velocity at any impulse.

During compression the impulse of the normal contact force does work $W_n(p_c)$ on the rigid bodies that surround the small deforming region — this work equals the internal energy of deformation absorbed in compressing the deformable region. An expression for this work is obtained by integrating the relative velocity at C when this is expressed as a function of normal impulse p,

$$W_c \equiv W_3(p_c) = \int_0^{p_c} v(p)\,\mathrm{d}p \tag{5}$$

This is just the kinetic energy of normal relative motion that is lost during compression. During the succeeding phase of restitution the rigid bodies regain some of this kinetic energy of normal relative motion due to the work $W_n(p_f)-W_n(p_c)$ done by contact forces.

$$W_f - W_c \equiv W_3(p_f) - W_3(p_c) = \int_{p_c}^{p_f} v(p)\,\mathrm{d}p \tag{6}$$

This work comes from and is equal to the elastic strain energy released during restitution.

These expressions for work done by the contact force during separate parts of the collision (period) are used to express the part of the initial kinetic energy of normal relative motion that is lost due to hysteresis of contact force. Expressions (5) and (6) give the part of this transformed energy that is irreversible and this can be used to define an *energetic coefficient of restitution*, e_*.

> Definition — *The square of the coefficient of restitution e_*^2 is the negative of the ratio of the elastic strain energy released during restitution to the internal energy of deformation absorbed during compression,*
>
> $$e_*^2 = -\frac{W_3(p_f) - W_3(p_c)}{W_3(p_c)} \tag{7}$$

This coefficient has values in the range $0 \le e_* \le 1$ where 0 implies a perfectly plastic collision (i.e. no final separation so that none of the initial kinetic energy of normal relative motion is recovered) while a value of 1 implies a perfectly elastic collision (i.e. no loss of kinetic energy of normal relative motion).

EXAMPLE PROBLEMS

Example 1: Compound pendulum colliding against rough, inelastic half-space

A compound pendulum is composed of a uniform slender bar of length L and mass M which pivots freely around a frictionless pin O. The pendulum is rotating with initial angular velocity $\boldsymbol{\omega}_0 = -\dot{\theta}_0 \mathbf{n}_2$ before the tip strikes a rough half-space at contact point C. The position of C relative to O is $\mathbf{r}_C = -r_1\mathbf{n}_1 - r_3\mathbf{n}_3$ where the unit vector $\mathbf{n}_1$ is parallel to the tangent plane and $\mathbf{n}_3$ is the common normal direction as shown in Fig. 3. The pendulum has radius of gyration $k_r = (L^2/3)^{1/2}$ for O and at the contact point the energy loss due to irreversible internal deformation is related to the energetic coefficient of restitution e_*. Find the ratio of terminal to incident angular velocities $\dot{\theta}_f / \dot{\theta}_0$ assuming that tangential compliance is negligible and that friction at C is represented by Coulomb's law with a coefficient of friction μ.

Solution

kinetic energy: $T = (1/2)Mk_r^2\dot{\theta}^2 = (1/2)M\dot{q}_1^2, \qquad \dot{q}_1 \equiv k_r\dot{\theta}$

velocity of C: $\mathbf{V}_C = \mathbf{V}_O + \boldsymbol{\omega} \times \mathbf{r}_C = (r_3\dot{q}_1 / k_r)\mathbf{n}_1 - (r_1\dot{q}_1 / k_r)\mathbf{n}_3$

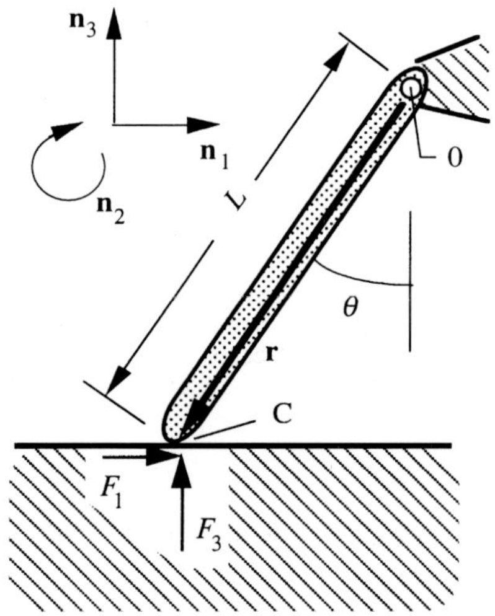

Fig. 3 Compound pendulum at inclination angle θ when it strikes a rough, inelastic half-space.

The solution of this problem employs the normal component of impulse $p \equiv p_3$ as an independent variable. (Since the normal force is compressive, the normal impulse p increases monotonously during the very brief period of collision.) The differential of impulse at contact point C, $d\mathbf{p} = dp_1\mathbf{n}_1 + dp_3\mathbf{n}_3$ has components that are related by the Amontons-Coulomb law of friction;

$$\begin{aligned} \mathrm{d}p_1 &= \mu\,\mathrm{d}p, && \mathbf{V}_\mathrm{C}\cdot\mathbf{n}_1 < 0 \\ -\mu\,\mathrm{d}p < \mathrm{d}p_1 &< \mu\,\mathrm{d}p, && \mathbf{V}_\mathrm{C}\cdot\mathbf{n}_1 = 0 \\ \mathrm{d}p_1 &= -\mu\,\mathrm{d}p, && \mathbf{V}_\mathrm{C}\cdot\mathbf{n}_1 > 0 \end{aligned}$$

Hence the differential of generalized impulse is obtained as

$$\mathrm{d}\Pi_1 = (r_3\,\mathrm{d}p_1 / k_r) - (r_1\,\mathrm{d}p_3 / k_r)$$

$$= \begin{cases} -k_r^{-1}(r_1 + \mu r_3)\,\mathrm{d}p, & \mathbf{V}_\mathrm{C}\cdot\mathbf{n}_1 < 0 \\ -k_r^{-1}(r_1 - \mu r_3)\,\mathrm{d}p, & \mathbf{V}_\mathrm{C}\cdot\mathbf{n}_1 > 0 \end{cases}$$

Equations of motion are obtained from (4) after recognizing that slip reverses in direction at impulse p_c simultaneous with the transition from compression to restitution. After integration, one obtains the following generalized speeds as a function of normal impulse p;

$$\begin{aligned} \dot{q}_1(p) &= \dot{q}_1(0) - (k_r M)^{-1}(r_1 + \mu r_3)p, && p < p_c \\ \dot{q}_1(p) &= -(k_r M)^{-1}(r_1 - \mu r_3)p, && p_c < p < p_f \end{aligned}$$

Notice that slip reversal requires $\mu < \bar{\mu} \equiv r_1 / r_3$; otherwise the pendulum sticks in the compressed configuration and $\dot{q}_1 = 0$ for $p > p_c$.

The normal impulse for compression p_c is obtained from a condition that for rate-independent material properties, the normal velocity vanishes simultaneously with termination of the compression period, $0 = \mathbf{n}_3 \cdot \mathbf{V}_C(p_c)$; hence

$$p_c = k_r(r_1 + \mu r_3)^{-1} M\dot{q}_1(0)$$

In two-body collision the normal contact force does work that transforms the kinetic energy of relative motion into internal energy of deformation during compression and subsequently restores a part of this kinetic energy during restitution. The square of the energetic coefficient of restitution is defined as the negative of the ratio of the work done by the normal contact force during restitution to the work done during compression, Stronge (1990). This coefficient relates the terminal normal impulse p_f to the normal impulse for compression p_c.

work of normal impulse during compression $W_3(p_c)$,

$$W_3(p_c) = \int_0^{p_c} -\frac{r_1}{k_r}\left[\dot{q}_1(0) - \frac{(r_1 + \mu r_3)}{k_r M} p\right] \mathrm{d}p = -\frac{r_1^2 + \mu r_3}{2k_r^2 M} p_c^2$$

work of normal impulse during restitution $W_3(p_f) - W_3(p_c)$,

$$W_3(p_f) - W_3(p_c) = \int_{p_c}^{p_f} \frac{r_1}{k_r}\left[\frac{(r_1 - \mu r_3)}{k_r M} p\right] \mathrm{d}p = \frac{r_1^2 - \mu r_3}{2k_r^2 M}(p_f^2 - p_c^2)$$

energetic coefficient of restitution,

$$e_*^2 = -\frac{W_3(p_f) - W_3(p_c)}{W_3(p_c)} = \frac{(r_1^2 - \mu r_1 r_3)}{(r_1^2 + \mu r_1 r_3)}\left[\frac{p_f^2}{p_c^2} - 1\right]$$

terminal impulse p_f as function of angle of inclination $\theta = \cot^{-1}(r_3 / r_1)$,

$$\frac{p_f}{p_c} = 1 + e_* \sqrt{\frac{r_1^2 + \mu r_1 r_3}{r_1^2 - \mu r_1 r_3}} = 1 + e_* \sqrt{\frac{1 + \mu \cot\theta}{1 - \mu \cot\theta}}$$

ratio of final to initial angular speed,

$$\frac{\dot{\theta}_f}{\dot{\theta}_0} = \frac{\dot{q}_1(p_f)}{\dot{q}_1(0)} = -e_* \sqrt{\frac{1 - \cot\theta}{1 + \cot\theta}}$$

Figure 4 illustrates the ratio of angular speeds as a function of the inclination angle θ of the pendulum at impact. With the energetic coefficient of restitution, the result shows the effect of energy dissipated by friction even if the bodies are elastic. At small angles of eccentricity the work done by friction is large in comparison with the work done by the normal component of contact force. At a sufficiently small angle of inclination θ there is no rebound (i.e. the contact sticks). Terminal stick occurs if the coefficient of friction is sufficiently large, $\mu \geq \bar{\mu} = \tan\theta$ (Stronge, 1991).

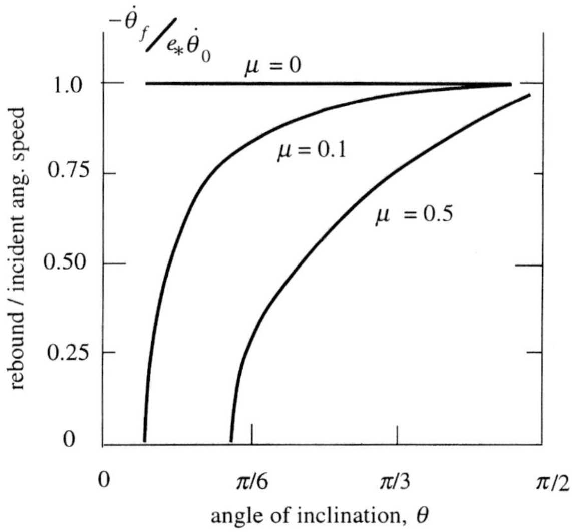

Fig. 4 Effect of friction coefficient μ on the ratio of angular speed of pendulum at rebound $\dot{\theta}_f$ to speed at incidence $\dot{\theta}_0$ as function of the angle of inclination θ.

A more complex problem of impact at the tip of a double compound pendulum was proposed by T.R. Kane (1985). The double pendulum is swinging when the tip strikes against a rough inelastic half-space. This problem has generated renewed interest in analytical methods for representing impact with friction (see Hurmuzlu & Marghitu, 1994).

Example 2: Double Pendulum colliding against rough, inelastic half-space
Two identical uniform rods OB and BC are joined at ends B by a frictionless joint in order to form a double pendulum; the other end of OB is suspended from a frictionless hinge at O as shown in Fig. 5. When the free end C of rod BC strikes against a rough half-space, the rods have angles of inclination from vertical denoted by θ_1 and θ_2 and angular speeds $\dot{\theta}_1$ and $\dot{\theta}_2$ respectively. Denote the coefficient of friction between C and the half-space by μ and the energetic coefficient of restitution at the same location by e_*. Assume the motion is planar.

Solution After defining generalized speeds, $\dot{q}_1 = L\dot{\theta}_1$, $\dot{q}_2 = L\dot{\theta}_2 / 2$ the kinetic energy of system can be expressed as

$$\begin{aligned} T &= (ML^2/6)[4\dot{\theta}_1^2 + \dot{\theta}_2^2 + 3\dot{\theta}_1\dot{\theta}_2 \cos(\theta_2 - \theta_1)] \\ &= (M/3)[2\dot{q}_1^2 + 2\dot{q}_2^2 + 3\dot{q}_1\dot{q}_2 \cos(\theta_2 - \theta_1)] \end{aligned}$$

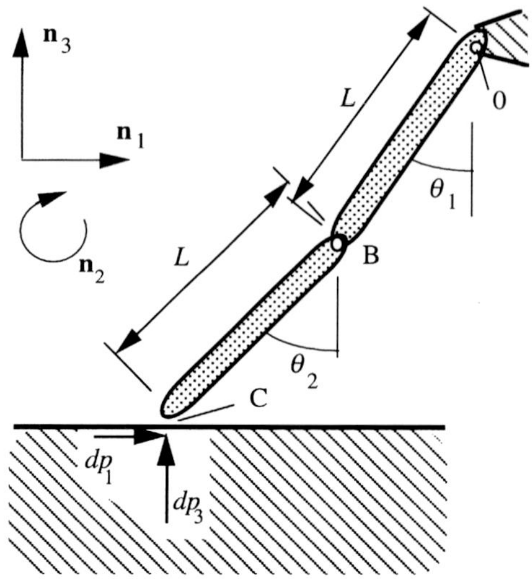

Fig. 5 Double pendulum striking rough inelastic half-space.

so that the generalized momenta,

$$\begin{aligned}\partial T / \partial \dot{q}_1 &= (M/3)[4\dot{q}_1 + 3\dot{q}_2 \cos(\theta_2 - \theta_1)] \\ \partial T / \partial \dot{q}_2 &= (M/3)[4\dot{q}_2 + 3\dot{q}_1 \cos(\theta_2 - \theta_1)]\end{aligned} \tag{a}$$

Velocity of contact point $\mathbf{V}_C$ with $s\theta_i \equiv \sin\theta_i$ and $c\theta_i \equiv \cos\theta_i$;

$$\begin{aligned}\mathbf{V}_C &= \dot{\theta}_1(-Lc\theta_1\mathbf{n}_1 + Ls\theta_1\mathbf{n}_3) + \dot{\theta}_2(-Lc\theta_2\mathbf{n}_1 + Ls\theta_2\mathbf{n}_3) \\ &= -(\dot{q}_1 c\theta_1 + 2\dot{q}_2 c\theta_2)\mathbf{n}_1 + (\dot{q}_1 s\theta_1 + 2\dot{q}_2 s\theta_2)\mathbf{n}_3\end{aligned} \tag{b}$$

Differential of generalized impulse $d\Pi_i$ for increment of impulse $d\mathbf{p} = dp_1\mathbf{n}_1 + dp_3\mathbf{n}_3$.

$$d\Pi_1 = -dp_1 c\theta_1 + dp_3 s\theta_1, \qquad d\Pi_2 = -2dp_1 c\theta_2 + 2dp_3 s\theta_2$$

Initially slip is in direction $\mathbf{n}_1$ so Coulomb's law gives $dp_1 = -\mu\, dp_3 \equiv -\mu dp$ and Eq. (4) results in equations of motion;

$$\begin{bmatrix} 4 & 3c(\theta_2 - \theta_1) \\ 3c(\theta_2 - \theta_1) & 4 \end{bmatrix} \begin{Bmatrix} d\dot{q}_1 / dp \\ d\dot{q}_2 / dp \end{Bmatrix} = \frac{3}{M} \begin{Bmatrix} \mu c\theta_1 + s\theta_1 \\ 2\mu c\theta_2 + 2s\theta_2 \end{Bmatrix} \equiv \begin{Bmatrix} b_1 \\ b_2 \end{Bmatrix}$$

After solving for the differentials and then integrating with initial conditions $\dot{q}_i(0)$,

$$\begin{Bmatrix} \dot{q}_1 \\ \dot{q}_2 \end{Bmatrix} = \begin{Bmatrix} \dot{q}_1(0) \\ \dot{q}_2(0) \end{Bmatrix} + \frac{1}{\Delta} \begin{Bmatrix} 4b_1 - 3b_2 \cos(\theta_2 - \theta_1) \\ 4b_2 - 3b_1 \cos(\theta_2 - \theta_1) \end{Bmatrix} p, \qquad p < p_s \tag{c}$$

$$\Delta = 16 - 9\cos^2(\theta_2 - \theta_1)$$

where at impulse p_s the tangential speed vanishes $\mathbf{V}_C \cdot \mathbf{n}_1 = \dot{q}_1 c\theta_1 + 2\dot{q}_2 c\theta_2 = 0$.

$$p_s = \frac{-[\dot{q}_1(0)c\theta_1 + 2\dot{q}_2(0)c\theta_2]\Delta}{[4b_1 - 3b_2 c(\theta_2 - \theta_1)]c\theta_1 + 2[4b_2 - 3b_1 c(\theta_2 - \theta_1)]c\theta_2} \tag{d}$$

while at impulse p_c the normal component of relative velocity vanishes, $\mathbf{V}_C \cdot \mathbf{n}_3 = \dot{q}_1 s\theta_1 + 2\dot{q}_2 s\theta_2 = 0$.

$$p_c = \frac{-[\dot{q}_1(0)s\theta_1 + 2\dot{q}_2(0)s\theta_2]\Delta}{[4b_1 - 3b_2 c(\theta_2 - \theta_1)]s\theta_1 + 2[4b_2 - 3b_1 c(\theta_2 - \theta_1)]s\theta_2} \tag{e}$$

If C slides in the positive direction, during compression the normal component of impulse does work $W_3(p_c)$ equal to,

$$W_3(p_c) = \{\dot{q}_1(0)s\theta_1 + 2\dot{q}_2(0)s\theta_2\}p_c + \{[4b_1 - 3b_2 c(\theta_2 - \theta_1)]s\theta_1 + 2[4b_2 - 3b_1 c(\theta_2 - \theta_1)]s\theta_2\}\frac{p_c^2}{2\Delta} \tag{f}$$

After initial sliding is brought to a halt, if sliding resumes it occurs in direction $-\mathbf{n}_1$ so that coefficients b_1 and b_2 transform to

$$\begin{Bmatrix} \bar{b}_1 \\ \bar{b}_2 \end{Bmatrix} \equiv \frac{3}{M}\begin{Bmatrix} -\mu c\theta_1 + s\theta_1 \\ -2\mu c\theta_2 + 2s\theta_2 \end{Bmatrix}$$

For impulse applied after slip is halted $p_s < p < p_f$ the critical coefficient of friction for stick $\bar{\mu}$ prevents the resumption of sliding; this coefficient is obtained from $\mathbf{V}_C \cdot \mathbf{n}_1 = \dot{q}_1 c\theta_1 + 2\dot{q}_2 c\theta_2 = 0$ with $\dot{q}_i(\bar{b}_1, \bar{b}_2)$,

$$\bar{\mu} = \frac{8s\theta_2 c\theta_2 + 2s\theta_1 c\theta_1 - 3s(\theta_2 + \theta_1)c(\theta_2 - \theta_1)}{8c^2\theta_2 + 2c^2\theta_1 - 6c\theta_1 c\theta_2 c(\theta_2 - \theta_1)} \tag{g}$$

For some specific initial values, $\theta_1 = \pi/9$, $\theta_2 = \pi/6$, $\dot{\theta}_1 = -0.1$ rad s^{-1}, $\dot{\theta}_2 = -0.2$ rad s^{-1}, $e_* = 0.5$, Table 8.1 contains results for this double pendulum obtained with an energetic coefficient of restitution at C.

Table 8.1 Result of double pendulum striking rough half-space

coeff. friction μ	initial velocity (rad s^{-1}) $\dot{\theta}_1(0)$	$\dot{\theta}_2(0)$	rel. imp. (slip = 0) p_s/p_c	rel. imp. (separation) p_f/p_c	final velocity (rad s^{-1}) $\dot{\theta}_1(p_f)$	$\dot{\theta}_2(p_f)$	final dir. slip (+/–)	final normal vel. $V_3(p_f)/V_3(0)$	final kin. energy T_f/T_0
0.0	–0.1	–0.2	1.32	1.50	–0.230	+0.292	–	–0.500	.707
0.2	–0.1	–0.2	1.30	1.52	–0.214	+0.259	–	–0.420	.621
0.5	–0.1	–0.2	1.29	1.56	–0.199	+0.223	–	–0.330	.552
0.7	–0.1	–0.2	1.28	1.58	–0.193	+0.210	stick	–0.280	.527

Configuration gives a coefficient of friction for stick $\bar{\mu} = 0.62$

CONCLUSION

Analyses of 'rigid' body impact problems with friction and possible changes in direction of slip during contact require differential equations of relative motion in order to correctly account for changes in tangential components of contact forces. If the tangential velocity is relatively small and the impact configuration is non-collinear (or unbalanced), these tangential forces are not proportional to the change in the normal component of force. The formulation developed in this paper is obtained from the principle of virtual power; this formulation is efficient for analyzing multi-body systems with perfect or workless constraints. If the connections between bodies have compliances that are similar in magnitude to that at the impact point however, it is necessary to explicitly include these compliances in modelling the system.

REFERENCES

Battle, J.A. (1996) "Rough balanced collisions". *ASME J. Appl. Mech.* **63**, 168-172.

Bahar, L. (1994) "On use of quasi-velocities in impulsive motion", *Int. J. Engr. Sci.* **32**, 1669-1686.

Drazetic, P., Level, P., Canaple, B. & Mongenie, P. (1996) "Impact on planar kinematic chain of rigid bodies: application to movement of anthropomorphic dummy in crash" Int. J. Impact Engng. **18**(5), 505-516.

Han, I. & Gilmore, B.J. (1993) "Multibody impact motion with friction — simulation and experimental validation" *ASME J. Design* **115**, 412-419.

Hurmuzlu, Y. & Marghitu, D. (1994) "Rigid body collisions of planar kinematic chains with multiple contact points" *Int. J. Robotics Research* 13(1), 82-92.

Kane, T. & Levinson, D.A. (1985) *Dynamics: Theory and Application*, McGraw-Hill, New York.

Pereira, M.S. & Nikravesh, P. (1996) "Impact dynamics of systems with frictional contact using joint coordinates and canonical equations of motion" *Nonlinear Dynamics* **9**, 53-71.

Souchet, R. (1993) "Analytical dynamics of rigid body impulsive motions". *Int. J. Engng. Sci.* **31**, 85-92.

Stronge, W. (1990) "Rigid body collisions with friction", *Proc. Roy. Soc. Lond.* A**431**, 169-181.

Stronge, W. (1991) "Friction in collisions — resolution of a paradox", *J. Appl. Phys.* **69**(2), 610-612.

Synge, J.L. & Griffith, B.A. (1959) *Principles of Mechanics*. McGraw-Hill, New York.

Wittenburg, J. (1977) *Dynamics of System of Rigid Bodies*, Teubner, Stuttgart.

Zhao, W. (1999) "Kinetostatics and analysis methods for the impact problem", *Eur. J. Mech. A/Solids* **18**, 319-329.

Challenges of finite element simulations of vehicle crashes

A ESKANDARIAN, G BAHOUTH, D MARZOUGUI, and **C D KAN**
FHWA/NHTSA National Crash Analysis Center, The George Washington University, Ashburn, Virginia, USA

The ability to model and analyze vehicle crashes accurately and efficiently is a necessity for automotive designers and safety engineers. Computer simulation of vehicle crashes using the latest in finite element methods has progressed rapidly during the past decade. Dynamic explicit finite element codes are widely used to model and simulate vehicle crashes, biomechanics of occupant injuries, and safety performance of barriers and roadside hardware. The field of crashworthiness, which previously dealt primarily with experimental impacts and full-scale vehicle crash tests, today, includes improvements in fundamentals as well as applications of finite element methods. Development of new element formulations, improved contact algorithms and material constitutive relationships, computational efficiency and parallel processing are among many other emerging topics.

This paper reviews some of the latest challenges of vehicle and occupant modeling using DYNA family of computer programs. Issues such as new elements dealing with fracture and dynamic crack propagation, various aspects of occupant mechanics, integration of optimization methods with non-linear impact dynamics, barrier/roadside hardware design challenges, and reliable and efficient computational methods are reviewed. Each concern is supported briefly by case studies. The current status and possible future research directions are discussed.

INTRODUCTION

Today, high performance computing technology and advanced finite element codes such as LS-DYNA, make it possible to analyze even the most complex structures within a reasonable time. Analysis of virtual prototypes and design concepts greatly reduce product development times and lead to an overall increase in product quality [Eskandarian, 98]. Traditionally, product design has included concept development followed by manufacturing of a prototype for testing and verification. During the development of an automobile, safety performance of prototypes is assessed using crash testing. Cost estimates for a single vehicle crash test range from one hundred thousand to over a million dollars depending on stage of development.

Over the complete vehicle design cycle, a manufacturer may conduct hundreds of these tests during the development of a single platform. With these cost estimates in mind, it becomes apparent that the use of analysis tools like FEM reduce the number of required crash tests saving auto makers millions of dollars in development costs.

Current simulation technology is capable of accurately representing a large number of conditions typical to a crash test. For example, frontal crash of a vehicle involves interaction between a barrier surface and the car structure. During an impact, load bearing members will deform as the vehicle structure and the occupant inside are decelerated. For typical structural designs and impact conditions, behavior of body components can be easily characterized and modeled using the finite element method. Further, modeling techniques are used to represent the behavior of interior components, restraints (airbags, seatbelts, etc.) and even the occupant. When addition complexities such as advanced material usage (composites, aluminum, etc.), material fracture and failure or atypical impact conditions exist, analysis tools often fall short. Listing all challenges of vehicle crash simulation is obviously beyond the scope of any single article. This paper identifies a number of critical issues in FE simulation which are currently addressed at the GW TRI. These are among the most critical and timely requirements of effective and accurate crashworthiness simulation as a true predictive tool.

VEHICLE FINITE ELEMENT MODELS

Currently, research is being conducted in the area of crashworthiness through the development and use of vehicle finite element models. Specialized techniques for part digitization and material characterization are used to create highly detailed vehicle models. To date, seven vehicle models across a broad range of vehicle classes have been created at the FHWA/NHTSA National Crash Analysis Center (NCAC) for use in a number of safety studies. Figure 1 shows an example of a single vehicle model and some relevant specifications [Eskandarian, 1997, Zaouk, 1998].

Model Specifications:

325	Parts
285,824	Nodes
2860	Solid Elems.
63	Beam Elems.
269,249	Shell Elems.
4,150	Spotwelds
675	Rigid Body Constraints

Figure 1- Dodge Neon Finite Element Model [Zaouk, 1998]

Although reliability and good accuracy of today's vehicle finite element models makes them valuable for crashworthiness studies, many important issues remain unresolved.

For solving dynamic problems, both implicit and explicit time integration schemes can be implemented in the finite element solution process. Due to the nature of impact problems involving contact, large deformation, and material non-linearity, it is most suitable to use explicit methods to solve this class of problems. While computational resources are more effectively used in explicit methods, it is still necessary to select a mesh-dependent small time step in order to achieve stability in the time integration. Therefore, it is nearly impossible to perform this analysis using fully integrated elements. In an effort to reduce the overall clock time for simulations of such models, reduced element formulations are often employed. The most commonly used element types in dynamic explicit FE codes include the Belytschko-Tsay Shell element and the Huighes-Liu formulations. These element types assume a constant stress across the face of the shell through a single integration point found at the center of the element. A consequence of using these element types is the introduction of hourglass deformations. This is controlled in explicit codes through several artificial numerical schemes. It is critical to choose the appropriate element formulation and hourglass control type when using reduced integration.

Another critical issue in vehicle analysis is connections between various components. Spotwelds are commonly used. This connection type assumes a rigid link between a number of specified points. Performance of weld models is known to be highly dependent on a number of conditions. These conditions include the following:

1. Element mesh size and location of element intersections adjacent to spotwelds have a significant effect on weld behavior. Using a course element mesh artificially stiffens parts. When a spotweld is attached to one of these less compliant regions, premature failure of spotwelds often takes place. It was found that a finite element mesh similar in size to the physical dimension of the spotweld leads to the most accurate behavior of welds. Also, it was found that spotweld behavior improves as the angle of connection between the weld and adjacent element is as normal to the mesh surface as possible [Hogg, 1998].
2. Based on current testing methods, spotweld failure properties are not well characterized. This situation prohibits true validation of spotweld models through comparison with real world data. In addition, large variation in manufacturing of the welds (variation in tooling) lead to highly scattered weld behaviors.

In recent years, a number of new materials have been introduced to vehicles and, as a result, finite element models. Some of these materials include plastics, foams and composites. The behavior of these materials is complex and implementation of their constitutive equations is involved. Several research studies are currently being conducted to incorporate these models into today's explicit finite element codes.

INTEGRATED OCCUPANT-AIRBAG INTERACTION MODELING

Currently, airbags are implemented in a number of finite element interior and vehicle models. For frontal cases with normal seating, these airbag models yield accurate results based on important assumptions. Normal or "in-position" seating is a requirement for full inflation of the airbag before any occupant interaction occurs. If the occupant contacts the bag during the unfolding process, unrealistic loading to the occupant will take place. Only during the final resting state or fully inflated state will interactive forces be realistic. When conditions change, interactions between occupants and the airbag systems lead to greatly varied restraint effectiveness during simulation [Noureddin 1998, Digges 1998]. Within actual airbag systems, deployment is a complex phenomenon where gases flow into a folded, confined space. The way which gases interact with the bag system depends on the direction of the gas jets, the rate of gas flow and folding patterns of the bag. These parameters, how they are modeled and their influence on airbag effectiveness require detailed investigation.

Within LS-DYNA, airbag models use a "control volume" assumption. The volume refers to the region surrounded by the airbag material or control surface. Green's theorem is used to calculate the volume of the bag versus time. An equation of state is then used to relate pressure, gas density and specific internal energy of the gas based on the 'Gamma Law Gas Equation of State.' Subsequently, the ideal gas law is used to characterize the relationship between that single pressure value for the whole bag volume and the gas temperature to calculate volume during bag inflation. A number of typical LS-DYNA input parameters, which contribute to the above relationship, are shown below [Hallquist, 1991].

Table 1- Airbag Input Parameters Available in LS-DYNA

Gas characteristics	Heat Capacities (C_v & C_P)
	Temperature vs. Time
	Molecular Weight (ie. gas composition)
Inflator characteristics	Mass flow vs. Time
	Tank Pressure vs. Time
Exhaust Parameters	Inflator Tank Orifice Size
	Bag Porosity and Venting

This method allows the folded airbag to inflate with an accurate volume of gas enclosed and arrive at a realistic final pressure. If an occupant involved in a frontal impact is seated away from the bag during deployment, the steps through which the airbag inflates will not effect the occupant in any way. Conversely, if there is interaction between the bag and occupant at any point during inflation, the region of the bag contacted and it's internal pressure will drastically effect occupant loading. This issue becomes critically important for out-of-position seating conditions, late deployments and non-frontal cases.

Preliminary studies indicate that interactions between the occupant and airbag during inflation significantly influences dummy loading mechanisms and subsequent deployment behavior of the bag. This behavior takes place during simulations using constant pressure assumptions like that found in LS-DYNA. Figure 2 shows results from an ongoing study where a single airbag is deployed into an obstructing pole. As the location and distance of the pole relative to the bag changes, the resulting variation in pole forces may be seen. When related to interaction of occupants seated near a deploying bag, these results indicate a need to further investigate options to reduce additional occupant loading [Bedewi, 1996].

In order to develop robust and accurate airbag representations for integration into complete vehicle systems, investigation of advanced airbag parameters and investigation of the capabilities of other simulation tools must be conducted. The use of coupled analysis techniques will most likely improve results of future airbag analysis. Coupled Lagrangian (like LS-DYNA) and Eularian techniques combine deformable and translational elements of an airbag surface with an Eularian mesh representing enclosed and ambient gases. The

Eularian mesh remains fixed in space but characterizes the flow of materials like gas within its boundaries over time. During inflation, pressure changes caused by the flow of gases through the stationary mesh will impose loads and deformation to the surface of the airbag modeled using Lagrangian techniques.

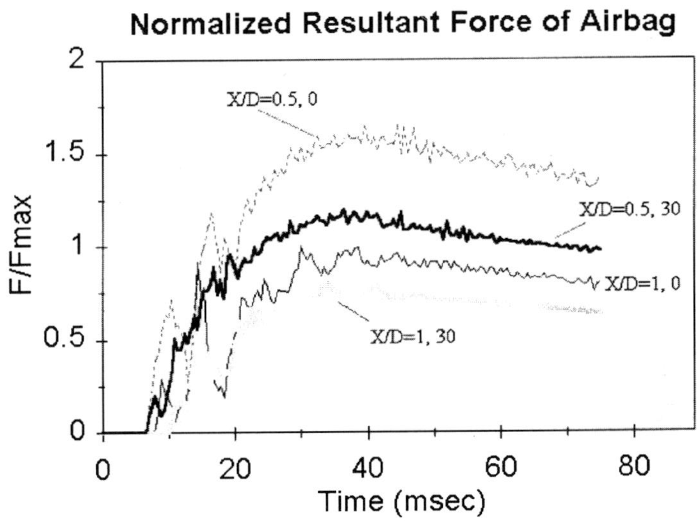

Figure 2- Contact force changes as a function of barrier distance and location

FAILURE MODELING AND FRACTURE

Several engineering designs use material failure and fracture to improve the crashworthiness and safety of structures. In these designs, failure mechanisms are introduced purposefully to cause the structure to respond in a more predictable and safe manner under impact loads. An example of such cases is in roadside hardware designs where structures are designed to break away upon impact without causing major damage to the vehicle and hence lowering the risk of injuries to the occupants [Eskandarian, 1999; Eskandarian, 1996; Marzougui, 1999]. A second example is in automotive safety designs where new materials, such as composites, are used in new vehicles to absorb more energy though fracture and reduce the severity of the crash. Also in conventional materials failure-inducing design elements like holes and slots are incorporated to increase crash pulse absorption. These problems can not be solved numerically unless accurate prediction of the failure phenomenon is achieved.

Current nonlinear explicit finite element codes have a major deficiency when modeling material failure and fracture. All explicit programs have simple material failure models that are based on element deletion. The stresses or strains for each element are checked at each time step and if an element reaches a certain critical stress or strain level, the element is considered to have failed. The stresses in the element are set to zero, which is equivalent to removing that element from the model for the rest of the calculations. In real life, failure occurs through the development and propagation of sharp cracks. This phenomenon can not be adequately predicted by this simple failure model. Deleting the failed element creates a hole the size of the element in the model instead of a sharp crack. Consequently, the stress concentration due to the sharp crack is excluded and the stresses around the failed element become inaccurate. Figure 3 shows a typical fracture mechanics problem. A plate with a center crack is subjected to normal loading. An explicit finite element program was used to predict the failure propagation in the plate. The figure shows the predicted results at different stages of the propagation process. It can be seen from the figure that as soon as the first element fails the stress concentration at the crack tip is reduced. The finite element predictions beyond this point are no longer accurate since the stress concentration and the stresses around the crack dictate how and when the crack should propagate, the failure behavior is not correctly captured.

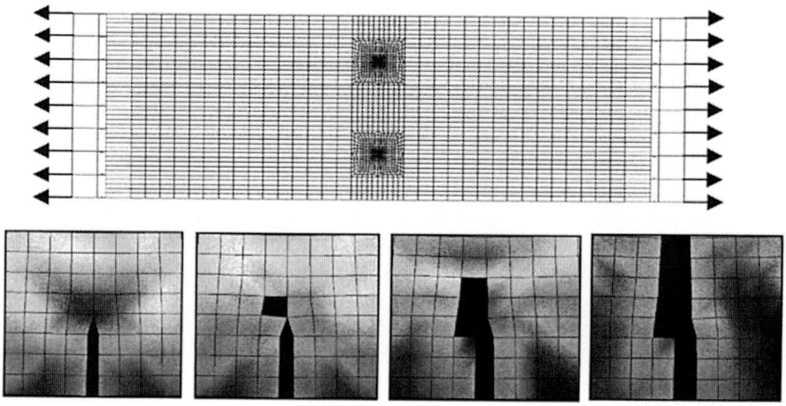

Figure 3: Fracture Prediction of a Center Crack Plate - Mesh 1

Using a finer mesh improves the prediction of the failure process but at the cost of an extremely larger model size. It was found that for this example a mesh size in the order of 0.1mm has to be used to get a reasonable approximation of the failure process. Such a small mesh may be adequate to solve small problems such us the center crack problem however it is impractical for the majority of engineering problems where the whole structure or a sub-component of the structure need to be analyzed. Consequently there is a need for a new failure model.

Fracture mechanics theories have great potential for accurately predicting the failure phenomenon for the general nonlinear three-dimensional dynamic problem. The field of fracture mechanics has seen significant progress and growth in the past few decades. Thanks to pioneering work of several researchers such as Inglis, Griffith, Irwin, Orowan, Westergaard, Dugdale, Barenblatt, Wells, and Rice; this field has become a widely recognized and greatly employed engineering discipline. Fracture mechanics concepts, which are based on conventional strength of material theories, incorporate the effects of cracks in the analysis. Several concepts have been introduced and found to be adequate for a variety of fracture problems. Griffith introduced the energy balance approach that is applicable for brittle materials. Irwin proposed the energy release rate and stress intensity factor approaches, which are suitable for linear elastic fracture mechanics problems. For elastic-plastic problems, Wells introduced the crack tip opening displacement (CTOD) approach, Rice introduced the J integral approach, DeKoning introduced the crack tip opening angle (CTOA) approach, and Lee introduced the plastic energy approach. Each of these approaches has been proven to accurately predict the fracture process within their limitations.

A new failure model based on fracture mechanics theories has been developed and implemented in a nonlinear explicit dynamic finite element program [Marzougui, 1998]. Fracture analyses introduce two additional unknowns to the governing dynamics equations of the system, namely the amount and direction of the crack extension. As is the case with any system of equations, a unique solution to the system can be determined only if the number of unknowns is equal to the number of equations. Therefore, additional equations are needed to solve the fracture problem. Two equations are needed: one that governs the amount of crack propagation and the other dictates the direction at which the crack should propagate. The new failure model uses the crack tip opening angle criterion theory to determine the amount of crack extension. This theory was introduced by DeKoning [DeKoning, 1975]. This criterion simply states that the crack will grow if the crack angle at the tip of the crack reaches a critical crack tip opening angle ($CTOA_{cr}$). The $CTOA_{cr}$ can be considered a material constant that does not depend on the loading and geometry of the crack system. This theory has been supported by several experimental tests and is widely accepted. The crack extension direction is determined based on the principal stress criterion theory. This theory is also well established and used by many fracture mechanics researchers. It states that the crack will propagate in the direction normal to the maximum principal stress direction. In the new failure model, the crack propagation phenomenon is simulated by splitting the elements along the crack extension direction. This method is more accurate than deleting the element since the crack tip remains sharp during the entire crack propagation process therefore conserving the stress concentration at the crack tip.

The new failure model has been tested using several fracture mechanics examples. The fist example is the same as the one shown in Figure 3. A center crack plate is subjected to a normal loading. Figure 4 shows the new failure model predictions at different stages of the propagation process. It can be seen from the figure that the stress concentration at the crack tip is maintained throughout the whole fracture process. Figure 5 shows a plot of the stress versus crack length from the simulation and experimental test. The figure shows that the two simulation results match the experimental data. This is highlighted by the fact that the onset of the crack propagation occurs at the same stress levels for both the experiment and simulations. In addition, the simulations produce results identical to the experimental data for the rest of the crack propagation process. Several other verification of the new model can be found at Marzougui, 1998.

Full implementation of this method to work with all existing contact models is one of remaining challenges. Furthermore, much work is needed in the area of crack initiation. The future FE methods for crash simulation need to predict crack initiation with modest mesh sizes.

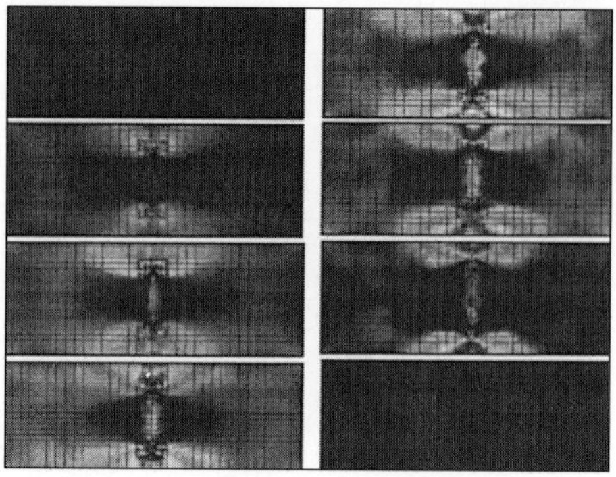

Figure 4: Fracture Prediction of a Center Crack Plate – New Failure Model

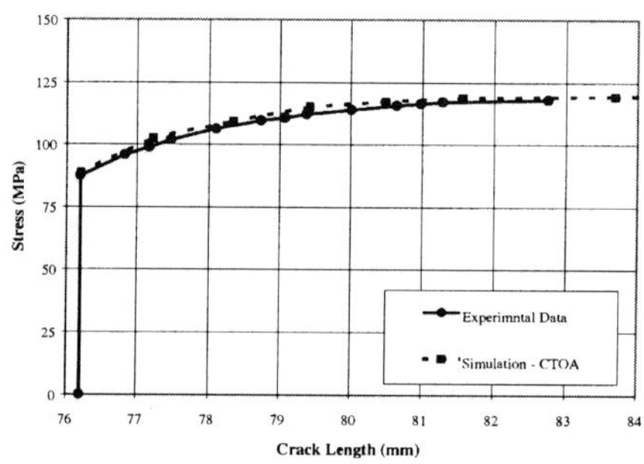

Figure 5: Experimental to FE Simulation Comparisons of Stress vs. Crack Length.

Optimization in Crashworthiness Modeling

In spite of the recent developments in the computer simulation technology, the high cost of the structural analyses to calculate crash responses still present a significant challenge to crashworthiness design. This leads to the requirement for optimization techniques for structural design under impact loading. Since optimizing algorithms are computationally intensive due to their iterative nature, additional research is needed to devise methods that converge on an acceptable design at less iterations.

Traditional crashworthiness relies on engineer's experience and intuition. It is a trial and error search for the best design. Design variables are changed one at a time followed by the analysis of the structure to check if the improvement in the performance of the structure is possible. The process continues until the design goals are achieved. If there are many design variables or if design objectives are in conflict, it will be difficult to decide on how to change these variables for further improvements. The decision process in those cases will exceed human capability and can be cumbersome. Since trial and error searches rely on an engineer's experience and intuition, the design process can not be automated. It must be done *manually*. As a result, the trial and error approach may result in an unnecessarily large number of FEM runs to reach an acceptable design. Therefore, a more systematic approach is necessary to automate the design process.

Mathematical programming or optimization algorithms can be of great help to reduce user interactivity. In other words, the design process can be easily automated by formulating it as an optimization problem.

The standard formulation of an optimization problem is given as to find the set of n design variable values $\mathbf{x} \in R^n$ that will minimize the objective function:

$$y_0(\mathbf{x}) \qquad (1)$$

subjected to the inequality constraints:

$$y_j(\mathbf{x}) \leq 0 \qquad (j = 1,\ldots,n_c) \qquad (2)$$

within the design space:

$$x_{li} \leq x_i \leq x_{ui} \qquad (i = 1,\ldots,n) \qquad (3)$$

where $\mathbf{x} = \{x_i\}\,(i = 1,\ldots,n)$ is the vector of design variables, the side-constraints x_{li} and x_{ui} define the lower and upper bounds of the i-th design variable. Side constraints describe the design space, i.e. the region in which optimum is searched. By formulating the design problem in the form of equations (1)-(3), the design process is systematized based on logic. Optimization, as shown above, will somewhat decrease the number of full FEM analysis required for design. Its advantages will be more pronounced when designing using a large number of variables.

Although *optimization* described above is an effective design tool, it does not sufficiently eliminate the high computational cost of the FEM. The construction and evaluation of the objective function $y_0(x)$ and constraints $y_j(x)$ in equations (1)-(2) and solution of the optimization problem by a suitable algorithm must still use a large number of full FEM analyses.

In order to alleviate the high computational cost of optimization several approximate analyses techniques replacing full FEM analysis have been developed. Approximate analyses techniques may include the replacement of the detailed FEM model of the structure with a crude FEM model, a lumped-mass model or an approximate model often in the form of a polynomial function. Since first and second choices require detailed information about the model at hand, third option, approximate functional models, is of most interest today. Functional models that are used to replace the full FEM analysis are often described by the term "approximation concepts" or "approximation methods". These are used very commonly to replace the objective function and constraints and thereby full FEM analysis with linear or quadratic polynomial functions.

First, limited number of full FEM analyses are carried out. Then linear, quadratic or other approximation models are fitted to the results of FEM analyses. These approximation models then constitute the response of the structure and replace the original objective function and constraints and thereby full FEM analyses. To create linear and quadratic approximation models for the objective or constraint, (n+1) and (n+1)(n+2)/2 numbers of full FEM analyses are needed, respectively where n is the number of design variables. Higher order approximation models require larger number of full FEM analyses in their construction, hence, loosing their advantages.

Usually crash responses that correspond to objective function and constraints are highly nonlinear. The use of linear or quadratic model may not give good approximations to those responses. Therefore, successive linear or quadratic approximations are used to approximate highly nonlinear crash responses. This approach is called as *successive approximate optimization* or *sequential approximate optimization*. In this approach, the original optimization problem given by equations (1)-(3) is replaced by successive approximate optimization sub-problems:

Minimize: $\tilde{y}_0^{(k)}(\mathbf{x})$ (4)

Subject to constraints: $\tilde{y}_j^{(k)}(\mathbf{x}) \leq 0 \qquad (j = 1,\ldots,n_c)$ (5)

Within the updated design space:

$$x_{li}^{(k)} \leq x_i \leq x_{ui}^{(k)} \qquad (j = 1,\ldots,n) \tag{6}$$

where

$$x_{li}^{(k)} \geq x_{li},\; x_{ui}^{(k)} \leq x_{ui} \tag{7}$$

where the superscript k refers to current iteration number of the successive approximate optimization. The current move limits $x_{li}^{(k)}$ and $x_{ui}^{(k)}$ define a sub-region of the original search region (i.e. design space) where the explicit functions $\tilde{y}_j^{(k)}(\mathbf{x})$ $(j = 0,\ldots,n_c)$ can be considered as adequate approximations of the initial implicit functions $y_j(\mathbf{x})$ $(j = 0,\ldots,n_c)$.

The above technique is currently being investigated and verified at GW-TRI for impact and crash problems. These methods are essential for cost effective structural designs.

Alternative optimization schemes involve varying internal parameters of the FE model during an execution based on the present state of the structure and the FE run. This involves internal iteration over variables of interest for design (along with optimizing method) and will require extensive changes in the FE codes.

High Performance Computing

As the size and complexity of finite element models for crash simulation has increased exponentially over the past five years, the need for faster computing is inevitable. High-performance computer platforms that are low-cost and easy to use are required in order to have reasonable turn-around time to solve crash simulation models. While traditional vector supercomputer architectures have continued to improve steadily in performance, the growth in the performance of microprocessors has proceeded at a far more rapid rate. The price-performance ratio of vector supercomputers lags far behind than that of today's microprocessor machines. However, individual microprocessors do not have the processing power to solve today's largest numerical simulation problems. Massively Parallel Processing (MPP) architecture computers connect a large number of small, relatively inexpensive processors together, and use the entire bank of processors together to solve a problem. This approach results in machines with aggregate CPU, I/O and memory bandwidth performance often matching or exceeding the performance of a traditional vector supercomputer, but at a dramatically reduced cost.

With respect to crash codes, there are a number of factors that have been impediments to the successful deployment of MPP systems in production environments. The conversion of existing vector codes to a form, which runs efficiently on an MPP system has proven an enormous task, one of similar complexity to re-writing the basic algorithms used in the codes. The MPP crash codes need to be at least as reliable as their serial or vector version counterparts.

While the DYNA based crash codes such as LS-DYNA's Symmetric Multiprocessor software are running reliably on shared-memory (SMP) parallel computers, the MPP versions of the code (running on distributed memory massively parallel platforms) still need further consideration. NCAC has been conducting a study to benchmark, verify, evaluate and validate different versions of this code on different computer hardware. The combination of the operating system, the new MPP code, and different computer platform creates a matrix of variability for performance evaluation.

Serial versions of the LS-DYNA have been in use at the NCAC on a variety of computer platforms including IBM SP1/2, Silicon Graphics Power Challenge, Silicon Graphics Origin 2000, HP Convex SP 1600, and HP V-class 2500. SMP Parallel version of LS-DYNA uses same algorithms as the serial code, and therefore offers identical and repeatable results. Because the order (sequence) of some parallel operations are inherently non-deterministic when implemented in parallel mode, an option is provided in the SMP versions of LS-DYNA where these operations are performed in a deterministic fashion. This option (the default) results in a small performance penalty, but ensures identical results every time the code is run.

The MPP Parallel version of LS-DYNA uses a domain-decomposition approach based on message passing to break a crash problem into smaller parts, and then perform the calculations on a distributed set of processors. Both the dynamics and the automatic contact detection are performed in parallel. This decomposition could result in inconsistent performance, i.e, different results could be obtained for the same problem when using different number of processors. Figure 5 shows speedup of a typical crash-impact simulation using MPP and SMP versions of LS-DYNA code.

As the number of CPU increases, the MPP version of the crash codes has much better performance. Additionally, in comparison with SMP version, the MPP codes have much better scalability with larger finite element models as illustrated in Figure 6.

Any contribution towards improvements on MPP FE crash codes needs to be closely conducted with the software vendors and requires detailed access to source code due to proprietary nature of the business. However, research can be conduced on performance evaluation of the developed software by the crashworthiness community. The three issues of the MPP code, namely, the consistency, reliability and repeatability are among the most critical challenges facing the users of parallel computing for crash simulation.

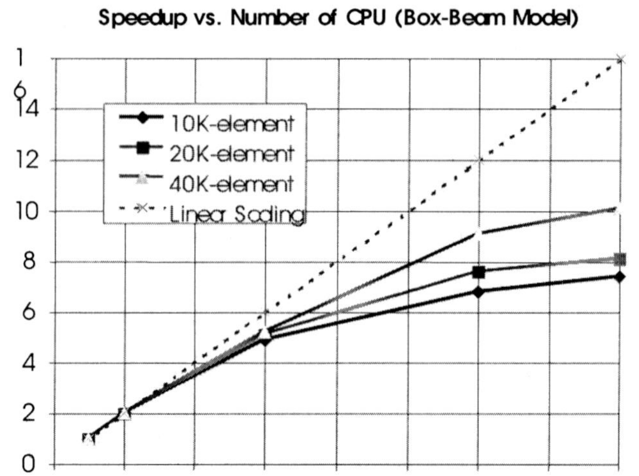

Figure 5- Speed up of MPP codes vs. SMP Codes (Taurus Model)

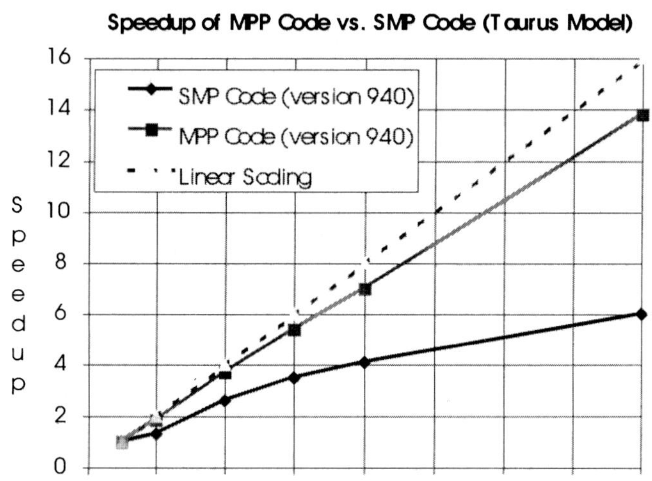

Figure 6- Speedup of vs. Number of CPU's (Box Beam Model)

CONCLUSIONS

The finite element simulation of vehicle crashes has so far proved to be a cost effective tool for crashworthiness design but with much room for improvements. Advances in dynamic nonlinear finite element methods has allowed certain level of accuracy and predictability in vehicle, occupant, and barrier impact simulations. However, many challenges and issues remain to be addressed in order to extend these methods to even higher levels of accuracy and dependency. It is only after such accomplishments that we can have a totally confident reliance on the simulation models for production and certification of vehicle safety.

Challenges are both in fundamentals of finite element and mechanics as well as modeling techniques. A few research areas with success potentials are described in this paper. In summary, for modeling, new materials characterization and development of constitutive relationships, joining techniques (spotwelds, bounding, etc.), and mesh compatibility optimization are among major areas of required development. In methodology, the issues concerning airbag and occupant interaction modeling highlight a present shortcoming. This introduces a fundamental requirement for coding developments, namely, the merging of Eulering and Lagrangian methods and their hybrid implementation to model this complex interaction phenomena. The need for the ability to model vehicle components and roadside barrier failure determines another important area of research. Sophisticated material fracture models and crack initiation methods need to be developed and incorporated in FE simulations.

For crashworthiness design and analysis efficiency, new optimization methods are an absolute necessity. Present manual analysis iterations are not cost effective and are bound to failure in complex problems. New optimization schemes need to be explored for the highly non-linear crash response problem. Finally, research efforts are needed in the computational aspects of crash simulations. A promising focus area is further development of DYNA family of codes on MPP platforms. Research in this field concerns three aress: The changes in the crash code itself, which is primarily accomplished by the software code vendors; Advancement in the computer hardware technology; And evaluation and validation of the developed codes which are the concern of the research and user community. From a user perspective, the issue with MPP code is the accuracy, consistency, and repeatability. The MPP version of the code needs to be evaluated to ensure it obtains identical results to serial or vector code.

Although certain level of success has been reported in this paper for each of the mentioned areas, much research remains to be conducted to fully address each problem. The present progress warrants much more success in this field in both the near and far future.

ACKNOWLEDGEMENT

This paper is the result of dedicated research of many staff scientists, faculty, and students at The FHWA/NHTSA National Crash Analysis Center and Center for Intelligent Systems Research at The George Washington Transportation Research Institute. The research presented here is partially sponsored by Federal Highway Administration of US Department of Transportation.

REFERENCES

Bedewi, N.E., Omar, T.A., and Eskandarian, A., (1994), "Effect of Mesh Density Variation in Vehicle Crashworthiness Finite Element Modeling", Proceedings of the ASME Winter Annual Meeting, Transportation Safety Session, Chicago, Ill, November 6-11.

Bedewi, N.E., Kan, C.D., Summers, S., Ragland, C. (1995), "Evaluation of Car-to-Car Frontal Offset Impact Finite Element Models Using Full Scale Crash Data," Issues in Automotive Safety Technology, SAE Pblication SP-1072, pp. 212-219, February.

Bedewi, N.E., Marzougui, D., and Motevalli, V. (1996), "Evaluation of Parameters Effecting Simulation of Airbag Deployment and Interaction with Occupants", International Journal of Crashworthiness, Vol. 1, No. 4.

Belytschko, Ted (1988), "On Computational Methods for Crashworthiness," *Proceedings of the 7th International Conference on Vehicle Structural Mechanics,* SAE, Detroit, pp.93-102.

DeKoning, A.U.(1975), "A Contribution to the Analysis of Slow Crack Growth", Rep. NLR MP75035, *National Aerospace Laboratory (NLR)*, The Netherlands.

Digges, K.H., Noureddine, A., Eskandarian, A., and Bedewi, N.E., (1998), "Effect of Occupant Position and Airbag Inflation Parameters on Driver Injury Measures", Proceedings of *Society of Automotive Engineers (SAE) International Conference and Exposition*, Detroit, MI, Feb.

Eskandarian, A., (1998) "Safety and Simulation", an invited article, *Testing Technology International*, UIP UK & International Press, pp. 94-98.

Eskandarian, A., Marzougui, D. and Bedewi, N.E., (1997), "Finite Element Model and Validation of a Surrogate Crash Test Vehicle for Impacts With Roadside Objects", *International Journal of Crashworthiness Research,* Vol. 2, No. 3, pp. 239-257.

Eskandarian, A., Marzougui, D., and Bedewi, N.E., (1999), "Impact Finite Element Analysis of Breakaway Sign Support Mechanism", *ASCE Journal of Transportation Engineering,*, Vol. 126 No.2, pp. 143-153.

Eskandarian, A., Marzougui, D. and Bedewi, N.E., (1996), "Failure Analysis of Highway Small Sign Support Systems in Crashes Using Impact Finite Element Models", Proceedings of *the 29th International Symposium On Automotive Technology and Automation (ISATA), Road and Vehicle Safety*, June 3-6, Florence Italy, pp. 395-402.

Hallquist, J.O. (1991), *LS-DYNA3D Theoretical Manual,* Livermore Software Technology Corporation, LSTC Report 1018.

Hallquist, J.O., Stillman, D.W., Lin, T.L. (1992), *LS-DYNA3D Users Manual,* Livermore Software Technology Corporation, LSTC Report 1007, Rev. 2.

Hogg, M. (1998), "Spotweld Behavior and Applications to Finite Element Analysis," Masters Thesis, The George Washington Unversity.

Marzougui, D., Kan, C.D., and Eskandarian, A., (1999), "Finite Element Simulation and Analysis of Portable Concrete Barriers Using LS-DYNA", Proceedings of *LS-DYNA Users Conference, Gothenburg Sweden,* June 14-15, pp. I.19-I.25.

Marzougui, D (1998), "Implementation of a Fracture Failure Model to a Three-Dimensional Dynamic Finite Element code (DYNA3D)", Doctoral Dissertation, Department of Civil,

Mechanical, and Environmental Engineering, The GeorgeWashington University, Washington, DC.

Marzougui, D., Eskandarian, A., and Bedewi, N.E., , (1999), “Analysis and Evaluation of a Redesigned 3”x3” Slipbase Sign Support System Using Finite Element Simulations”, *International Journal of Crashworthiness Research,* Vol. 4, No. 1, pp. 7-16.

Miller, L., Bedewi, N., Chu, R. (1995)., “Performance Benchmarking of LS-DYNA3D for Vehicle Impact Simulation on the Silicon Graphics POWER CHALLENGE” *Presented at the High Performance Computing Asia 95,* Taiwan, October.

Noureddine, A., Digges, K., Eskandarian, A., and Bedewi, N.E., (1998), “Analysis of Airbag Depowering and Related Parameters in Out of Position Environment”, *International Journal of Crashworthiness Research,* Vol. 3. No. 4, pp. 237-248.

Omar, T., Eskandarian, A, and Bedewi, N.E., (1999), “Artificial Neural Networks for Modeling Dynamics of Impacting Bodies and Vehicles”, I. Mech E. *Journal of Multi-Body Dynamics.*

Omar, T., Bedewi, E., Kan, C.D., and Eskandarian, A., (1999.), “Major Parameters Affecting Nonlinear Finite Element Simulations of Vehicle Crashes”, Proceedings of *ASME International Mechanical Engineering Congress and Exposition, Nashville, TN*, Nov. 14-19.

Omar, T., Eskandarian, A., and Bedewi, N.E., (1998), “Vehicle Crash Modeling Using Recurrent Neural Networks”, *Mathematical and Computer Modeling*, Pergamon, Vol. 28, No. 9, pp. 31-42.

Omar, T.A., Kan, C.D., and Bedewi, N.E., (1996), “Crush Behavior of Spot Welded Hat Section Load Bearing Components with Material Comparison”, ASME Winter Annual Meeting, Atlanta, GA. November. ASME Publication: Crashworthiness and Occupant Protection in Transportation Systems, AMD-Vol. 218, pp.65-78.

Omar, T.A., Kan, C.D., and Bedewi, N.E.(1996),“Non-linear Finite Element Analysis of Box Beam Crush Buckling: Experimental Validation and Material Comparison,” 29th International Symposium on Automotive Technology and Automation, Florence, Italy.

Phen, R.L., Dowdy, M. W., Ebbeler, D. H., Kim, E-H,. Moore, N. R., VanZandt, T. R. (1998), “Advanced Air Bag Technology Assessment. Final Report”. Jet Propulsion Laboratory, California Institute of Technology. Pasadena, California, April.

Schinke, H., Zaouk, A., KAN, C.D. (1995), “Vehicle Finite Element Development of a Chevy C1500 Truck with Varying Application,” *FHWA/NHTSA National Crash Analysis Center Internal Report.*

Simha, K.R.Y., Fourney, W.L., Baker, D.B., and Dick, R.D.(1986), "Dynamic Photoelastic Investigation of Two Pressurized Cracks Approaching One Another", *Engineering Fracture Mechanics*, Vol. 23, pp. 237-49.

Zaouk, A., Bedewi, N.E., Kan, C.D., Marzougui, D. (1996), “Validation of a Non-linear Finite Element Vehicle Model Using Multiple Impact Data,” ASME Winter Annual Congress and Exposition, Atlanta, GA. November 1996, ASME Publication: Crashworthiness and Occupant Protection in Transportation Systems, AMD-Vol. 218, pp.91-106.

Free vibrations of flexible thin rotating discs

H R HAMIDZADEH
Mechanical Engineering Department, South Dakota State University, USA

ABSTRACT

Analytical methods are adopted to study the transverse and in-plane vibrations of rotating discs. The disc is assumed to be isotropic and rotating under steady state conditions. For the geometrically non-linear transverse vibration, the effects of lateral displacement amplitude and rotating speed on natural frequencies are determined. For the in-plane vibration, a linear model is considered and the variation of natural frequencies versus rotating speeds for different modes are computed. The mathematical model includes the effects of radial, tangential, centripetal, and coriolis accelerations. Validity of these procedures is verified by comparing some of the computed results with those previously established for certain cases.

1. INTRODUCTION

The earliest attention to the vibration of a spinning disc was provoked for the investigation of failure of simple turbine blades. In most of these analyses the periodic motions were investigated only in its linear form. In this connection, works of Lamb and Southwell (1), which became classical, must be mentioned. While the linear case has attracted much attention, nonlinear dynamical analyses have scarcely been discussed. Because of the recent wide application of high-speed thin rotating discs, one should realize the possibility of large vibration amplitudes that they may be subjected to. Therefore it is vital to investigate nonlinear vibration of these systems. Extensive literature review on linear vibration of rotating discs can be found in a paper by Parker (2). He analyzed a spinning disc spindle system and presented governing equations.

The first non-linear analysis of transverse vibration in asymmetric spinning disc system is due to Nowinski (3). He studied nonlinear transverse vibration of a spinning isotropic disc using von Karman field equations. Large amplitude vibrations of the spinning disc were analyzed. The study was limited to demonstrating dependency of natural frequency of vibration on amplitudes only for a case of two nodal diameters. However, the aforementioned von Karman theory for rotating discs does not include the coriolis inertia caused by rotation, and the membrane force contribution on the force balance is considered only in the transverse direction. Advani and Bulkely (4) analyzed nonlinear transverse vibrations in spinning membrane discs. Two exact solutions were obtained to nonlinear equations governing transverse motion of spinning circular membrane discs. Malhotra et al (5) considered nonlinear finite amplitude vibration of a flexible spinning disc. The transverse vibration of a high speed spinning disc which is clamped at the inner radius and rotating with time-varying spin rate was examined in a fixed space frame of reference. Raman and Mote (6) studied nonlinear oscillations of circular plates near critical speed subjected to space-fixed transverse force. The analysis included averaged Hamiltonian for damped as well as undamped disc rotation. They also investigated forward and backward travelling waves. Boulabal and Crandall (7) experimentally demonstrated the presence of stationary waves in rotating discs. Based on Nowinski's analytical method Hamidzadeh, et al. (8) presented numerical results for natural frequencies, mode shape, and modal stresses for thin spinning discs.

On the in-plane vibration of discs, Bhuta and Jones (9) considered the axisymmetric planar vibration of a solid disc and found that the effect of rotation was to lower the natural frequencies. Doby (10) investigated the elastic stability of a Coriolis-coupled oscillation for a rotating disc. Burdess et al. (11) investigated the general in-plane response of a solid rotating disc (clamping ratio of zero). Properties of the forward and backward travelling circumferential waves were discussed. Chen and Jhu (12, 13) studied the in-plane vibration of a spinning annular disc and investigated the effects of clamping ratio on the natural frequencies and stability of the disc. Hamidzadeh and Dehghani (14) and Hamidzadeh and Wang (15) have also presented analytical solution to this problem and provided results for variation of dimensionless natural frequencies versus rotational speeds for several modes.

This paper adopts Nowinski's (4) approach and provides natural frequencies for transverse vibration of spinning disc with large amplitudes. The method outlined here assumes a typical disc with small and uniform thickness, elastic in nature, rotating with constant angular velocity, and having negligible in-plane vibration. It is also assumed that the vibration is controlled both by the flexural stiffness of the disc and by the tensions induced due to centrifugal forces. For the in-plane vibration linear equations of motion for a rotating disc is derived based on the two-dimensional theory of elasto-dynamics. The mathematical model is reduced to a wave propagation problem and time dependent and time independent modes are considered. Modal displacements and stresses are formulated and computational analysis is then made to obtain the natural frequencies of the system.

2. NON-LINEAR TRANSVERSE VIBRATION

The transverse vibration of flat and thin elastic spinning disc is considered. The disc rotates about its axis of symmetry with a constant angular velocity ω. In the analysis henceforth, the transverse deflections of the disc are assumed to be large in comparison with thickness. A typical disc and its transverse deflection are shown in Figure 1. Although a geometric non-linearity creeps into the system due to large deflections, it has been documented that Hook's law remains valid in its isotropic form. Since displacement amplitudes of in-plane vibration are very small in comparision with that of transverse vibration, the inertia terms in foregoing equations are ignored. Considering that the non-linearity in this problem is associated with large transverse displacement, the following non-linear strain-displacement relations are assumed.

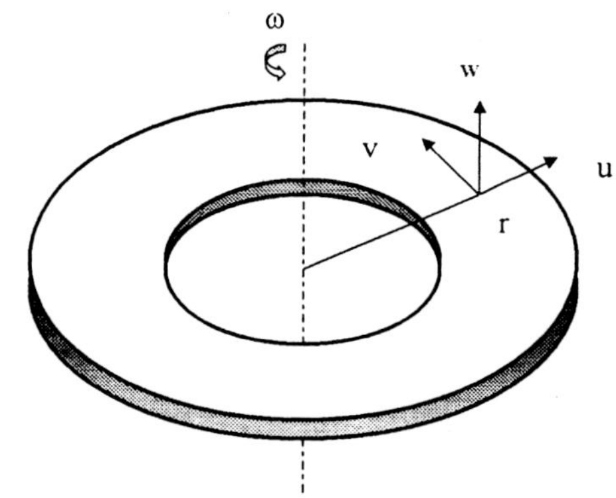

Figure 1. Rotating annular disc

$$\varepsilon_r = \frac{\partial u}{\partial r} + \frac{1}{2}\left(\frac{\partial w}{\partial r}\right)^2 \tag{1.a}$$

$$\varepsilon_\theta = \frac{u}{r} + \frac{1}{r}\frac{\partial v}{\partial \theta} + \frac{1}{2r^2}\left(\frac{\partial w}{\partial \theta}\right) \tag{1.b}$$

$$\gamma_{r\theta} = \frac{1}{r}\frac{\partial u}{\partial \theta} + \frac{\partial v}{\partial r} - \frac{v}{r} + \frac{1}{r}\frac{\partial w}{\partial r}\frac{\partial w}{\partial \theta} \tag{1.c}$$

Where ε_r, ε_θ and $\gamma_{r\theta}$ are radial, hoop and shear strains respectively. Also u, v and w are the displacement in cylindrical coordinates. The above strains consist of linear and non-linear components. The non-linear governing equation for the transverse vibration of a rotating disc can be obtained by transforming the von Karman equation and supplementing it with the body and inertia forces. Assuming free vibration, the governing equation in the polar coordinate system will become:

$$[\frac{D}{h}\nabla^4 w + \rho\frac{\partial^2 w}{\partial t^2}] = +\frac{\partial^2 w}{\partial r^2}\left(\frac{1}{r}\frac{\partial \phi}{\partial r} + \frac{1}{r^2}\frac{\partial^2 \phi}{\partial \theta^2}\right) + \left(\frac{1}{r}\frac{\partial w}{\partial r} + \frac{1}{r^2}\frac{\partial^2 w}{\partial \theta^2}\right)\frac{\partial^2 \phi}{\partial r^2} - 2\frac{\partial}{\partial r}\left(\frac{1}{r}\frac{\partial w}{\partial \theta}\right)\frac{\partial}{\partial r}\left(\frac{1}{r}\frac{\partial \phi}{\partial \theta}\right) - \frac{1}{2}\rho\omega^2 r^2\nabla^2 w - \rho\omega^2 r\frac{\partial w}{\partial r} \tag{2}$$

Where $D = \frac{Eh^3}{12(1-v^2)}$ is the bending rigidity of the disc, ∇^4 is biharmonic operator, and ∇^2 is the two dimensional Laplacian operator in polar coordinates. The required compatibility equation in terms of stress function and the transverse displacement is expressed as:

$$\frac{1}{E}\left[\nabla^4\phi - 2(1-\nu)\omega^2\rho\right] = -\frac{1}{r}\left(\frac{\partial w}{\partial r} + \frac{\partial^2 w}{r\partial\theta^2}\right)\frac{\partial^2 w}{\partial r^2} + \frac{1}{r^2}\left(\frac{\partial^2 w}{\partial r\partial\theta}\right)^2 - \frac{2}{r^2}\frac{\partial^2 w}{\partial r\partial\theta}\frac{\partial w}{\partial\theta} + \frac{1}{r^4}\left(\frac{\partial w}{\partial\theta}\right)^2 \tag{3}$$

The above equation in conjunction with equation (2) should be solved to obtain the displacement function 'w' and the stress function 'ϕ'. Thus, the problem reduces to the integration of two nonlinear equations (2) and (3) along with the boundary conditions of the system. An assumed solution for the displacement of the disc has to be incorporated into the analysis. It has been well documented by Prescot (16), that the radial profile of deflection of the disc surface for various modes of vibration assumes a separable form of power series. In the absence of nodal circles for any number of nodal diameters, the assumed displacement function reduces to the following deflection pattern:

$$w(r,\theta,t) = A_0\tau(t)r^n\cos(n\theta+\psi) \tag{4}$$

Where $w(r,\theta,t)$ is the deflection of the disc in polar coordinates, 'ψ' is the phase constant, $\tau(t)$ is a time function describing variation of 'w' with respect to time, and 'n' is the number of nodal diameters. As presented by Nowinski (3), the stress function 'ϕ' can be determined by substituting equation (4) into (3) and solving for 'ϕ'. Upon further simplification the stress function becomes:

$$\phi=\kappa[\frac{a_1}{c_1}]+\frac{1-v}{32}\omega^2\rho r^4+Ar^2+[Cr^{2n}+Dr^{2(n+1)}]\cos2(n\theta+\varphi) \tag{5}$$

To calculate unknown coefficients of A, B and C, two boundary conditions need to be satisfied. These boundary conditions require that the radial and tangential stresses on the outer radius of the disc to be zero. To satisfy the boundary stresses it is required that values of B and C to be zero. According to Nowinski (3), by applying Galerkin's method and substituting equations (4), (5), into equation (2) and integrating over the disc boundary it results in the following second order nonlinear time equation.

$$\frac{d^2\tau}{d\tau^2}+\alpha\tau+\beta\tau^3=o \tag{6}$$

In the absence of nodal circles it was shown by Hamidzadeh et al. (8) that the coefficients of the above nonlinear differential equation are:

$$\alpha=-\frac{\omega^2}{2s}s_4 \tag{7.a}$$

$$\beta=-\frac{EA_o^2a^{2(n-2)}}{\rho s}\frac{1}{2}s_5 \tag{7.b}$$

The solution to the nonlinear equation (6) is a Jacobian Elliptical function represented by:

$$\tau(t)=cn(\omega^*t,k) \tag{8.a}$$

where:

$$\omega^*=\sqrt{(\alpha+\beta)} \tag{8.b}$$

$$k^2=\frac{\beta}{2(\alpha+\beta)} \tag{8.c}$$

It should be noted that $cn(\omega^*t,k)$ is a periodic function which has the period of $4K/\omega^*$, and $K=F(k, \pi/2)$ is the complete elleptical integral of the first kind.

For the ease of analysis and presentation of results, the following dimensionless parameters are introduced:

$$w^*=\frac{w}{h} \qquad \text{(for deflection)} \tag{9.a}$$

$$\Omega_1 = (\omega a / \sqrt{\frac{E}{\rho}}) \text{ (for speed)} \tag{9.b}$$

$$\Omega_2 = 2\pi / (\frac{4K}{\omega^{*} a}\sqrt{\frac{E}{\rho}}) \text{ (for natural frequency)} \tag{9.c}$$

Moreover, the dimensionless stresses are given by:

$$\sigma_r^{*} = \frac{\sigma_r}{E}\frac{a^2}{h^2} \quad \text{(for radial stress} \tag{9.d}$$

$$\sigma_\theta^{*} = \frac{\sigma_\theta}{E}\frac{a^2}{h^2} \quad \text{(for hoop stress)} \tag{9.e}$$

The dimensionless amplitude of radial stresse due to nonlinear vibrationis.

$$\sigma^{*}_{r-\text{nonlin}} = \left[n\frac{a_1}{c_1}\left\{\left(\frac{r}{a}\right)^{2(n-1)} - 1\right\}\left(\frac{w}{h}\right)^2 \right] \tag{9.f}$$

and the dimensionless amplitude of hoop stresses due to nonlinear vibration can be written as:

$$\sigma^{*}_{\theta-\text{nonlin}} = \left[n\frac{a_1}{c_1}\left\{(2n-1)\left(\frac{r}{a}\right)^{2(n-1)} - 1\right\}\left(\frac{w}{h}\right)^2 \right] \tag{9.g}$$

where (r/a) is the radius ratio.

4. IN-PLANE VIBRATION

The disc material is assumed to be homogeneous, isotropic, and elastic and it is rotating with a constant angular speed. Two-dimensional theory of elasticity is used to define the stress and train in polar coordinates. These relationships are then implemented into the dynamic equilibrium equations to obtain the general equations of motion. The linear equations of motion may be given by:

$$c_1^2\nabla^2\Delta - \ddot{\Delta} + \omega^2\Delta + 2\omega\dot{\psi} = -2\omega^2 \tag{10.a}$$

$$c_2^2\nabla^2\psi - \ddot{\psi} + \omega^2\psi - 2\omega\dot{\Delta} = 0 \tag{10.b}$$

Wherep is the mass density of the medium, and

$$\Delta = \frac{\partial u}{\partial r} + \frac{u}{r} + \frac{1}{r}\frac{\partial v}{\partial \theta} \tag{11.a}$$

$$\psi = \frac{\partial v}{\partial r} + \frac{v}{r} - \frac{1}{r}\frac{\partial u}{\partial \theta} \tag{11.b}$$

$$2\mu = E/(1+\nu) \tag{11.c}$$

$$\gamma = \nu E / (1-\nu^2) \tag{11.d}$$

$$c_1^2 = (\gamma + 2\mu)/\rho \tag{11.e}$$

$$c_2^2 = \mu / \rho \tag{11.f}$$

Since no external forces are acting on the disc, vibration of the disc is solely due to the rotation of the disc. Therefore, the following solutions for equations (10) can be assumed

$$\Delta = \delta_0(r) + \Delta_n(r)e^{i(n\theta - pt)} \tag{12.a}$$

$$\psi = i\Psi_n(r)e^{i(n\theta - pt)} \tag{12.b}$$

Where $\delta_0(r)$ is a time independent function, and $\Delta_n(r)$ and $\Psi_n(r)$ are time dependent functions and can be given in terms of Bessel functions of the first and second kind J_n and Y_n by:

$$\Delta_n(r) = B_n J_n(r^* x_1^*) + C_n Y_n(r^* x_1^*) + D_n J_n(r^* x_2^*) + E_n Y_n(r^* x_2^*) \tag{13.a}$$

$$\Psi_n(r) = t_{1n} B_n J_n(r^* x_1^*) + t_{1n} C_n Y_n(r^* x_1^*) + t_{2n} D_n J_n(r^* x_2^*) + t_{2n} E_n Y_n(r^* x_2^*) \tag{13.b}$$

The time dependent equations play an important role in determining the natural frequencies of the system. Assuming that the radial and tangential displacements are related to time by:

$$u(r,\theta,t) = U_n(r)e^{i(n\theta - pt)} \tag{14.a}$$

$$v(r,\theta,t) = iV_n(r)e^{i(n\theta - pt)} \tag{14.b}$$

Since no external forces are acting on the disc, the stress distribution in the disc is caused solely by the rotation of the disc. The three stresses that occur due to the rotation of the disc are the radial, shear, and hoop stresses. The modal radial and shear stress are:

$$\sigma_r(r,\theta,t) = \sigma_{rn}(r)e^{i(n\theta + pt)} \tag{15.a}$$

$$\tau_{r\theta}(r,\theta,t) = i\tau_{r\theta n}(r)e^{i(n\theta + pt)} \tag{15.b}$$

Before obtaining the solutions for the time dependent equations, it is convenient to introduce the following non-dimensional variables:

Ω_1	$= \omega\, b / c_1$	(16.a)	Ω_2	$= p\, b / c_1$	(16.b)
U^*	$= U_n / b$	(16.c)	V^*	$= V_n / b$	(16.d)
σ_r^*	$= \sigma_r / \mu$	(16.e)	$\tau_{r\theta}^*$	$= \tau_{r\theta} / \mu$	(16.f)
r^*	$= r / b$	(16.g)	ζ	$= a / b$	(16.h)
x	$= x / b$	(16.i)	q	$= c_1 / c_2$	(16.j)
t_1	$= [\Psi_n(r) / \Delta_n(r)]_1$	(16.k)	t_2	$= [\Psi_n(r) / \Delta_n(r)]_2$	(16.l)

Substituting equations (13) and (14) into equations (11) and rearranging, the result yields the modal solution for the non-dimesional radial and tangential displacements. Similarly modal stresses can be obtained in terms of modal displacements and Ψ_n and Δ_n. These modal displacements and stresses are:

$$U_n^* = m_1^*[\,[x_1^* J_n'(r^* x_1^*) + \frac{nt_{1n}}{q^2 r^*} J_n(r^* x_1^*)]B_n + [x_1^* Y_n'(r^* x_1^*) + \frac{nt_{1n}}{q^2 r^*} Y_n(r^* x_1^*)]C_n +$$

$$[x_2^* J_n'(r^* x_2^*) + \frac{nt_{2n}}{q^2 r^*} J_n(r^* x_2^*)]D_n + [x_2^* Y_n'(r^* x_2^*) + \frac{nt_{2n}}{q^2 r^*} Y_n(r^* x_2^*)]E_n\;] +$$

$$m_2^*[\,[\frac{n}{r^*} J_n(r^* x_1^*) + \frac{t_{1n}}{q^2} x_1^* J_n'(r^* x_1^*)]B_n + [\frac{n}{r^*} Y_n(r^* x_1^*) + \frac{t_{1n}}{q^2} x_1^* Y_n'(r^* x_1^*)]C_n +$$

$$[\frac{n}{r^*} J_n(r^* x_2^*) + \frac{t_{2n}}{q^2} x_2^* J_n'(r^* x_2^*)]D_n + [\frac{n}{r^*} Y_n(r^* x_2^*) + \frac{t_{2n}}{q^2} x_1^* Y_n'(r^* x_2^*)]E_n\;] \tag{17.a}$$

$$V_n^* = m_2^*[\,[x_1^* J_n'(r^* x_1^*) + \frac{nt_{1n}}{q^2 r^*} J_n(r^* x_1^*)]B_n + [x_1^* Y_n'(r^* x_1^*) + \frac{nt_{1n}}{q^2 r^*} Y_n(r^* x_1^*)]C_n +$$

$$[x_2^* J_n'(r^* x_2^*) + \frac{nt_{2n}}{q^2 r^*} J_n(r^* x_2^*)]D_n + [x_2^* Y_n'(r^* x_2^*) + \frac{nt_{2n}}{q^2 r^*} Y_n(r^* x_2^*)]E_n\;] +$$

$$m_1^*[\,[\frac{n}{r^*} J_n(r^* x_1^*) + \frac{t_{1n}}{q^2} x_1^* J_n'(r^* x_1^*)]B_n + [\frac{n}{r^*} Y_n(r^* x_1^*) + \frac{t_{1n}}{q^2} x_1^* Y_n'(r^* x_1^*)]C_n +$$

$$[\frac{n}{r^*} J_n(r^* x_2^*) + \frac{t_{2n}}{q^2} x_2^* J_n'(r^* x_2^*)]D_n + [\frac{n}{r^*} Y_n(r^* x_2^*) + \frac{t_{2n}}{q^2} x_1^* Y_n'(r^* x_2^*)]E_n\;] \tag{17.b}$$

where

$$m_1^* = -\frac{\Omega_1^{*2} + \Omega_2^{*2}}{(\Omega_1^{*2} - \Omega_2^{*2})^2} \qquad and \qquad m_2^* = -\frac{2\Omega_1^*\Omega_2^*}{(\Omega_1^{*2} - \Omega_2^{*2})^2} \tag{18}$$

$$\sigma_{rn}^* = \frac{\gamma}{\mu}[B_n J_n(r^* x_1^*) + C_n Y_n(r^* x_1^*) + D_n J_n(r^* x_2^*) + E_n Y_n(r^* x_2^*)] +$$

$$2[s_1^* x_1^{*2} J_n''(r^* x_1^*) + s_2^* \frac{n}{r^*} x_1^* J_n'(r^* x_1^*) - s_2^* \frac{n}{r^{*2}} J_n(r^* x_1^*)]B_n +$$

$$2[s_1^* x_1^{*2} Y_n''(r^* x_1^*) + s_2^* \frac{n}{r^*} x_1^* Y_n'(r^* x_1^*) - s_2^* \frac{n}{r^{*2}} Y_n(r^* x_1^*)]C_n +$$

$$2[s_3^* x_2^{*2} J_n''(r^* x_2^*) + s_4^* \frac{n}{r^*} x_2^* J_n'(r^* x_2^*) - s_4^* \frac{n}{r^{*2}} J_n(r^* x_2^*)]D_n +$$

$$2[s_3^* x_2^{*2} Y_n''(r^* x_2^*) + s_4^* \frac{n}{r^*} x_2^* Y_n'(r^* x_2^*) - s_4^* \frac{n}{r^{*2}} Y_n(r^* x_2^*)]E_n \tag{19.a}$$

$$\tau_{r\theta}^* = 2[s_2^* x_1^{*2} J_n''(r^* x_1^*) + s_1^* \frac{n}{r^*} x_1^* J_n'(r^* x_1^*) - s_1^* \frac{n}{r^{*2}} J_n(r^* x_1^*)]B_n +$$

$$2[s_2^* x_1^{*2} Y_n''(r^* x_1^*) + s_1^* \frac{n}{r^*} x_1^* Y_n'(r^* x_1^*) - s_1^* \frac{n}{r^{*2}} Y_n(r^* x_1^*)]C_n +$$

$$2[s_4^* x_2^{*2} J_n''(r^* x_2^*) + s_3^* \frac{n}{r^*} x_2^* J_n'(r^* x_2^*) - s_3^* \frac{n}{r^{*2}} J_n(r^* x_2^*)]D_n +$$

$$2[s_4^* x_2^{*2} Y_n''(r^* x_2^*) + s_3^* \frac{n}{r^*} x_2^* Y_n'(r^* x_2^*) - s_3^* \frac{n}{r^{*2}} Y_n(r^* x_2^*)]E_n -$$

$$[t_{1n} B_n J_n(r^* x_1^*) + t_{1n} C_n Y_n(r^* x_1^*) + t_{2n} D_n J_n(r^* x_2^*) + t_{2n} E_n Y_n(r^* x_2^*)] \tag{19.b}$$

Where expressions for s_1, s_2, s_3, and s_4 are given by Hamidzadeh and Dehghani (1999). Rearranging equations (17) and (19), the following equation is obtained.

$$\begin{Bmatrix} U_n^*(r) \\ V_n^*(r) \\ \sigma_{rn}^*(r) \\ \tau_{r\theta n}^*(r) \end{Bmatrix} = \begin{bmatrix} a_{11} & a_{12} & a_{13} & a_{14} \\ a_{21} & a_{22} & a_{23} & a_{24} \\ a_{31} & a_{32} & a_{33} & a_{34} \\ a_{41} & a_{42} & a_{43} & a_{44} \end{bmatrix} \begin{Bmatrix} B_n \\ C_n \\ D_n \\ E_n \end{Bmatrix} \tag{20.a}$$

or

$$\left\{U_n^*(r), V_n^*(r), \sigma_{rn}^*(r), \tau_{r\theta n}^*(r)\right\}^T = [A_n(r)]\{B_n, C_n, D_n, E_n\}^T \tag{20.b}$$

Expressions for elements of A_n are given in Hamidzadeh and Dehghani (1999). To determine the modal information the boundary conditions must be satisfied. These boundary conditions are:

radial displacement is zero at r = a: $U_n(a) = 0$
tangential displacement is zero at r = a: $V_n(a) = 0$
radial stress is zero at r = b: $\sigma_{rn}(b) = 0$
shear stress is zero at r = b: $\tau_{r\theta n}(b) = 0$

Satisfying the inner and outer boundary conditions using equation (19.b) and combining them one can relate displacements and stresses at the boundaries.

$$\left\{U_n^*(b), V_n^*(b), 0, 0\right\}^T = [T_n(b)][T_n(a)]^{-1}\left\{0, 0, \sigma_{rn}^*(a), \tau_{r\theta n}^*(a)\right\}^T \tag{21}$$

Assuming that the product of $[T_n(b)][T_n(a)]^{-1}$ is given by:

$$[T_n(b)][T_n(a)]^{-1} = \begin{bmatrix} d_{11} & d_{12} & d_{13} & d_{14} \\ d_{21} & d_{22} & d_{23} & d_{24} \\ d_{31} & d_{32} & d_{33} & d_{34} \\ d_{41} & d_{42} & d_{43} & d_{44} \end{bmatrix} \tag{22}$$

Then the frequency equation can be presented as:

$$\begin{vmatrix} d_{33} & d_{34} \\ d_{43} & d_{44} \end{vmatrix} = 0 \tag{23}$$

In the above equation, the frequency, p, determines the type of wave occurring. If the frequency is negative, $p < 0$, then a forward travelling wave is induced. This wave is in the direction of positive rotation, θ. Backward waves occur in the direction of negative rotation, $-\theta$, and when the frequency is positive, $p > 0$. It should be noted that both of these waves do not generally travel at the same speed.

5. RESULTS AND DISCUSSIONS

The presented results for the nonlinear transverse vibration are for a disc with thickness ratio of h/a = 0.004 and the Poisson ratio of ν= 0.3. The dimensionless natural frequencies for the nonlinear vibration of rotating discs are also computed. Variations of dimensionless frequencies for different numbers of nodal diameters versus a wide range of rotating speeds is provided in Figure 2 for a dimensionless deflection ratio of w/h = 5. Results reveal that the natural frequency of nonlinear vibration is independent of amplitude when n=1, and increasingly dependent on amplitude for n=2 to 6 with amplitude being more effective at lower speeds than higher ones. Figure 3 depicts the effect of relative amplitude ratios of w/h on variation of the dimensionless frequency versus speed ratio for the mode with 4 nodal diameters. The results indicate that at higher speeds, dimensionless frequencies for different w/h approach the corresponding linear dimensionless frequency.

Comparison of the ratios of nonlinear to linear periods of vibration versus relative amplitudes for n=2 at different speeds are made with those of Nowinski (3) in Figure 4. As demonstrated in this Figure the validity of the present procedure is verified by an excellent agreement between these results. Computed results indicate that at lower relative amplitude and higher operating speed, the nonlinear results differ slightly from linear ones. However, at higher amplitude, the nonlinear periods decrease drastically in comparison with the period of linear vibration. This corroborates the fact that the frequency of vibration is getting higher and higher. A similar trend is observed for higher numbers of nodal diameters. Therefore, the linear theory can not provide an accurate estimation for modal frequencies for larger amplitudes of vibration at high speeds.

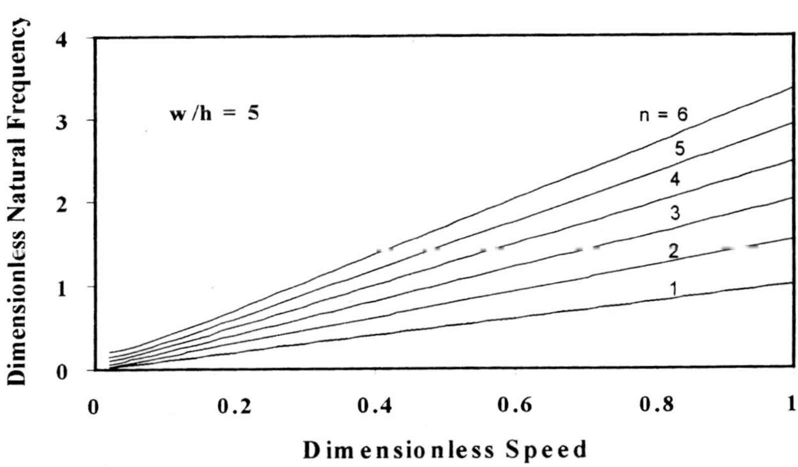

Figure 2: Variation of dimensionless natural frequencies versus Speed for amplitude Ratio of w/h = 5, and different number of nodal diameters.

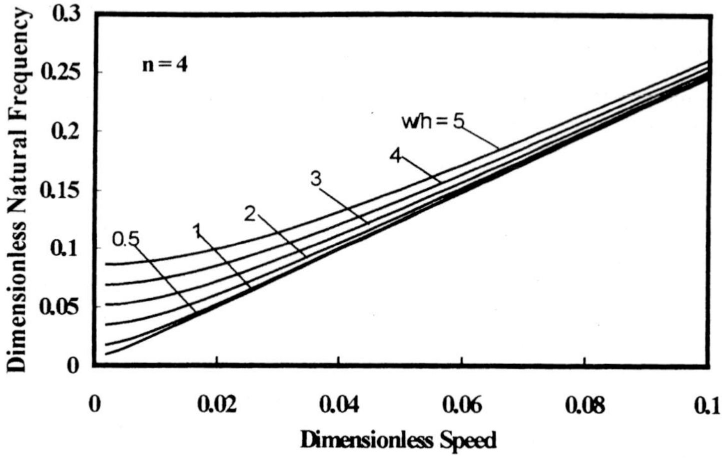

Figure 3: Variation of dimensionless natural frequencies versus Speed For n = 4, and different amplitude ratios.

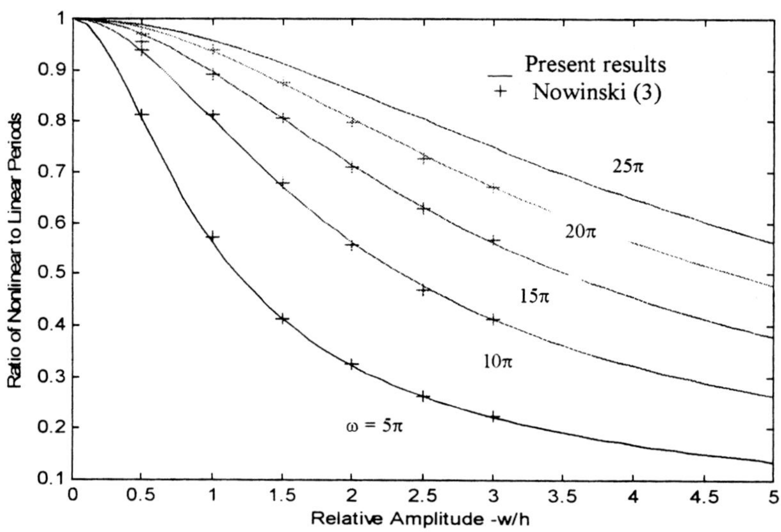

Figure 4: Variation of ratios of nonlinear to linear periods versus amplitude ratio at different speeds, for n=2.

For in-plane vibration, natural frequencies for various modes of discs with different clamping ratios are computed for Poisson's ratio of 0.3. Results of the analysis are compared very well with those provided by Burdess et al. (11) and Chen and Jhu (12). Figure 5 shows the comparison of the present dimensionless natural frequency with those of Burdess et al. (11) for a full disc and the mode with two nodal diameters. As illustrated, excellent comparison between these results is established. Referring to this figure, the forward wave frequency starts at a certain value when the speed is zero, and decreases as the speed increases. When the frequency approaches 0, the direction of the wave is reversed and becomes a backward wave. For the backward wave, the frequency increases until it reaches a maximum value of about 1.8, at a speed ratio of about 0.6. Further increase in speed causes the frequency to decrease. Burdess et al. (11) indicates that the two roots converge and become equal at a speed ratio of about 1.8. Present results also show this same speed ratio. Beyond this speed, the roots of equation (23) become imaginary.

For the clamp ratio of $\zeta = 0.3$ results for several modes indicate that the variation of the dimensionless frequency versus speed ratio are in good comparison with those of Chen and Jhu (12). The label $(m,n)_b$ in Figures 6 - 10 refers to the backward wave with m nodal circles and n nodal diameters. The subscript f refers to the forward wave.

6. CONCLUSION

Analytical procedures are presented to determine natural frequencies, mode shapes, and modal in-plane stresses for nonlinear transverse and in-plane vibrations of spinning discs. Nonlinear analysis indicates that modal parameters are highly dependent on the amplitudes of the transverse vibration. Also, in this case, it is concluded that at faster rotating speeds the significance of amplitude over natural frequency of vibration is greater. For the in-plane analysis, the results are in agreement with available results. It was observed that the effect of rotational speed on the natural frequency depended on the clamping ratio, mode of vibration, and the type of wave occurring.

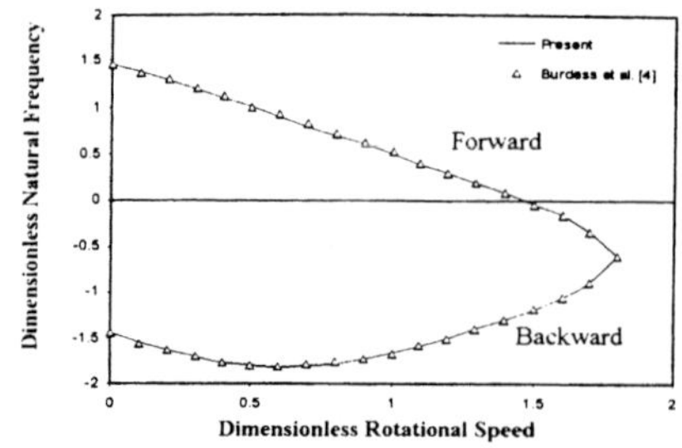

Figure 5. Comparison of the dimensionless natural frequency for the mode (0,2) and $\zeta = 0$.

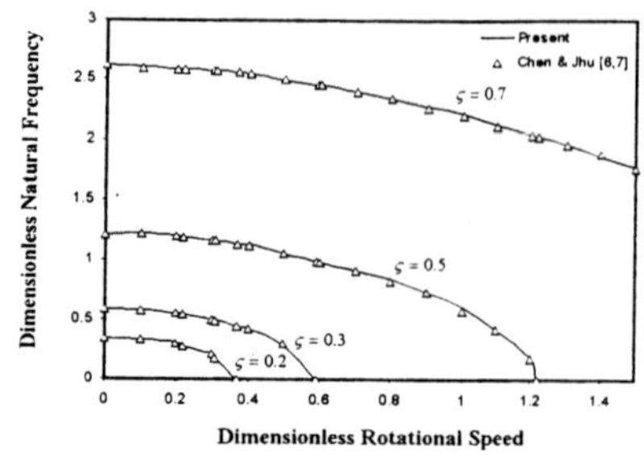

Figure 6. Comparison of the dimensionless natural frequencies for the mode (0,0) and different radius ratios.

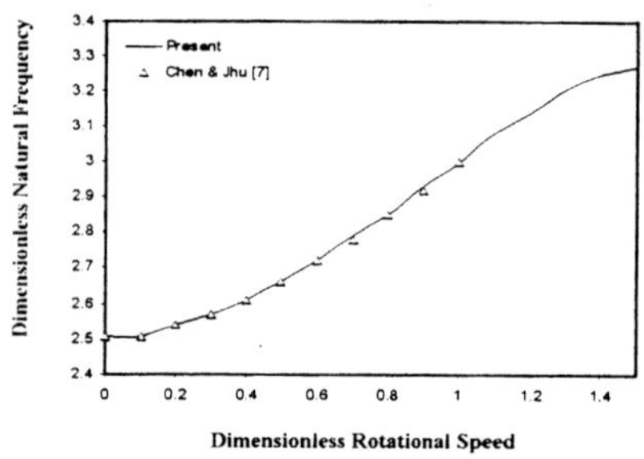

Figure 7. Comparison of the dimensionless natural frequencies for the mode (1,0) and $\zeta = 0.3$.

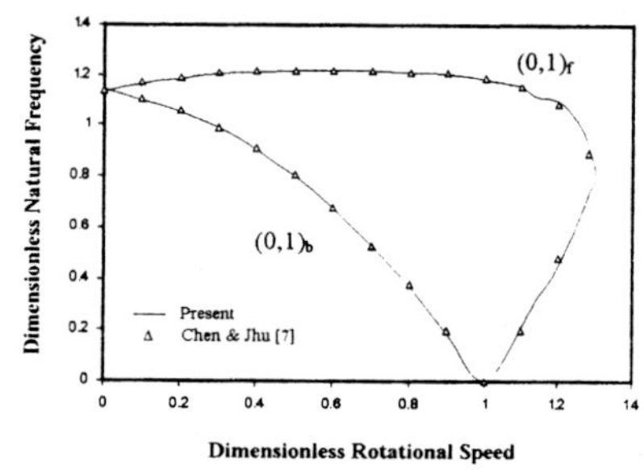

Figure 8. Comparison of the dimensionless natural frequency for the mode (0,1) and $\zeta = 0.3$.

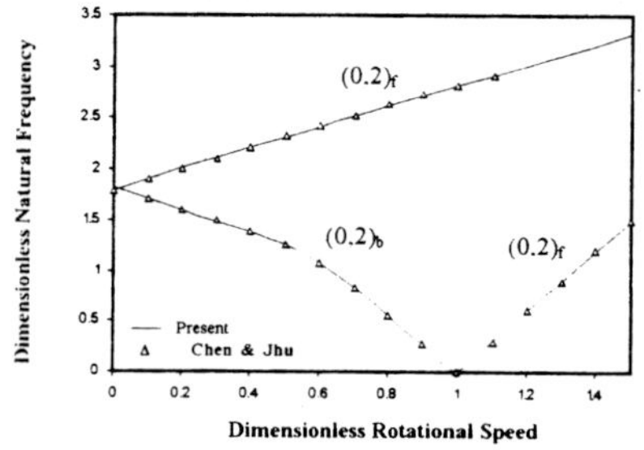

Figure 9. Comparison of the dimensionless natural frequencies for the mode (0.2) and $\zeta = 0.3$.

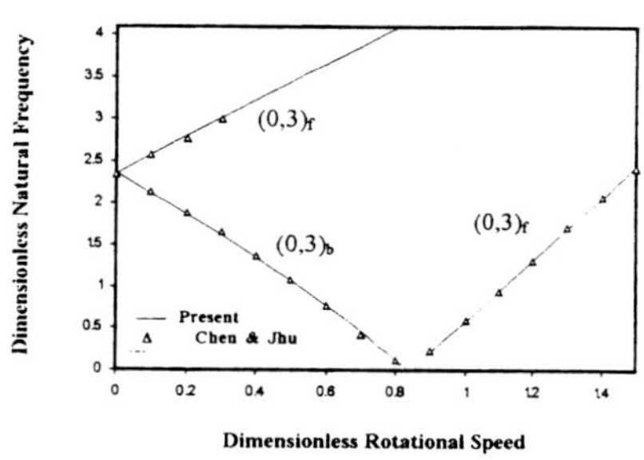

Figure 10. Comparison of the dimensionless natural frequency for the mode (0,3) and $\zeta = 0.3$.

7. REFERECNCES

Lamb, H. and Southwell, R. V., "The Vibrations of a Spinning Disc", Proceeding of the Royal Society, London, 99 (1922), 272-280.

2. Parker, R. G., "Modeling and Analysis of Spinning Disk-Spindle Vibration", ASME Design Engineering Technical Conferences, Proceedings of DETC'97(1997), 1-9.

Nowinski, J. L., "Nonlinear Transverse Vibrations of Spinning Disk", Journal of Applied Mechanics, 31(1964), 72-78.

4. Advani, S. H. and Bulkely, P. Z., "Nonlinear Transverse Vibrations and Waves in Spinning membrane Disks", International Journal of Nonlinear Mechanics, 4 (1969), 123-127.

5. Malhotra, N., Namachchivaya, N. S. and Whalen, T., "Finite Amplitude Dynamics of a Flexible Spinning Disk", ASME Design Engineering Technical Conferences, III, Part A (1995), 239-250.

Raman, A. and Mote Jr., C. D., "Nonlinear Oscillations of Circular Plates Near Critical Speed", Active/ Passive Vibration Control and Nonlinear Dynamics of Structures, DE-Vol. 95/AMD- 223, (1997), 171-183.

Boulabal, D. and Crandall S. H., "Self-Excited Harmonic and Solitary Waves in Spinning Disk," ASME Proceedings of DETC97, (1997), Vib-4093.

8. Hamidzadeh, H. R., Nepal, N. and Dehghani, M., "Transverse vibration of thin rotating disks—Nonlinear modal analysis,' ASME International Mechanical Engineering Congress and Exposition, DE-98, (1998), 219-225.

9. Bhuta, P. G. and Jones, J. P., "Symmetric planar vibrations of a rotating disk," *Journal of the Acoustical Society of America* **35**(7), (1963), 982-989..

10. Doby, R., "On the elastic stability of Coriolis-coupled oscillations of a rotating disk," *Journal of the Franklin Institute* **288**(3), (1969), 203-212.

11. Burdess, J. S. Wren, T. and Fawcett, J. N., "Plane stress vibrations in rotating discs," *Proceedings of Institute of Mechanical_Engineers* **201**, (1987), 37-44.

12. Chen J.S. and Jhu, J. L., "On the in-plane vibration and stability of a spinning annular disk," *Journal of Sound and Vibration* **195**(4), (1996), 585-593.

13. Chen J.S. and Jhu, J. L., "In-plane response of a rotating annular disk under fixed concentrated edge loads," *International Journal of Mechanical Sciences* **38**(12), (1996), 1285-1293

14. Hamidzadeh, H. R. and Dehghani, M., "Linear In-Plane Free Vibration of Rotating Disks," Proceedings of the ASME 17th Biennial Conference on Mechanical Vibration and Noise, DETC99, (1999), Vib-8146.

Hamidzadeh, H. R. and Wang, H., "In-plane free vibration and stability of annular rotating disks," (2000) submitted to the ASME Journal of Vibration and acoustics.

16.Prescot, J., "Applied Elasticity," Dover Publications, New York, (1961).

Study of sub-harmonic vibration of a tube roll using simulation model

J SOPANEN and **A MIKKOLA**
Department of Mechanical Engineering, Lappeenranta University Technology, Finland

ABSTRACT

The current paper highlights the capability of the commercial multi-body systems simulation software (ADAMS) to analyse the non-idealities of a tube roll of a paper machine. The flexibility of the roll is modelled by a utilising modal flexibility method. In this method the tube's modes and corresponding frequencies, obtained from the FE-model, are used for defining the flexibility behaviour of the tube. To ensure the validity of the simulation results, the theoretical results are compared with those obtained by measuring the real structure. The comparison shows that good agreement between the simulated and measured results is obtained.

1 BACKGROUND

The rolls of a rotating machine system include a number of non-idealities, such as an uneven mass distribution and unsymmetrical bearings. This kind of non-idealities are harmful because they cause sub-harmonic vibration that may lead to uncontrolled dynamic behavior. As a consequence excessive wearing or even fatal damage of the machine may take place. At the same time, competition on the market is forcing companies to shorten their product development cycles and reduce their product development costs. This, in turn, is forcing companies to minimise the number of traditional physical prototypes being used in the design phase. The drawbacks of physical prototypes are the costs and time associated with manufacturing unique components, the manual assembly of each prototype, the installation of measurement instruments and finally the measurements that have to be made under realistic working conditions.

Multibody systems simulation has proved to be an effective tool when analysing the dynamics of rotating systems [3]. The increase of computational capacity has improved the possibilities of simulating models accurate enough to describe the sub-harmonic vibration reliably. The paper introduces a way to use the commercial, generally available multibody systems simulation software (MBS) for analysing the sub-harmonic vibration of the tube roll.

The structural flexibility of the tube is modelled utilising the assumed mode method. The modes are defined using a detailed finite element model, which defines the mass distribution obtained by measuring the existing roll. The bearings and the support of the roll are also described in the model. To ensure the model matches the real phenomena, the verification is done using test data measured from an actual system.

2 THE THEORY OF ROLL SYSTEM MODELLING

The motion of a flexible roll consists of reference rotation and elastic deformation. The reference rotation is a rigid body motion whereas the elastic deformation can be seen as a vibration around the rigid body motion. Employing the multibody system software such as ADAMS the dynamic behaviour of a roll system can be solved.

A flexible body must be modelled using a specific approximation method, which reduces the partial differential equation that defines the structural deformation into a set of ordinary differential equations. Assumed modes method is one of the most commonly used approximation methods. This approach can be characterised as a distributed parameter method. When this method is used, the deformation of the flexible body can be obtained using a set of admissible functions. These functions are also known as assumed modes, which describe the deformation of the entire body. The following introduces the modelling premises used in this study as well as the most essential mathematical equations used in ADAMS.

2.1 The kinematics of flexible body

In the mathematical sense the body consists of particles whose locations are described using a local coordinate system. The local coordinate system is attached to the body and the linear deformation of the body is defined in respect to these coordinates. The local coordinate system can undergo large non-linear translation and rotation in respect to the unmoveable global coordinate system. Figure 1 introduces vectors that define the global position of a particle [10].

The global position of an arbitrary particle, P, on the body i can be expressed in the following form:

$$\mathbf{r}^i = \mathbf{R}^i + \mathbf{A}^i \bar{\mathbf{u}}^i = \mathbf{R}^i + \mathbf{A}^i \left(\bar{\mathbf{u}}_o^{\ i} + \bar{\mathbf{u}}_f^{\ i}\right) \tag{1}$$

where $\mathbf{R}^i$ is the position vector of the origin of a local coordinate system, $\mathbf{A}^i$ is a rotation matrix which describes the rotation of the local coordinate system in respect to the global coordinate system, $\bar{\mathbf{u}}^i$ is the position vector of a particle in the local coordinate system, $\bar{\mathbf{u}}_o^{\ i}$ is the position vector which defines the undeformed position of the particle and $\bar{\mathbf{u}}_f^{\ i}$ is the position vector which defines the deformation of the body.

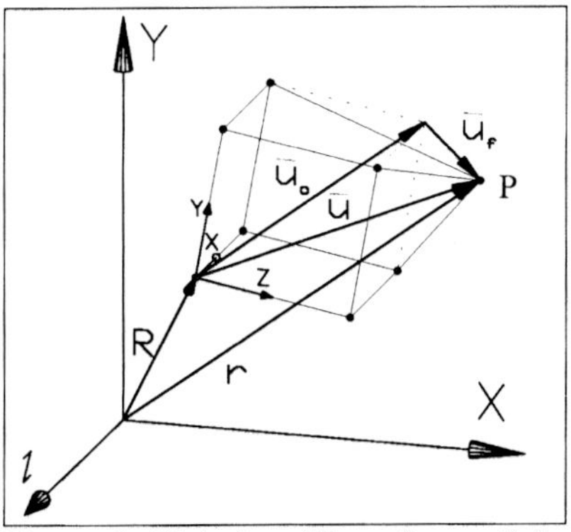

Figure 1. Global position of a particle.

Flexible bodies have an infinite number of degrees of freedom which defines the position of every particle of a body. Thus the components of vector $\bar{\mathbf{u}}_f^{\,i}$ can be expressed in the following form:

$$\bar{u}_{f1} = \sum_{m=1}^{\infty} a_m \cdot f_m\,,\quad \bar{u}_{f2} = \sum_{m=1}^{\infty} b_m \cdot g_m\,,\quad \bar{u}_{f3} = \sum_{m=1}^{\infty} c_m \cdot h_m \tag{2}$$

where a_m, b_m, c_m are the coordinates which are functions of time and f_m, g_m, h_m are the base functions. Because of the computational reason the deformation vector must be defined using a finite number of coordinates. This approximation can be carried out using the Rayleigh-Ritz method. Using matrix formulation this can be expressed using equation (3)

$$\bar{\mathbf{u}}_f^{\,i} \cong \sum_{i=1}^{n} \varphi_i\, p_i = \mathbf{\Phi p} \tag{3}$$

where $\mathbf{\Phi}$ is a (3 x n) shape matrix whose elements are base functions and $\mathbf{p}$ is the vector of modal coordinates. The shape matrix $\mathbf{\Phi}$ can be obtained using finite element method [10].

2.2 Solving the shape matrix using finite element method

By using finite element method the equation of the motion for an undamped system can be written as:

$$\mathbf{M}\ddot{\bar{\mathbf{u}}}_f + \mathbf{K}\bar{\mathbf{u}}_f = \mathbf{Q}_e \tag{4}$$

where $\mathbf{M}$ is the mass matrix, $\bar{\mathbf{u}}_f$ is the vector of nodal displacements, $\mathbf{K}$ is the stiffness matrix and $\mathbf{Q}_e$ is the vector of applied loads. In most engineering applications the structure is rather complex and the finite element model of the entire structure may contain so many degrees of freedom that it would be infeasible to perform a dynamic analysis based on the finite element equations for the complete system. For this reason methods like component mode synthesis (CMS) has been developed. The following is an overview of the Craig-Bampton method [4].

In CMS the structure is divided into interior and boundary degrees-of-freedom. The stiffness and mass matrices of the structure in a partitioned form are:

$$\mathbf{K} = \begin{bmatrix} \mathbf{K}^{BB} & \mathbf{K}^{BI} \\ \mathbf{K}^{IB} & \mathbf{K}^{II} \end{bmatrix}, \qquad \mathbf{M} = \begin{bmatrix} \mathbf{M}^{BB} & \mathbf{0} \\ \mathbf{0} & \mathbf{M}^{II} \end{bmatrix} \tag{5}$$

where superscripts B and I refer to boundary and interior, respectively. The component normal modes may be classified as fixed-interface normal modes, when the interface degrees-of-freedom are fixed and the normal modes are obtained by solving the following eigenvalue problem:

$$(\mathbf{K}^{II} - \omega^2 \mathbf{M}^{II})\mathbf{\Phi}^{N} = \mathbf{0} \tag{6}$$

where $\mathbf{\Phi}^{\mathrm{N}}$ is the matrix of constrained normal modes and ω^2 is an eigenvalue [4], [5].

Constraint modes are the mode shapes of the interior degrees-of-freedom due to the successive unit displacement of boundary degrees-of freedom. Constraint modes are obtained from the static force balance equation (7) by setting all the forces $\mathbf{F}^{I}$ at interior degrees-of-freedom to zero, which yields to equation (8).

$$\begin{Bmatrix} \mathbf{F}^{B} \\ \mathbf{F}^{I} \end{Bmatrix} = \begin{bmatrix} \mathbf{K}^{BB} & \mathbf{K}^{BI} \\ \mathbf{K}^{IB} & \mathbf{K}^{II} \end{bmatrix} \begin{Bmatrix} \boldsymbol{\delta}^{B} \\ \boldsymbol{\delta}^{I} \end{Bmatrix} \tag{7}$$

$$\boldsymbol{\delta}^{I} = -\mathbf{K}^{II^{-1}} \mathbf{K}^{IB} \boldsymbol{\delta}^{B} = \mathbf{\Phi}^{C} \boldsymbol{\delta}^{B} \tag{8}$$

where $\boldsymbol{\delta}^{I}$ and $\boldsymbol{\delta}^{B}$ are the physical displacements of interior and boundary nodes, respectively. The matrix $\mathbf{\Phi}^{C}$ is the matrix of constraint modes.

The coordinate transformation that relates the substructure final i.e. modal coordinates $\hat{\mathbf{p}}$ to the substructure initial i.e. physical coordinates is [4]:

$$\begin{bmatrix} \boldsymbol{\delta}^{B} \\ \boldsymbol{\delta}^{I} \end{bmatrix} \cong \begin{bmatrix} \mathbf{I} & \mathbf{0} \\ \mathbf{\Phi}^{N} & \mathbf{\Phi}^{C} \end{bmatrix} \begin{bmatrix} \hat{\mathbf{p}}^{C} \\ \hat{\mathbf{p}}^{N} \end{bmatrix} = \hat{\mathbf{\Phi}} \hat{\mathbf{p}} \tag{9}$$

The generalised stiffness matrix of the substructure is

$$\hat{\mathbf{K}} = \begin{bmatrix} \hat{\mathbf{K}}^{CC} & \mathbf{0} \\ \mathbf{0} & \hat{\mathbf{K}}^{NN} \end{bmatrix} \tag{10}$$

where $\hat{\mathbf{K}}^{CC} = \mathbf{K}^{BB} + \mathbf{K}^{BI} \mathbf{\Phi}^{C}$ and $\hat{\mathbf{K}}^{NN} = \mathbf{\Phi}^{N^{T}} \mathbf{K}^{II} \mathbf{\Phi}^{N}$. Correspondingly, the generalised mass matrix for the substructure is

$$\hat{\mathbf{M}} = \begin{bmatrix} \hat{\mathbf{M}}^{CC} & \hat{\mathbf{M}}^{NC} \\ \hat{\mathbf{M}}^{NC} & \hat{\mathbf{M}}^{NN} \end{bmatrix} \tag{11}$$

where $\hat{\mathbf{M}}^{CC} = \mathbf{M}^{BB} + \mathbf{\Phi}^{C^T}\mathbf{M}^{II}\,\mathbf{\Phi}^{C}$, $\hat{\mathbf{M}}^{CN} = \mathbf{M}^{BB} + \mathbf{\Phi}^{C^T}\mathbf{M}^{II}\,\mathbf{\Phi}^{N}$ and $\hat{\mathbf{M}}^{NN} = \mathbf{\Phi}^{N^T}\mathbf{M}^{II}\,\mathbf{\Phi}^{N}$

ADAMS slightly modifies the Craig-Bampton method because it adds six non-linear rigid body coordinates for each flexible body. Therefore, the rigid body motion that is normally contained in Craig-Bampton modes must be removed. By orthonormalizing the Craig-Bampton modes, any rigid body modes can be identified and removed. Matrices can be orthonormalized by solving a new eigenvalue problem for the component interface:

$$(\hat{\mathbf{K}} - \hat{\omega}^2\,\hat{\mathbf{M}})\,\hat{\mathbf{p}} = \mathbf{0} \tag{12}$$

The transformation from modal coordinates $\hat{\mathbf{p}}$ to orthogonal modal coordinates $\mathbf{p}$ is defined as $\mathbf{N}\,\mathbf{p} = \hat{\mathbf{p}}$. Thus, the modal superposition formula (3) becomes:

$$\bar{\mathbf{u}}_f \cong \hat{\mathbf{\Phi}}\hat{\mathbf{p}} = \hat{\mathbf{\Phi}}\,\mathbf{N}\,\mathbf{p} = \sum_{j=1}^{n} \boldsymbol{\varphi}_j\, p_j \;= \mathbf{\Phi}\,\mathbf{p} \tag{13}$$

As a result of the orthonormalization procedure the stiffness and mass matrices become diagonal. i.e. [9]:

$$\mathbf{K} = \mathbf{\Phi}^T\hat{\mathbf{K}}\,\mathbf{\Phi} = \begin{bmatrix} \hat{\omega}_1^{\,2} & & 0 \\ & \ddots & \\ 0 & & \hat{\omega}_n^{\,2} \end{bmatrix}, \qquad \mathbf{M} = \mathbf{\Phi}^T\hat{\mathbf{M}}\,\mathbf{\Phi} = \mathbf{I} \tag{14}$$

A useful side-effect of this orthonormalization procedure is that all modes have an associated frequency so that their frequency contribution to the dynamic system can be identified [1].

2.3 The system's equations of motion

2.3.1 Lagrangian equation

The dynamics of a multibody system can be calculated using either the Lagrangian or the Newton-Euler method. The Newton-Euler method can be defined as a "force balance" approach whereas the Lagrangian method is an "energy-based" approach to dynamics. Both methods have advocates, but the mechanism will, however, have the same principle equations of motion independent of the chosen method. A number of computer orientated textbooks on the Newton-Euler [7], [8] and Lagrangian methods [10] provide the basics required for the multibody system analysis.

A number of general-purpose computer applications for multibody dynamics are based on the Lagrangian method . The Lagrangian method relies upon Lagrange's equations which can be derived using the concepts of generalised coordinates, virtual work and generalised force while employing D'Alembert's principle. When the constraint equations are taken into account by Lagrange multipliers the Lagrangian equation can be written in the following form:

$$\frac{d}{dt}\left(\frac{\partial L}{\partial \dot{\mathbf{q}}_i}\right)-\left(\frac{\partial L}{\partial \mathbf{q}_i}\right)+\mathbf{C}_q{}^T\boldsymbol{\lambda}=\mathbf{Q}_e \tag{15}$$

where L is the Lagrangian, $\mathbf{C}_q$ is the Jacobian matrix calculated using generalised coordinates $\mathbf{q}$, $\boldsymbol{\lambda}$ is the vector of Lagrange multipliers and $\mathbf{Q}_e$ is the vector of non-conservative generalised forces.

Substituting kinetic energy as well as potential energy into the Lagrangian equation (15) gives the equations of the constrained motion of a flexible multibody system:

$$\mathbf{M}\ddot{\mathbf{q}}+\dot{\mathbf{M}}\dot{\mathbf{q}}-\frac{1}{2}\left[\frac{\partial \mathbf{M}}{\partial \mathbf{q}}\dot{\mathbf{q}}\right]^T\dot{\mathbf{q}}+\mathbf{K}\mathbf{q}+\mathbf{f}_g+\mathbf{D}\dot{\mathbf{q}}+\mathbf{C}_q{}^T\boldsymbol{\lambda}=\mathbf{Q}_e \tag{16}$$

where $\mathbf{D}$ is the damping matrix and $\mathbf{f}_g$ is gravitational forces. Using the following expressions

$$\mathbf{Q}_v=-\dot{\mathbf{M}}\dot{\mathbf{q}}+\frac{1}{2}\left[\frac{\partial \mathbf{M}}{\partial \mathbf{q}}\dot{\mathbf{q}}\right]^T\dot{\mathbf{q}}\,,\qquad \mathbf{Q}_E=\mathbf{Q}_e-\mathbf{K}\mathbf{q}-\mathbf{f}_g-\mathbf{D}\dot{\mathbf{q}} \tag{17}$$

The equation of constrained motion (16) takes the form:

$$\mathbf{M}\ddot{\mathbf{q}}+\mathbf{C}_q{}^T\boldsymbol{\lambda}=\mathbf{Q}_E+\mathbf{Q}_v \tag{18}$$

where $\mathbf{Q}_v$ is a quadratic velocity vector which describes the gyroscopic and coriolis forces. Quadratic velocity vector is the non-linear function of the system's generalised coordinates and velocities. The force vector $\mathbf{Q}_E$ includes the descriptions of both conservative and non-conservative forces. Dividing generalised coordinates to the components that describe rigid body motion and deformation the equation of motion can be expressed in the following form:

$$\begin{bmatrix}\mathbf{m}_{rr} & \mathbf{m}_{rp}\\ \mathbf{m}_{pr} & \mathbf{m}_{pp}\end{bmatrix}\begin{bmatrix}\ddot{\mathbf{q}}_r\\ \ddot{\mathbf{p}}\end{bmatrix}+\begin{bmatrix}(\mathbf{C}_q)_{q_r}{}^T\\ (\mathbf{C}_q)_p{}^T\end{bmatrix}\cdot\boldsymbol{\lambda}=\begin{bmatrix}(\mathbf{Q}_E)_{q_r}\\ (\mathbf{Q}_E)_p\end{bmatrix}+\begin{bmatrix}(\mathbf{Q}_v)_{q_r}\\ (\mathbf{Q}_v)_p\end{bmatrix} \tag{19}$$

3 STUDIED STRUCTURE AND SIMULATION MODEL

3.1 Studied structure

The studied structure is a paper machine's tube roll which is located at the laboratory of Machine Design at Helsinki University of Technology. The tube roll is supported with hard bearing type balancing machine. The roll has been used in the development of the measuring and analysis methods of rotor dynamics. The measuring equipment comprises a PC based data acquisition system, four laser sensors with amplifiers, connection panel and a guide bar in which the sensors have been installed. The throw of the roll is measured by using laser sensors whose function is based on the movement of the intensity maximum of the diffuse reflection of the laser beam with the detector. From the measured bend line, the throw and its harmonic components and resonance sections are analysed. From the throw signal the sizes of the components, the mutual relations and phase angles and changes are perceived as a function of speed.

3.2 Simulation model

Both FEA and MBS software were used in the construction of the simulation model. The structural flexibility of the roll is imported from a detailed FE-model to the dynamics simulation model. The stiffness properties of the pedestal of the balancing machine are determined by using a FE-model and the mass properties are exported to ADAMS. In ADAMS the separate bodies are connected by using constraints and forces.

3.2.1 FE-model of the roll

The FE-model of the roll is made using ANSYS finite element programme (Figure 2). The shafts and end disks of the roll are modelled using 8-node structural solid elements (SOLID45). The shell is modelled using 4-node shell elements (SHELL63). Attachment nodes are stiffened using stiff beam elements (BEAM4) of which modulus of elasticity is 100-times larger than that of steel. In ADAMS, the forces and constraints are attached to one node so unstiffened attachment location gives corrupted results. The beam elements divide the bearing load on multiple nodes.

In the FE-model the wall thickness of the shell is in accordance with the measured results and because of that the mass and stiffness distribution of the roll is taken into account in the simulation model. The thickness of the shell element SHELL63 can be defined at four corner nodes [2]. The wall thickness of the shell is measured so that the measuring points have the same location than the nodal points of the FE-model. A program which uses both the measurement results and the node and element data of the FE-model as input was written using MATLAB software. The program prints the ANSYS commands for changing the thickness of the elements to a file.

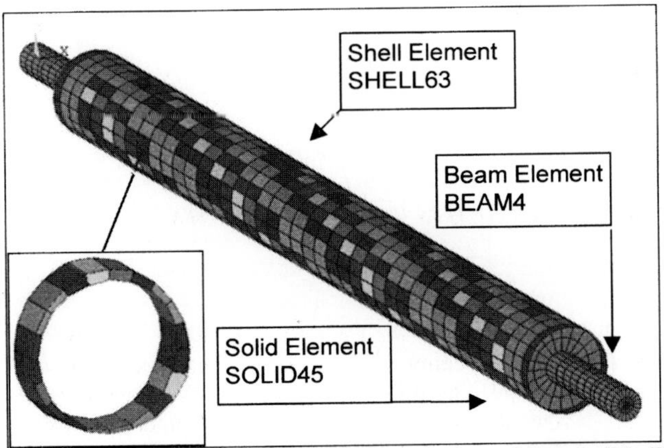

Figure 2. The FE-model of the roll. Thickness variation of the shell is illustrated by different colours. In the section the error is 25-fold.

3.2.2 Dynamics simulation model

The flexible roll and the rigid pedestals of the balancing machine are imported to ADAMS (Figure 3). The same balancing masses as in the real roll are attached to the roll. The pedestal of the balancing machine is attached to ground with spring-damper forces, where horizontal and vertical stiffness and damping constants are: $K_x = 73.80 \cdot 10^6$ N/m, $K_y = 146.42 \cdot 10^6$ N/m, $C_x = 23.1$ Ns/m and $C_y = 32.5$ Ns/m. The stiffness constants are obtained using the FE-model of the pedestal and the damping constants are equal to 6 % of critical damping.

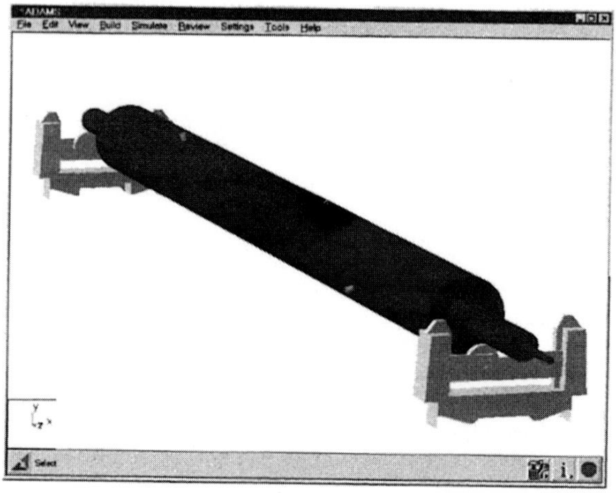

Figure 3. The simulation model in ADAMS.

3.2.2.1 Modelling of shafts and support rollers' roundness error

The measured roundness profile of shaft necks and support rollers is analysed with FFT. As a result the amplitudes and phase angles of the harmonic components are obtained. The roundness profile can be presented as Fourier cosine series by using equation (20).

$$H(\alpha) = \sum_{k=1}^{4} c_k \cos(k \cdot \alpha + \phi_k), \tag{20}$$

where c_k is the amplitude and ϕ_k is the phase angle of the kth harmonics and α is the rotation angle of the roll. In the simulation model attention is paid to the harmonic components of only 1st - 4th order because the amplitudes of higher components are insignificantly small.

The roundness errors of the support rollers are distinctly smaller than those of the shafts necks. Furthermore, the impulses which come from them are not repeated similarly on every rotation of the roll because the diameter of the shaft necks is 125 mm and the diameter of the support rollers is 115 mm. It is not possible to model the throw of the support rollers exactly because there may be differences of the amount of one tenth millimetre in the diameters of the rollers. In that case the rotation angle of one support roller will not change in the same relation with other support rollers and the rotation angle of the roll. Therefore it is decided to model the roundness error of the support rollers without the phase angles. The measuring accuracy of the roundness is ± 1 μm for the shaft and ± 0.2 μm for the support rollers. The error estimate for the phase angles of the harmonic components is ± 20 degrees.

Table 1. Roundness errors of the rolls shaft and support rollers.

SHAFT OF THE ROLL					
Driven Side			Service Side		
k	Amplitude c_k [μm]	Phase ϕ_k [rad]	k	Amplitude c_k [μm]	Phase ϕ_k [rad]
1	31.89	0.6021	1	43.38	5.2360
2	2.75	0.0349	2	4.05	0.0873
3	0.50	1.4312	3	0.35	1.1170
4	0.20	1.3963	4	0.25	0.2443
SUPPORT ROLLERS					
Driven Side			Service Side		
k	Amplitude c_k [μm]		k	Amplitude c_k [μm]	
	Front Roll	Rear Roll		Front Roll	Rear Roll
1	2.50	4.00	1	2.00	2.00
2	0.20	0.25	2	0.15	0.15
3	0.10	-	3	-	-

3.2.2.2 Contact force between shaft and support rollers

In order to model impulses which come from the roundness errors of the shaft and support rollers the force-deflection relationship of the contacting parts must be known. This problem can be solved using Hertzian theory related to the contact between elastic solids [6]. In this case the type of contact is cylinder-cylinder and the contact area is rectangular. The force-deflection relationship can be solved from equations (21) and (22):

$$b = \sqrt{\frac{4 \cdot F}{\pi L_1} \cdot \frac{(1-\nu^2)}{E} \cdot \left(\frac{1}{D_1} + \frac{1}{D_2} \right)^{-1}} \tag{21}$$

where b is the semiwidth of the contact, F is the force, L_1 is the length of the cylinder, ν and E are the Poisson's ratio and the modulus of elasticity of the material and D_1 and D_2 are the diameters of the cylinders. The total deflection between the cylinders is:

$$y = \frac{2(1-\nu^2)F}{\pi L_1 E} \cdot \left(\frac{2}{3} + \ln \frac{4 D_1 D_2}{b^2} \right) \tag{22}$$

The contact force can not be solved directly from the above equations. If one wishes to obtain an exact solution, an iteration scheme must be applied. However, in the case of time-integration where force must be solved at each time step the iteration of forces is not computationally efficient. Therefore, the force-deflection relationship is calculated in MATLAB program using equations (21) and (22). Resulted data points are fitted by least squares method to the equation $F = k_c \cdot y^e$, where k_c is the stiffness constant and e is the exponent of the force-deflection relationship. For steel cylinders, whose dimensions are D_1=125 mm, D_2=115 mm and L_1=25 mm, the fitting results in values $k_c = 899.6 \cdot 10^6$ N/m and $e = 1.07220$.

4 RESULTS

The simulation results are compared with those obtained by measuring the real structure. A vertical throw is measured both from the simulation model and from the real structure from the middle and from a 700 mm distance from the driven side and service side. The sampling frequency is 1 kHz which is about tenfold compared to the highest interesting frequency. The results are compared by analysing the measured results of the throw with FFT in the frequency domain.

In the verification the subharmonic behaviour of the tube roll is examined at rotational frequencies from 7.2 Hz to 20.0 Hz. Experiments are made as series of constant speed measurements where speed is increased by steps of 0.2 Hz. From the constant speed step a sample of 8192 points is separated and weighted with the Hanning window to reduce the truncation error of the FFT. The FFT conversion of the measured results is made in the measuring software. In the measuring the limit frequency of the low-pass filtering is 10 times the rotational frequency of the roller. The phase lag of the low pass filter is linear so it is easy to correct the delay in an analysis. The FFT conversion of the simulation results is made in the MATLAB program from which the spectrum maps are printed. The following values are used as damping ratios of the roller in the simulation:

- 0,5 % of critical damping for all modes with frequency lower than 250 Hz
- 1 % of critical damping for modes with frequency in the 250 - 1070 Hz range
- 100 % critical damping for modes with frequency above 1070 Hz.

The error estimate of the measured amplitude is ± 5 μm. The measured rotational frequency is about 0.15 Hz lower than the real rotational frequency. The error is caused by the pulse sensor that is used in the measurement and it is perceived from the frequency of the first harmonic component of the spectrum map. The measured and simulated amplitudes and frequencies of sub-harmonic resonance peaks are shown in Table 2. The measured spectrum map is presented in Figure 4 and the simulated one in Figure 5.

Table 2. Values of the sub-harmonic resonance peaks.

k	Measured				Simulated			
	Speed [Hz]	Amplitude [μm]			Speed [Hz]	Amplitude [μm]		
		Middle	Driven side	Service side		Middle	Driven side	Service side
2	15.95	104	55.3	51.9	16.2	243.2	120.4	123.5
3	10.75	5.1	1.5	1.4	10.8	25.6	12.9	12.8
4	8.15	13.5	7.8	6.5	8.2	21.4	10.5	10.6

By comparing the harmonic components of the first order it can be seen that the roll in the simulation model is not so well balanced than the real roll. Only the wall thickness variation of the shell is modelled to the simulation model. In the real roll there are other non-idealities such as the waviness of the shell and the initial curvature of the roll. In order to model the mass distribution correctly, more accurate information about the geometry of the roll would be needed. Then the non-idealities of the roll could be modelled in the finite element model by moving nodes. Because of this, the same balancing masses as in the real roll do not balance the roll of the simulation model.

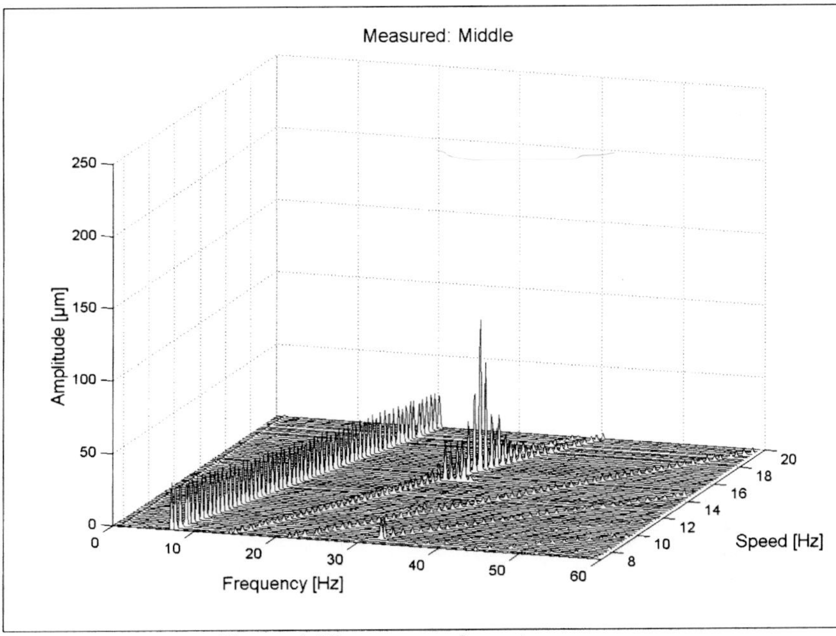

Figure 4. The measured spectrum map.

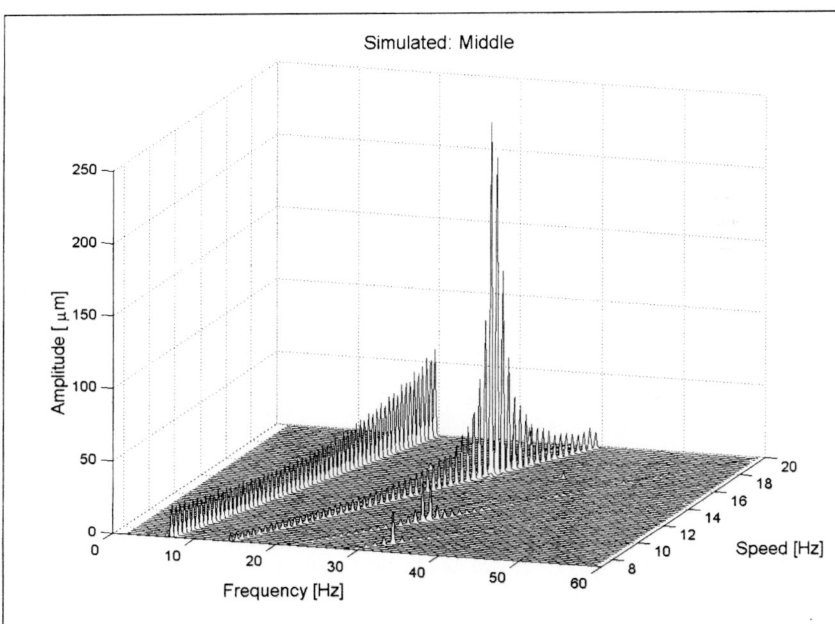

Figure 5. The simulated spectrum map

The frequencies of the resonance peaks in the simulation results are slightly larger than those of the measured results. According to the results the critical frequency of the roll system is about 0.5 Hz higher in the simulation model. The largest difference between the simulated and measured results is in the amplitude of the half-critical resonance peak. Probably the reason for this is the wrong damping of the roll-balancing machine system. In the real balancing machine there are several bolted joints whose friction damping is not modelled. Another possible reason for large amplitude is too large impulse force from the support of the roll. The stiffness coefficient of the contact is large and so minor roundness errors affect greatly the contact force. Therefore the measuring accuracy of the roundness of the shaft and support rollers may be too low.

5 CONCLUSIONS

The dynamic behaviour of one rotor system was studied using commercial multibody simulation software. The structural flexibility of studied tube roll was imported from the detailed finite element model to the MBS software. The modelled non-idealities of the roll were the thickness variation of the shell and the roundness errors of shaft necks. To ensure the validity of the simulation results, the theoretical results were compared with those obtained by measuring the real structure.

The comparison between the measured and simulated results shows that good correlation is achieved. The results indicate that the used method is suitable for an analysis of the rotor dynamics and for the modelling of the non-idealities of a tube roll. With the simulation model the sub-harmonic rotational frequencies are retrieved and the critical speed of the roll corresponds well to the measured results.

REFERENCES

[1] ADAMS 9.1 Online Documentation. Mechanical Dynamics, Inc. 1998.
[2] ANSYS 5.4 Elements Reference. Online Help. SAS IP, Inc. 1998.
[3] Brown, M. A. and Shabana, A. A. Application of Multibody Methodology to Rotating Shaft Problems. Journal of Sound and Vibration, Vol. 204, No. 3, 1997, pp. 439-457.
[4] Craig, R. R., Bampton, M. C. C. Coupling of Substructures for Dynamic Analyses. AIAA Journal, Vol. 6, No. 7, 1968, pp. 1313-1319.
[5] Craig, R. R. Structural Dynamics: An Introduction to Computer Methods. John Wiley & Sons, New York, 1981.
[6] Hamrock, B. J. Fundamentals of Fluid Film Lubrication. McGraw-Hill, New York, 1994.
[7] Haug, E. J. Computer-Aided Kinematics and Dynamics of Mechanical Systems, Vol. 1: Basic Methods, Allyn and Bacon, Massachusetts, USA 1989.
[8] Nikravesh, E. P. Computer-Aided Analysis of Mechanical Systems, Prentice-Hall, New Jersey, 1988.
[9] Óttarsson, G. Modal Flexibility Method in ADAMS/FLEX. MDI Technical Paper, 1998. Available at: http://support.adams.com/kb/faq.asp?ID=kb7247.dasp.
[10] Shabana, A. Dynamics of Multibody Systems. John Wiley & Sons, New York, 1998.

Non-collocated tracking control of a rotating Euler–Bernoulli beam attached to a rigid body

C-F J KUO and **C-H LIU**
Intelligence Control and Simulation Laboratory, Department of Textile and Polymer Engineering, National Taiwan University of Science and Technology, Taipei, Taiwan

Most of the distributed parameter system designs as shown in the literature are based on colocated control. It has been known for such control, the stable operation is easy to achieve at low frequencies. Oppositely, when one wants to control the flexural structure by noncolocated actuator and sensor, the problem of achieving stability and fast trajectory is difficult. In this paper, the noncolocated control for distributed parameter system is illustrated by application to feedback control of a rotating Euler-Bernoulli beam attached to a rigid body. Within this paper, a completely solution strategy including design based model, system dynamics and control due to such rotating Euler-Bernoulli beam is presented. The assumed modes method is used to obtain the system numerical solution. A control scheme, which uses a realizable actuator and sensor without involving truncation the higher–frequency modes, has shown that good stability and efficient tracking property can be achieved.
Keywords:

1. Introduction

It has been known form the previous literature (Canon *et al*, 1983; Gevarter, 1970) that if a flexible structure is controlled by locating the sensor exactly at the actuator it will control, then stable operation is easy to achieve. Oppositely, when one wants to control a flexure structure with noncolocated actuator and sensor, the problem of tracking property and stability is severe. At the same time, for the convenience of control design and simulation, many control system design methods for a flexible structure are either based-on reduced-order models (Szary *et al*, 1992; Canon *et al*, 1984) or require distributed actuators (Crawley, 1986; Miller, 1986). As a result, controller designs based on reduced-order models can destabilized the truncated modes due to the residual uncontrolled modes. On the other hand, the use of distributed actuators are rarely available in reality.

In this paper, the control problem of the flexural structure is studied again. The proposed systematically control system analysis and design method, which uses an actuator and a sensor without involving truncation of the higher frequency modes for good tracking property, is based on modal suppression concept. The performance, which includes in the presence of the linear and nonlinear dynamics, can be effectively shown through the computer simulation.

2. MATHEMATICAL MODEL

Consider a rotating Euler-Bernoulli beam is constrained to move in the horizontal plane (no gravity effect). It is actuated by a DC motor with a torque input at the hub as shown (Zhu, 1988; Akulenko *et al*, 1983) in Fig. 1. The Hamilton principle (Meirovitch, 1990) can be stated as:

$$\int_{t_1}^{t_2} \delta(T_k - V)dt + \int_{t_1}^{t_2} \delta\omega_{nc} dt = 0 \qquad (1)$$

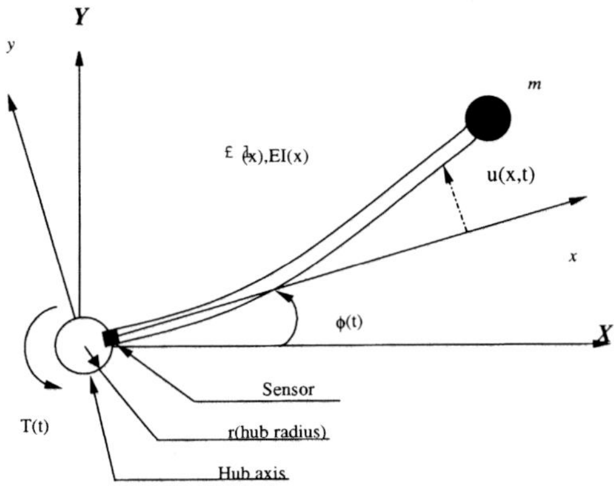

Fig. 1 Schematic diagram of the Euler-Bernoulli beam attached to a rigid body

where T_k is the total kinetic energy, V is the potential energy, ω_{nc} is the work by nonconservative force.
For this model:

$$T_k = \frac{1}{2}\int_0^{\ell} \rho(z)\left[u^2\left(\frac{d\phi}{dt}\right)^2 + z^2\left(\frac{d\phi}{dt}\right)^2 + \left(\frac{\partial u(z,t)}{\partial t}\right)^2 + \right.$$

$$\left. 2z\frac{d\phi}{dt}\frac{\partial u(z,t)}{\partial t}\right]dz + \frac{1}{2}m\left[u^2(r+\ell,t)\left(\frac{d\phi}{dt}\right)^2 + (r+\ell)^2\left(\frac{d\phi}{dt}\right)^2\right.$$

$$\left. + \left(\frac{\partial u(r+\ell,t)}{\partial t}\right)^2 + 2(r+\ell)\frac{d\phi}{dt}\frac{\partial u(r+\ell,t)}{\partial t}\right] + \frac{1}{2}I_H\left(\frac{d\phi}{dt}\right)^2 \qquad (2)$$

$$V = \frac{1}{2}\int_0^{\ell} EI\left(\frac{\partial^2 u(z,t)}{\partial x^2}\right)^2 dz \qquad (3)$$

$$\omega_{nc} = -\frac{1}{2}\int_0^{\ell} C(z,t)\left(\frac{\partial u(x,t)}{\partial x}\right)^2 dz + T\phi \qquad (4)$$

C(z,t) = the axial force due to the centrifugal force (Meirovitch, 1990)

$$= \int_x^{r+\ell} \rho(x)\left(\frac{d\phi}{dt}\right)^2 x dx + m(r+\ell)\left(\frac{d\phi}{dt}\right)^2 \qquad (5)$$

where u(x,t) is the link's transverse displacement relative to a relating frame fixed to the hub, $\phi(t)$ is the angle between the hub's frame and an inertia reference frame, EI is the bending stiffness, $\rho(x)$ is the mass density, m is the mass of the end effector, I_H is the moment of inertia of the hub, r is the hub's radius, ℓ is the length of the link, and T is the applied torque of the motor.

Substituting Eqs. (2)-(5) into Eq. (1), the following governing equation with the boundary conditions can be obtained.

$$EI\frac{\partial^4 u(z,t)}{\partial z^4} + \rho(z)\frac{\partial^2 u(z,t)}{\partial t^2} = -\rho(z)\frac{d^2\phi}{dt^2}z + \left(\frac{d\phi}{dt}\right)^2\left\{\frac{\partial^2 u(z,t)}{\partial z^2}\right.$$

$$\left.\int_z^{\ell}\rho(z)zdz - \left[\frac{\partial u(z,t)}{\partial z} - u(z,t)\right]\rho(z) + m(r+\ell)\left(\frac{d\phi}{dt}\right)^2\frac{\partial^2 u(z,t)}{\partial z^2}\right. \qquad (6)$$

$$T(t) = \int_r^{r+\ell}\rho(x)x^2\frac{d^2\phi}{dt^2}dx + m(r+\ell)^2\frac{d^2\phi}{dt^2} + \int_r^{r+\ell}\rho(x)x\frac{\partial^2 u}{\partial t^2}dx + m(r+\ell)\frac{\partial^2 u(r+\ell,t)}{\partial t^2} + I_H\frac{d^2\phi}{dt^2} \qquad (7)$$

$$u(r,t) = 0 \qquad (8)$$

$$\frac{\partial u(r,t)}{\partial x} = 0 \qquad (9)$$

$$EI\left.\frac{\partial^2 u}{\partial x^2}\right|_{x=r+\ell} = 0 \qquad (10)$$

$$EI\left.\frac{\partial^3 u}{\partial x^3}\right|_{x=r+\ell} = m\left[\frac{\partial^2 u}{\partial t^2} + x\frac{d^2\phi}{dt}\right]_{x=r+\ell} + m\dot{\phi}^2\left.\left(x\frac{\partial u}{\partial x} - u\right)\right|_{x=r+\ell} \qquad (11)$$

where z=x-r.

When $\dot{\phi}$ is small, the linearized equation with the boundary conditions are as follows:

$$EI\frac{\partial^4 u}{\partial z^4} + \rho\frac{\partial^2 u}{\partial t^2} = -\rho z\frac{d^2\phi}{dt^2} \qquad (12)$$

$$u(r,t) = 0 \qquad (13)$$

$$\frac{\partial u(r,t)}{\partial x} = 0 \qquad (14)$$

$$EI\frac{\partial^2 u}{\partial x^2}\bigg|_{x=r+\ell}=0 \quad (15)$$

$$EI\frac{\partial^3 u}{\partial x^3}\bigg|_{x=r+\ell}=m\left[\frac{\partial^2 u}{\partial t^2}+x\frac{d^2\phi}{dt}\right]_{x=r+\ell} \quad (16)$$

$$EI\frac{\partial^2 u}{\partial x^2}\bigg|_{x=r}+T-I_H\frac{d^2\phi}{dt^2}=0 \quad (17)$$

can be obtained.

3.Equation Discretization

Lagrange's equation may be stated as

$$\frac{d}{dt}\left(\frac{\partial L}{\partial \dot{q}_i}\right)-\frac{\partial L}{\partial q_i}=Q_i, i=1,2,...,n \quad (18)$$

where $L=T_k-V$, and Q_i is the generalized force.

By using assumed-modes method (Zhu, 1988; Meirovitch, 1990), let

$$u(x,t)=\sum_{i=1}^{n-1}\Psi_i(x)u_i(t) \quad (19)$$

where Ψ_I are functions of spatial coordinates, $u_i(t)$ are the generalized displacements. By choosing $q_1=\phi$, $q_i=u_i$, with the following boundary conditions:

$$u_i^2(x,t)=0 \quad (20)$$

$$u^2(x,t)=0 \quad (21)$$

$$u_i^2(r+\ell,t)=0 \quad (22)$$

$$u^2(r+\ell,t)=0 \quad (23)$$

Using virtual work

$$\delta\omega_{nc}=T\delta\phi+Q_i\delta u_i \quad (24)$$

to get the ith mode generalized force:

$$Q_i=T \quad (25)$$

$$Q_1 = -\frac{1}{2}\int_r^{r+\ell}\left[\frac{\rho\left((\ell+r)^2 - x^2\right)}{2} + m(r+\ell)\right]\dot{\phi}^2\Psi_j'(x)u_j(t)dx \qquad (26)$$

Then, the following compact form of the system equation can be obtained:

$$M\{\ddot{u}\} + K\{u\} + \dot{\phi}^2 D\{u\} = T(t)\{P\} \qquad (27)$$

where $M = M_{n\times n} = \lfloor m_{ij} \rfloor_{n\times n}$ = mass matrix

$K = K_{n\times n} = \lfloor k_{ij} \rfloor_{n\times n}$ = stiffness matrix

$D = D_{n\times n} = \lfloor d_{ij} \rfloor_{n\times n}$ = nonlinear term matrix

$$m_{11} = \int_r^{r+\ell} \rho x^2 dx + m(r+\ell)^2 + I_H \qquad (28)$$

$$m_{ij} = m_{ji} = \begin{cases} \int_r^{r+\ell} \rho x \Psi_{j-1}(x)dx + m(r+\ell)\Psi_{j-1}(r+\ell) \\ \text{when } i = 1, j = 2,3,\ldots,n \\ \int_r^{r+\ell} \rho \Psi_{i-1}(x)\Psi_{j-1}(x)dx + m\Psi_{i-1}(r+\ell)\Psi_{j-1}(r+\ell) \\ \text{when } i,j = 2,3\ldots,n \end{cases} \qquad (29)$$

$$k_{ij} = k_{ji} = \begin{cases} 0, \text{ when } i = 1, j = 1,2,3,\ldots,n \\ \int_r^{r+\ell} EI\Psi_{i-1}''(x)\Psi_{j-1}''(x)dx, \text{ when } i,j = 2,3,\ldots,n \end{cases} \qquad (30)$$

$$d_{ij} = d_{ji} = \begin{cases} 0, \text{ when } i = 1, j = 1,2,3,\ldots,n \\ \int_r^{r+\ell}\left\{m(r+\ell) + \frac{\rho\left[(r+\ell)^2 - x^2\right]}{2}\right\}\Psi_{i-1}'(x)\Psi_{j-1}'(x)dx - \\ \int_r^{r+\ell}\rho\Psi_{i-1}(x)\Psi_{j-1}(x)dx - m(r+\ell)\Psi_{i-1}(r+\ell)\Psi_{j-1}(r+\ell), \\ \text{when } i,j = 2,3,\ldots,n \end{cases} \qquad (31)$$

$$\{u\} = (\phi\ u_1\ u_2\ \ldots\ldots\ u_{n-1})^T \qquad (32)$$

$$\{P\} = (1\ 0\ 0\ \ldots\ 0)^T \qquad (33)$$

Equation (27) is the discretized form of equations (6)-(11). When $\dot{\phi}$ is small, the linearized form of Eq. (27) is:

$$M\{\ddot{u}\} + K\{u\} = T(t)\{P\} \qquad (34)$$

which is the discretization form of Eqs.(12)-(17)

4.OPEN LOOP TRANSFER FUNCTION

By taking Laplace transform of Eqs. (12) and using boundary conditions (13)-(17), the open loop transfer function from the angular velocity to control torque can be obtained as:

$$\frac{\dot{\phi}}{T}=\frac{sG}{EI\beta(\sin\beta\ell\cosh\beta-\cos\beta\ell\sinh\beta\ell+\dfrac{m\beta}{\rho}(\sin\beta\ell\sinh\beta(\ell-r)}+\sinh\beta\ell\sin(\ell-r))+s^2I_HG \quad (35)$$

where

$$G=\frac{m\beta}{2\rho}[\sin\beta\ell coh\beta(\ell-r)+\sin\beta(\ell-r)\cosh\beta\ell-\cos\beta\ell\sinh\beta(\ell-r)-\cos\beta(\ell-r)\sinh\beta\ell+\sin\beta(\ell-r) \quad (36)$$
$$\cos\beta\ell-\sin\beta\ell\cos\beta(\ell-r)+\sinh\beta\ell\cosh\beta(\ell-r)-\sinh\beta(\ell-r)\cosh\beta\ell]-\cos\beta\ell\cosh\beta\ell-1$$

When system physical parameters are (Qian *et al*, 1991): total length = 1.001 m, height = 0.0507m, thickness = 0.0032m, mass of the end effector = 0.4528 kg, mass density =0.4578362kg/m, Young's Modulus=$6.895\text{x}10^{10}\,\text{m}^4$, cross-sectional area moment of inertia = $0.1384448\text{x}10^{-9}\,\text{m}^4$, motor hub moment of inertia = $0.00044\ \text{kg}-\text{m}^2$, and motor hub radius=0.05m, The pole-zero plot of this transfer function is shown in Fig. 2. This is a nonminimum phase problem.

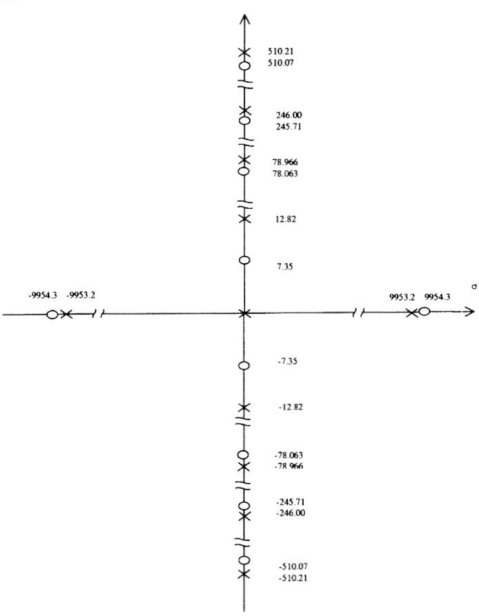

Fig. 2 Poles and zeros' location of the open loop transfer function

5.CONTROLLER DESIGN

Stabilization Theorem (Saeks *et al*,1992): For the feedback system of Fig.3, let the plant have a coprime fractional representation as

$$p(s)=\frac{n_p(s)}{d_p(s)},\ \text{where } u_p(s)n_p(s)+v_p(s)d_p=1,\ \text{for some stable } u_p(s) \text{ and } v_p(s).$$

Then for any stable w(s) such that $w(s)n_p(s)+v_p(s)$ is not identically zero the compensator;

$$c(s)=\frac{[-w(s)d_p(s)+u_p(s)]}{[w(s)n_p(s)+v_p(s)]}=\frac{n_c(s)}{d_c(s)} \qquad (37)$$

stabilizes the feedback system and yields a coprime fractional representation on p(s)c(s)= $[n_p(s)n_c(s)]/\ [d_p(s)d_c(s)]$. Conversely, every such stabilizing compensator is of this form for some stable w(s).

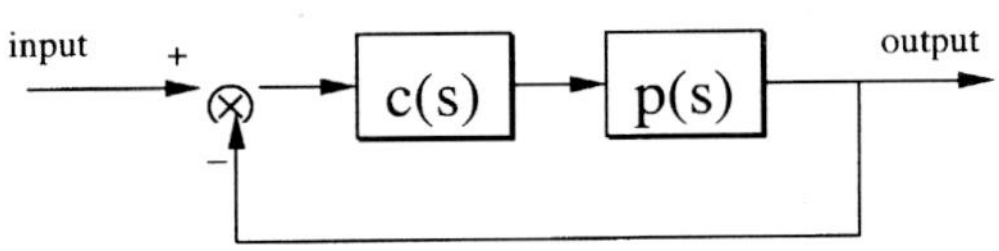

Fig.3 Basic control system

Design procedure ;
Step 1: Let

$$p(s)=\frac{n(S)}{d(s)}=\frac{s^2-z^2}{s(s^2-p^2)}=\frac{\dfrac{s-z}{(s+p)^2}}{\dfrac{s(s-p)}{(s+z)(s+p)}} \qquad (38)$$

Step 2:

$$c(s)=\frac{-w(s)d_p(s)+u_p(s)}{w(s)n_p(s)+v_p(s)}=\frac{n_c(s)}{d_c(s)}=\frac{\dfrac{As+B}{s+z}}{\dfrac{Ds+E}{(s+p)(s+z)}} \qquad (39)$$

For simplest c(s), let w(s)=0, then $u_p(s)=n_c(s)$, $v_p(s)=d_c(s)$

Step 3:

Because u(s)n(s)+v(s)d(s)=1, p=9953.2, z=9954.3,

From (39), $c(s)=\frac{-(s+1)(s+9953.2)}{s+9956}$ (40)

6.COMPUTER SIMULATION

For a reference input:

$$\phi_i(t) = \phi_f\left[3\left(\frac{t}{t_f}\right)^2 - 2\left(\frac{t}{t_f}\right)^3\right] \qquad (41)$$

Boundary conditions:

$$\phi_i(0) = \dot{\phi}_i(0) = 0$$

$$\phi_i(t_f) = \phi_f$$

$$\phi_f = \frac{1}{20}\pi$$

t_f=0.8sec, arrival time $\Psi_i(x-r) = \left[\cosh\beta_i(x-r) - \cos\beta_i(x-r) - \left(\frac{\cos\beta_i l + \cosh\beta_i l}{\sin\beta_i l + \sinh\beta_i l}\right)\sinh\beta_i(x-r) - \sin\beta_i(x-r)\right]$

=shape function (42)

The governing differential equation has been employed by modal analysis to represent the response as a superposition of the shape functions multiplied by corresponding time-dependent generalized coordinates as shown in Eq. (27). Figs. (4)-(6) indicates the transient response of the joint angle, endpoint displacement, and motor torque. Good tracking property and no steady state errors can be shown through the computer simulation.

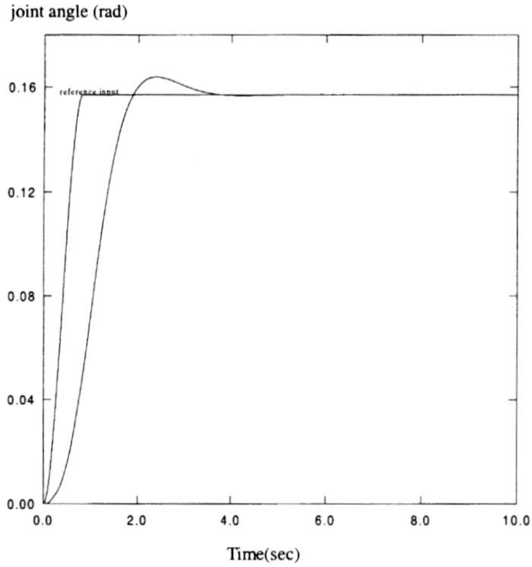

Fig. 4 the transient response of the joint angle

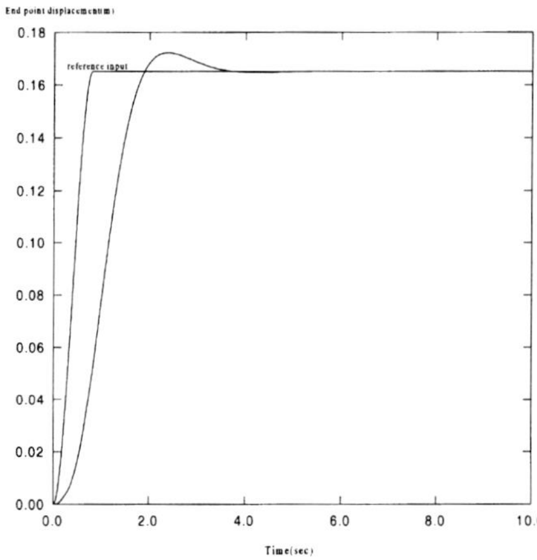

Fig. 5 The transient response of the end point displacement

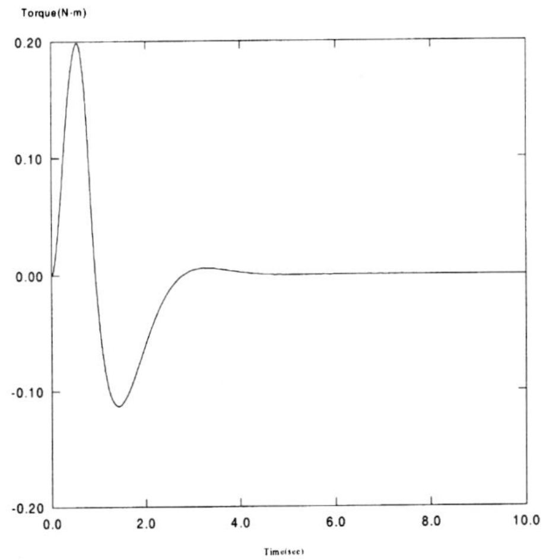

Fig. 6 The transient response of the control torque

7.CONCLUSIONS

In this paper, a systematic approach to design a controllers for a rotating Euler-Bernoulli beam attached to a rigid body is presented. It includes system dynamics and control. The transcendental transfer function from control torque to hub angle has been derived. This makes it possible to determine as many exact poles and zeros of the infinite dimensional open loop transfer function as desired and the poles and zeros pattern of the system can be obtained. The assumed modes method and fourth order Rung-Kutta method are used for computer simulation. It can be seen that the proposed controller not only can get good transient response characteristics, but also can eliminate the steady-state errors. Science joint angle rate can be easily measured, the control does not need state estimation.

ACKNOWLEDGEMENT

The authors gratefully acknowledge the support for the project provided by National Science Council, Taiwan, Republic of China (NSC84-2216E-011-019)

REFERENCES

Akulenko, L. D. and Bolotnik, N. N., 1983, 'On controller rotation of an elastic rod', PMM U.S.S.R.,46, 465-471.

Canon, R. H. and E. Schmitz, 1983, 'Precise control of flexural manuipulators', Journal of Robotics, Pre-print, 841-861.

Canon, R. H. and Schmitz, E., 1984, 'Initial experiments on the end point control of a flexible one-link robot', International Journal of Robotics Research 3, 62-75.

Crawley, E. F., 1986, 'Use of piezo-ceramics as distributed actuators in large space structures. Strsuctures', Structural Dynamics and Materials Conference, Orlando, FL, Aiaa Paper No.85-0626,126-133

Gevarter,W. B., 1970. 'Basic relations for control of flexible vehicles', AIAA Journal 4, 666-672.

Meirovitch, L., 1990, 'Dynamics and Control of Structures', John Wiley & Sons Inc.

Miller, D. W. and Crawley, E. FD. and Ward, B.A, 1986, 'Inertial actuator design for maximum passive and active energy dissipation in flexural space structures.. Structures', Structural Dynamics and Materials Conference, Orlando, FL, AIAA Paper No.85-0626,126-133

Qian, W. T. and Ma, C. H., 1991, 'Experiments on a flexible one-link manipulator. IEEE Pacific Rim Conference on Communications', Computers and Signal Processing 262-265.

Saeks, R. et al., 1992, 'Feedback system design: The single-variate case-part I. Circuits system signal process. 1 (19820 137-169. flexible robot arm', Active Control of Noise and Vibration ASME 38, 149-156.

Szary, M., Rong, Y., and Lee, F., 1992, 'Transverse vibration control in the free end of flexible robot arms. Active Control of Noise and Vibration', ASME, Dynamic Systems and Control Division 38, 271-274.

Zhu, W., 1988, 'Dynamical analysis and optimal control of a flexible robot arm. M.S. thesis', Department of Mechanical Engineering, Arizona State University.

Vehicle Dynamics

Concepts for the modelling of a passenger car

P LUGNER, M PLÖCHL, and **Ph HEINZL**
Institute of Mechanics, Vienna University of Technology, Austria

Abstract

Simplified problem oriented models, that can be analysed analytically provide basic understanding for the dynamic behaviour and are today used for the design of control systems. First 3D-Models, including kinematic and force nonlinearities, provided a comprehensive analysis till to the drive limits. Later Multibody System programs make it relatively simple to include a large number of design details. Easy accessability and good possibilities to illustrate and interpret results make them the first choice for most investigations. The ongoing integration of control features into the system sees the car model itself as one component of the whole only. The challange of the furture will be the balanced tuning of the model details of all components.

1. INTRODUCTION

With respect to the modelling and simulation of passenger car dynamics there is steady progress to include more details, nonlinearities, flexible parts, and more and more control components. This development is strongly supported by the availability of Multibody System (MBS) programs or programs especially tuned for control design.

A short retrospect shows that for nearly 50 years until about 1940 the main car development with respect to its dynamic behaviour like cornering was based on trial and error. The then introduced two-wheel vehicle model already provides essential insight into the principal behaviour – and today this model is often used for the controller design. The application of MBS-programs for car behaviour simulation spans the last 15 to 20 years only.

Today talking about the modelling of the dynamic behaviour of a passenger car inevitably leads to a splitting up into two different approaches. The first is the classical way of problem oriented modelling with the substitute for the real car as simple as possible. The second uses available programs describing the car as a very complex system applicable for most problems. While for the first approach different models (e.g. for vertical dynamics or cornering), relatively simple with few parameters has to be established or selected from literature, the second approach often starts with the structure and components and their interconnections and the main emphasis on determining the generally large number of parameters or force characteristics necessary for the problem to be investigated. Naturally there are also combinations of these two extremes but the polarisation is also driven by the field of application – the design of controllers for the car and the car model as plant in a control loop or as substitute in the first stage of car development.

Since the simple car model provides a better insight for the effects of special design features working with MBS-programs should be based on this experience. Also therefore some main features of the essential problem oriented models will be discussed.

2. SIMPLIFIED PROBLEM ORIENTED MODELS

Especially this modelling approach leads to separat models for three areas of investigation: lateral, longitudinal and vertical vehicle dynamics. Basically these models are described in such a way, that the results can be obtained in an analytical way – and they are common standards for the design and implementation of control features today.

2.1 Two-wheel Car Model

The problem to analyse the lateral dynamics of a car leads to a linearized, plane model, first introduced with its main characteristics about 1940, (1). Its range of application is limited to relatively low lateral accelerations (about $a_q \leq 4ms^{-2}$ for dry horizontal surface) and it is widely used for controller design today.

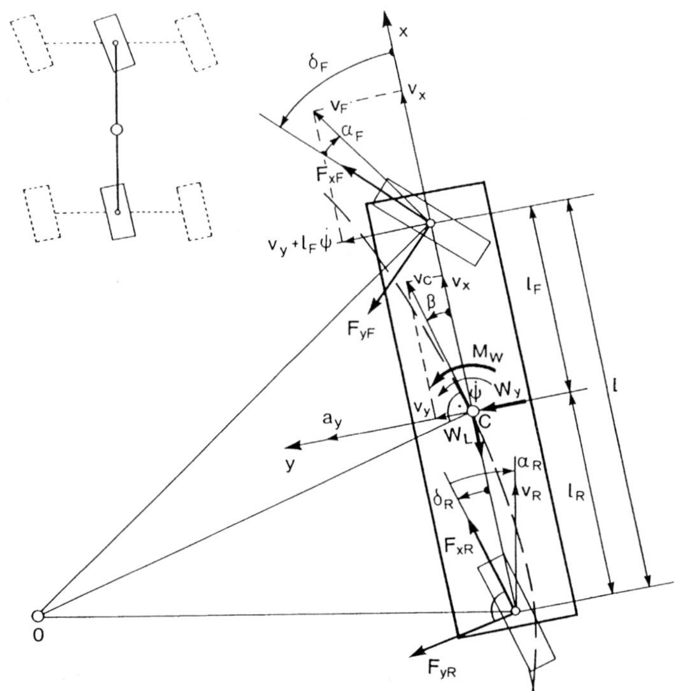

Fig.2.1: Two-wheel car model with additional rear wheel steering

The essential simplifications of this model, Fig.2.1, are:

- two wheels of an axle summarized into one massless substitutive wheel, that is always normal to the road surface,
- lateral tyre forces F_{yi} of a substitutive wheel can include tyre force characteristics and effects of steering and suspension compliance; approximated as linear functions of the side slip angles α_i (cornering stiffness C_i):

$$F_{yi} = C_i \alpha_i \qquad i = F, R \qquad (2.1),$$

- all angles assumed to be small and trigonometric functions are linearized,
- no influences of longitudinal tyre forces with respect to the lateral dynamics; aerodynamic forces W_L, W_y and the corresponding moment M_W often neglected.

Since no relative body motion is considered there are 2 DOF only: yaw and lateral motion of the CG (or side slip angle β of the car). This simplification also includes that there is no change in load distribution front to rear. The longitudinal velocity v_x (or v_c) acts as a system parameter.

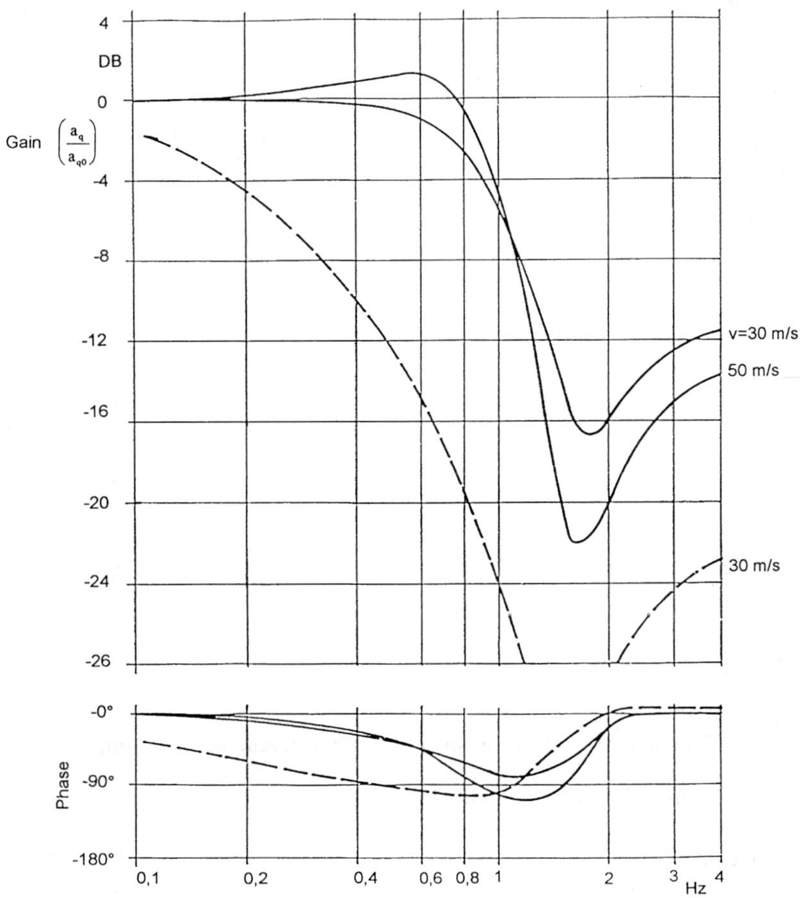

Fig.2.2 Frequency response of a medium size passenger car with oversteer (- -) and understeer (—) characteristics

The advantage of these simplifications is the possibility to formulate the essential equations of motion in an easy to handle analytical way, that is widely used, see e.g. (2,3,4). The application of this kind of modelling allows investigations in

- steady state cornering and steering behaviour,
- stability properties for small disturbances,
- effects of rear wheel steering,
- frequency and step input responses as far as lateral accelerations remain small,
- controller design,
- driver model development and driver – vehicle – roadway interaction.

As an example Fig.2.2 shows a characteristic frequency response for an understeering and an oversteering medium size passenger car, calculated with this kind of model.

2.2. Longitudinal Dynamics

For basic considerations the lateral and longitudinal dynamics can be separated and with respect to problems of braking, acceleration and drive train design correspondingly tuned modells can be established (see e.g. (5, 6)). So Fig.2.3 shows a plane car model for motions in longitudinal direction and a description of a drive train with braking system that can be connected to this car model.

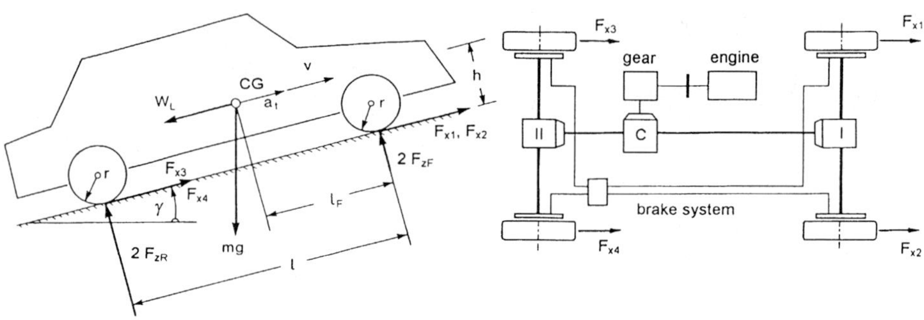

Fig.2.3: Vehicle model for longitudinal dynamics and possible drive train model for 4WD with 3 differentials (C, I, II)

The essential simplifications are

- no heave and pitch motion
- same normal tyre forces left and right

$$\begin{aligned} F_{zF} &= F_{z1} = F_{z2} \\ F_{zR} &= F_{z3} = F_{z4} \end{aligned} \qquad (2.2)$$

- longitudinal tyre slip neglected but longitudinal tyre forces limited by friction coefficient.

Though by neglecting the tyre slip no higher frequency transient motions can be investigated quite a number of problems are in the range of application of this model:

- braking and acceleration on grades,
- effects of different drive configurations for different friction conditions,
- controller design for drive train, and basic considerations for dynamic stability control,
- tuning of the braking system including ABS,
- necessary power and fuel consumption (if corresponding engine characteristics are included in the model).

As an example (5) the accelerating behaviour on a μ-split surface is shown in Fig.2.4. Instead of the longitundinal acceleration a_t a normalized a^x is used that also includes effects of a grade q, the aerodynamic drag W_L and rotating masses (factor ξ_R) with a_m^x being the maximum possible value:

$$a^x = \left(\frac{a_t}{g} + \frac{\sin\gamma}{(1+\xi_r)} + \frac{W_L}{mg(1+\xi_r)} \right) \qquad (2.2).$$

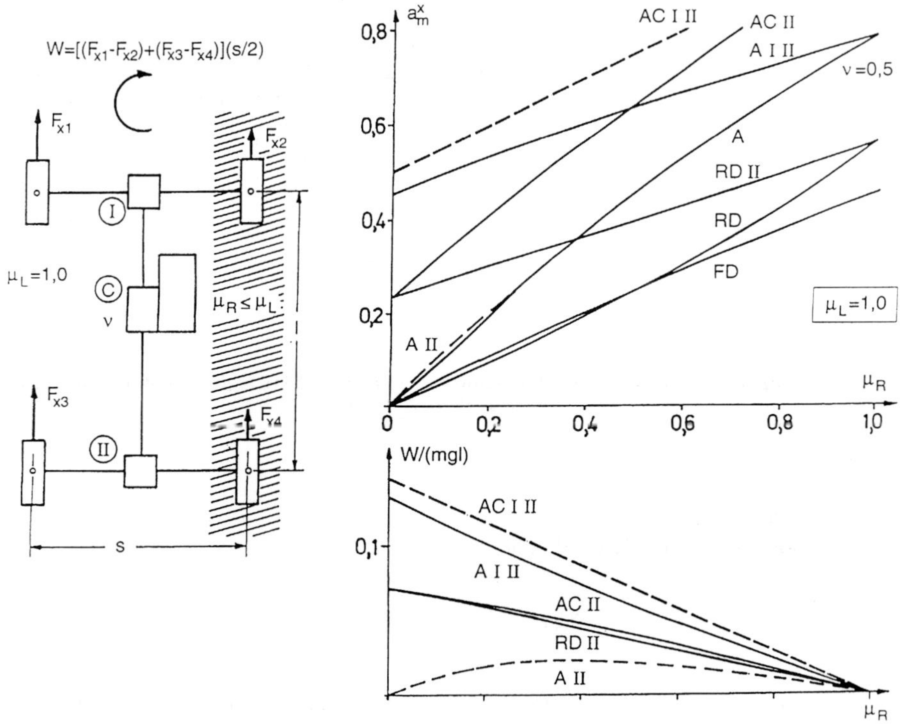

Fig.2.4: Maximum accelerating capability a_m^x and disturbing yaw moment W for different μ split conditions and drive configurations (A all wheel drive, R rear wheel drive, F front wheel drive, locked differential according to C, I, II)

Due to the differences in longitudinal tyre forces a disturbing yaw moment W occurs that depends on which differentials are locked and how large is the difference between the friction coefficients of right and left wheel track μ_R to μ_L. As can be seen by this Fig.2.4 the locking of all differentials and all wheel drive, marked AC I II provides the highest potential for a_m^x but also the largest yaw moments that need to be compensated by the driver.

2.3 Vertical Dynamics

To be able to get estimates for ride comfort and the dynamics of the normal tyre forces and thereby the road holding capability of the wheels (and the car) a plane model with sprung and unsprung masses with the road surface profile as system input is established, Fig.2.5, see e.g. (4,6,7). An even more simplified model, the so called "quater car model" takes into account one unsprung mass and the mass of a quater of the car body and its vertical motion only, see e.g. (8).

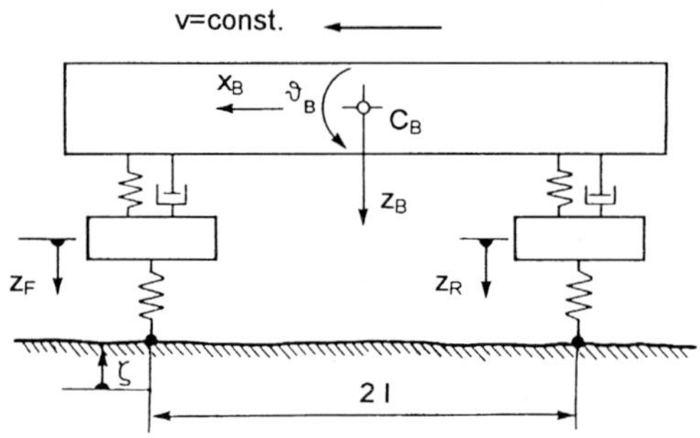

Fig.2.5: Vertical vehicle model

The essential simplifications are:

- linear, plane system,
- linear springs and dampers, no other suspension features,
- constant driving velocity especially with respect to the single track excitation,
- no external forces, continuous tyre-road contact,

and the consequently range of application

- vertical and pitch motion and corresponding accelerations,
- comfort and dynamic wheel load estimation and consequence for the tuning of the suspensions,
- controller design.

Since an analytical description of the vertical dynamics via transfer function is possible a model like shown in Fig.2.5 is especially useful to investigate random road inputs and the evaluation of stochastic quantities taking into account human sensitivity like weighted RMS-values $\bar{a}_{RMS}$ of vertical accelerations. As an example Fig.2.6 shows the influence of the position $\lambda = x_B / l$ on the car and the velocity for a medium good asphalt road on the $\bar{a}_{RMS}$ (4).

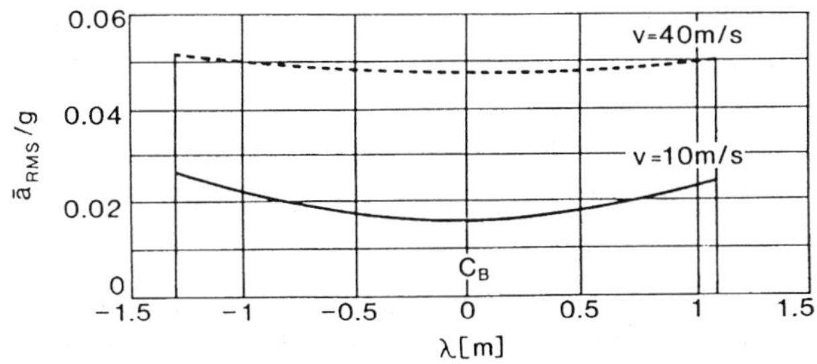

Fig.2.6: Weighted RMS value of vertical acceleration as function of the position on the car $\lambda = x_B / l$

An extension of this model may also be used to evaluate the reactions to the road itself (8).

3. NONLINEAR MODELS

Before a modelling with MBS-programs was available and easy to apply, some developments with respect to complex 3D-models extended the range of investigation of the Simplified Models and combined many of their features, see e.g. (9).

The nonlinearities employed included tyre characteristics, see Fig.3.1, (10), and features of the suspension systems. Since moreover in general drive train and steering system were also modelled, the car was able to move over an ondulated road (excitations with low frequency contents) with individual behaviour of car body and each wheel/suspension. Thereby the wheel-road contact was calculated for each wheel separately. Fig.3.2 indicates with the coordinate frames how to proceed from road fixed frame x_I, y_I, z_I via carbody (frame B), wheel system (frame 1) back to the road surface (indicated by z_I). The general motion and the wheel spin ω_1 of the presented left front wheel 1 delivers the slip quantities α, s_x and via the tyre deflexion $(r - \lambda)$ the normal tyre force (4).

Still simplifications are included

- non or simplified elastisities of suspension components and bushings,
- drive train and steering system modelled separately and directly connected to the wheels or suspension system,
- simplified tyre transient reaction by first order filter depending on travel distance of wheel,
- surface contour wavelength larger than tyre patch.

The range of application nearly spans all usually investigated manoeuvres. Only ride comfort and road holding capabilities cannot include higher frequencies in a proper way. Also effects with very fast transient tyre motions cannot be represented correctly.

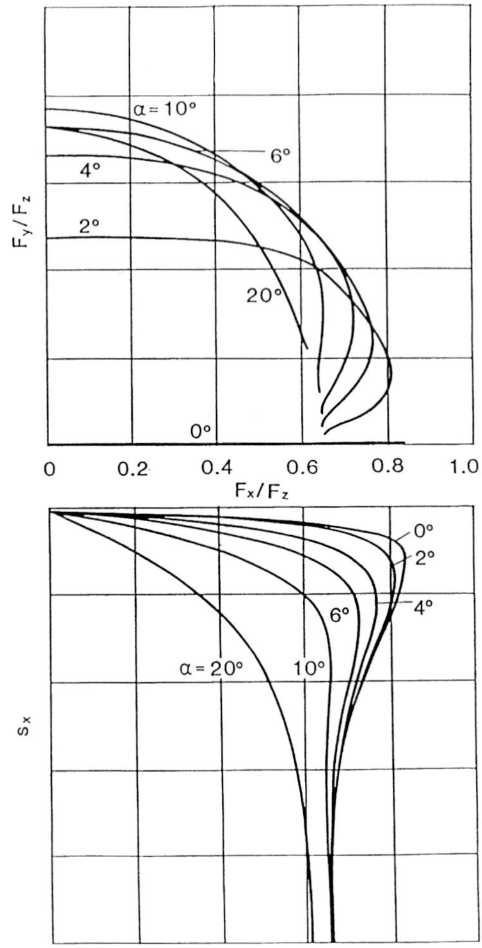

Fig.3.1: Typical tyre characteristics: steady state lateral F_y and longitudinal F_x tyre forces as functions of sideslip angle α, longitudinal slip s_x at constant normal force F_z; dry road surface

A disadvantage is the necessity of a larger number of tuned parameters and/or force characteristics are necessary. The example in Fig.3.3, (9), a braking out of steady-state cornering, demonstrates the possibilities of this approach in relation to the measurements.

By the measurements it can be seen that ideal conditions like assumed for the simulation never exist and especially for such an extreme manoeuvre differences measurement – simulation are unavoidable. On the other hand the potential of simulation and precalculation of even such an emergency braking can very well be noticed also!

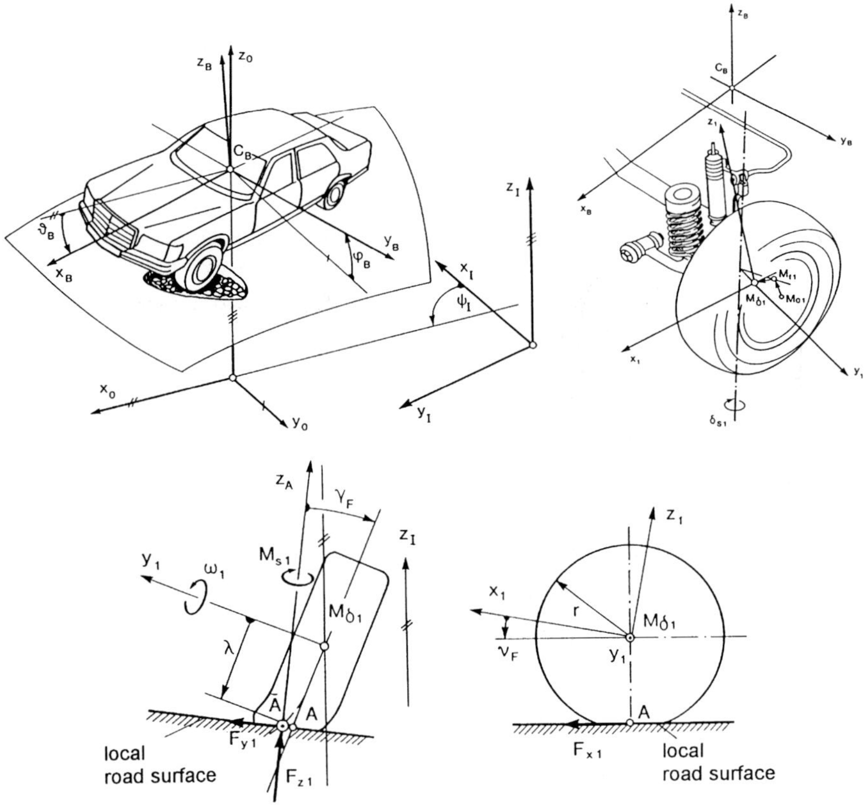

Fig.3.2: 3D-car model and wheel-road surface contact

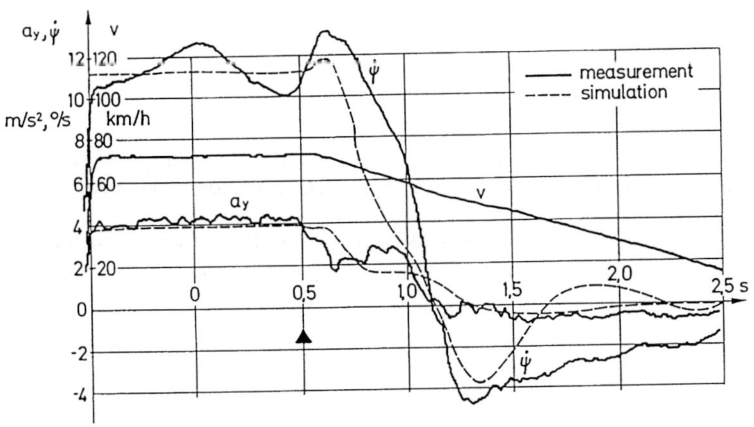

Fig.3.3: Emergency braking at steady-state cornering

4. MULTIBODY-SYSTEM PROGRAMS (MBS)

This approach allows to compose the vehicle as an assembly of rigid or even flexible bodies connected by springs, dampers or flexible joints or by kinematic constraints (11, 12). So in principle there is no limit to include an increasing number of components, nonlinearities and other details, see Fig.4.1.

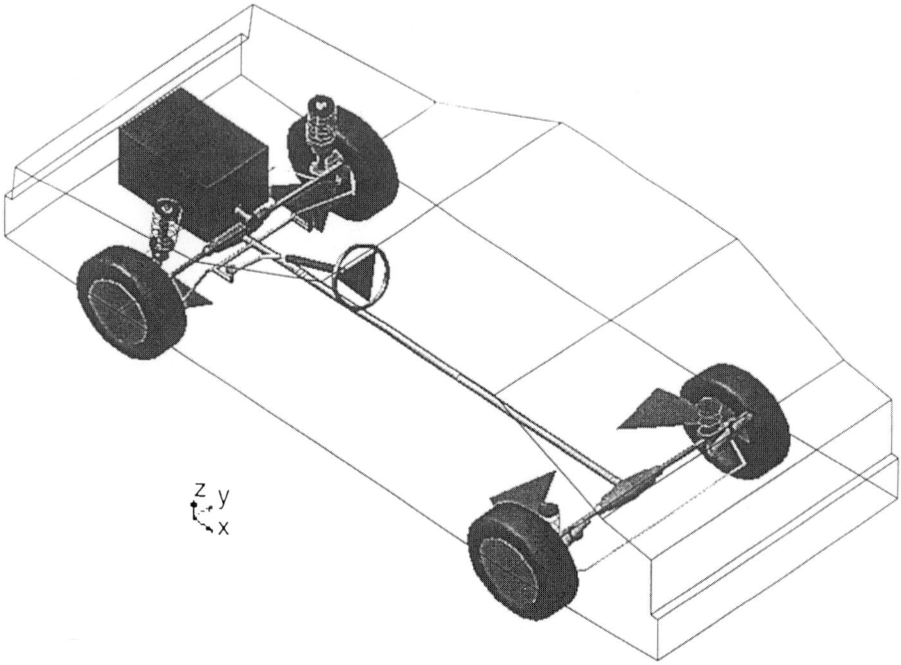

Fig.4.1: Passenger car modelled with ADAMS-car (13)

So it seems there are no real restricting simplifications or disadvantages despite maybe necessary computing time. Without going into the details of Multibody-System programs themselves – what would need the good knowledge of quite a lot of manuals – still problems remain:

- more bodies with more connections need more parameters and/or force characteristics, that are not always available,
- the balanced tuning of the modelling for a frequency range requires experience and practical knowledge,
- steady-state and transient tyre characteristics lacks behind detailed modelling of the vehicle at the moment,
- interpretation of results, especially higher frequency responses, should consider the uncertainties of parameters.

Though this features may impose the impression that the advantages with respect to nonlinear models are not convincing, taking into account the possibility to establish equations of motion by program, working with an interactive input surface and possible animation of results provide great improvements especially for "every day" applications. Problems may arise when information with respect to the program itself (source code) are necessary and the provided interfaces or input possiblities are not sufficient. This maybe the case for including control systems.

As an example Fig.4.2 shows the influences of changes in the front suspension properties. Though these changes of the lateral elasticity of the rear bushing of the wishbone of the outer wheel are extensive the vehicle reactions are not very different and especially the interpretation of the results may pose difficulties. Also in the second case neglecting the damping of the front wheel does not induce large qualitative changes of the results for this flat surface.

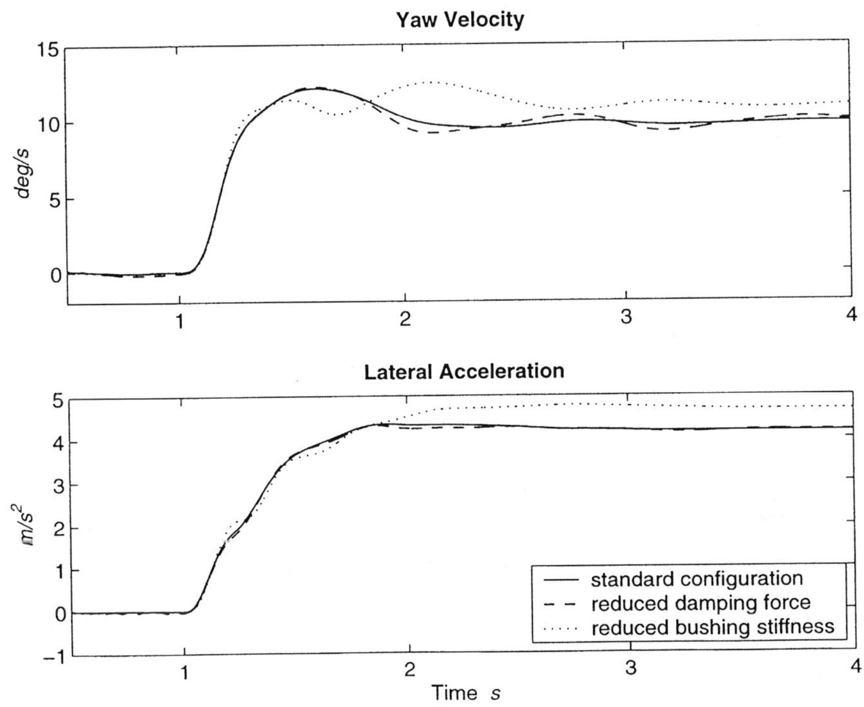

Fig.4.2: Step steering input at $v = 25km/s$

A concluding remark should be supported by Fig.4.3. Going into more and more details will be a fractal problem with increasing uncertainties of the properties of always smaller parts and their connections!

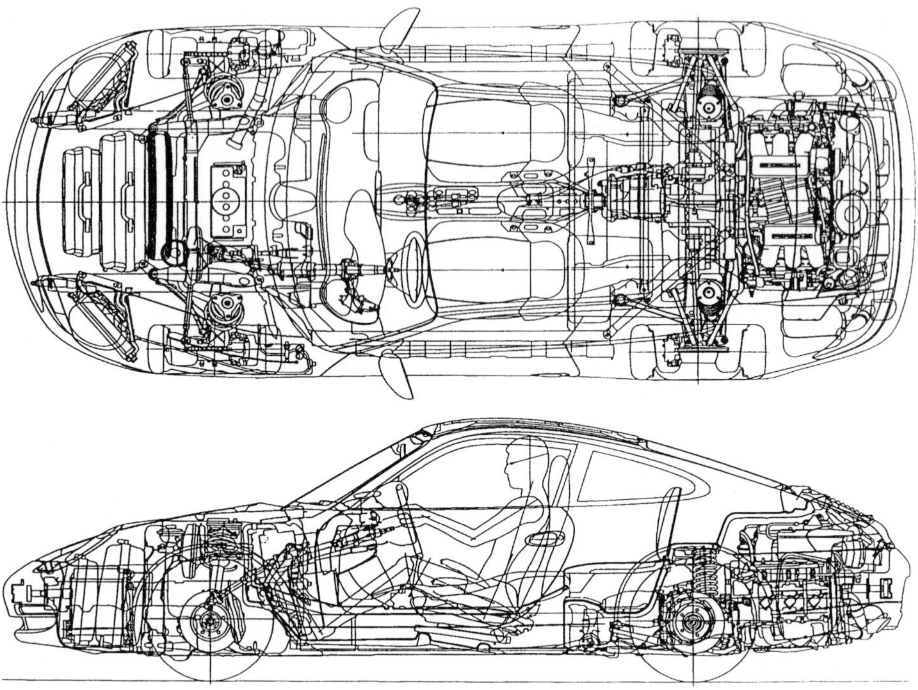

Fig.4.3.: Package view of Porsche 911 Carrera

5. MBS AND CONTROL

Today the simulation of the vehicle behaviour needs to include devices like electronic stability program (ESP), 4-wheel steering and more. Moreover closed-loop manoeuvres or keeping the vehicle on a predetermined road (with locally changing surface structures and conditions) are increasingly important. So the MBS-car model becomes only a part of control loops with active systems, driver models and road descriptions. Especially for the vehicle control the Simplified Models are an essential part for the design of the controllers.

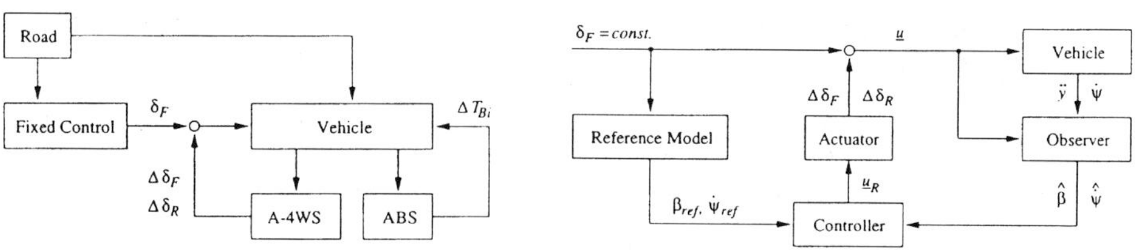

Fig.5.1: General system overview and detail of 4 WS feedback control loop

Block diagrams dominate the modelling of such systems like the example shown in Fig.5.1, (14), for a system without driver input (fixed control).

The system part "Vehicle" contains the full car model while "Reference Model" and "Observer" are based on the 2-Wheel-Model and provide information on side slip angle β and yaw velocity $\dot{\psi}$. "Controller" and "Actuator" with their dynamics deliver the correction steering angles $\Delta\delta_F, \Delta\delta_R$ for the front and rear wheels and are parts of the system. So the restrictions and problems of this modelling approach does not include the vehicle model only but also the other components:

- harmonization of the frequency range of the components and depth of detail,
- linear controller design only; implementation of nonlinear controllers just starts and thereby there are deficits in the control of vehicle limit behaviour,
- necessity for program connections of MBS and mechatronical programs.

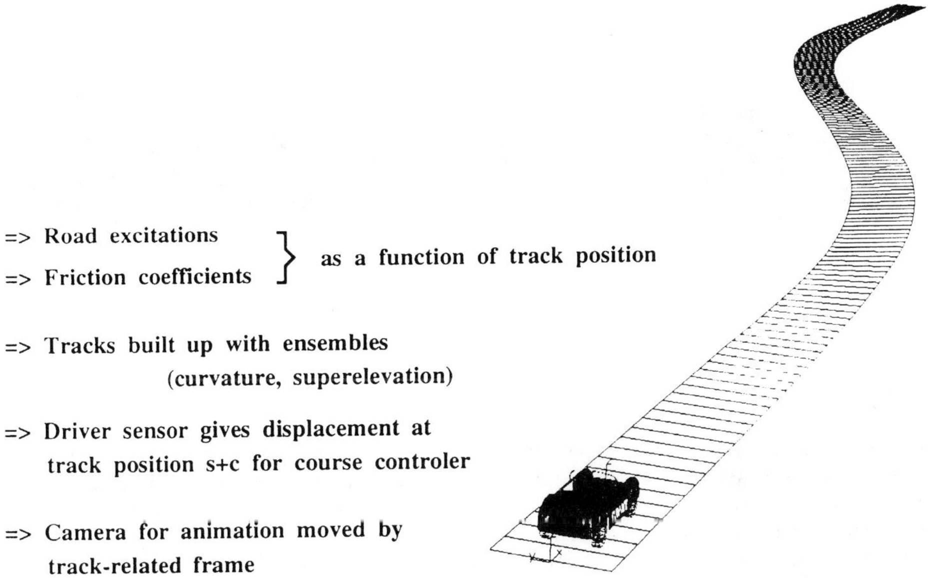

Fig.5.2: Simulation application with the use of MBS-program SIMPACK Automotive (15) for driving along a trajectory (pathlength s) with a lateral position c

For including more details the same holds like for the MBS-vehicle modelling: more details need more parameters and include more uncertainties. This should always be taken into account for the comparison of simulation and measurement of an individual car!

As an example Fig.5.2 indicates the possibilities of the simulation of a car running on a given track where the capability of the driver is essential. This symbolic figure should only present some clues of what can be done by specialized MBS-programs in connection with control features.

6. ONGOING AND FUTURE DEVELOPMENTS

Based on the complete system like shown in Fig.5.1 on the one hand improvements of the components will be considered and on the other hand an integration of the complete system car into the traffic surrounding is becoming essential.

The MBS-programs themselves show improvements in calculation time, user friendly software for input and result interpretation and animation. But also problems of one – sided contact with friction and vibro impact are beginning to being integrated, e.g. (16).

With respect to car modelling an integration of the overall dynamic behaviour and local stresses and information for construction (with FE-codes) is emerging. Hereby a model specialisation and a focusing on special problems is taking place comparable to a specified modelling at higher level of detail. Just at the beginning is the combination of FE-modelling and MBS-car modelling for crash analysation and accident reconstruction. Moreover the ongoing development of more sophisticated tyre models will make it possible to take into account short wave length street excitation in the near future, e.g. (17).

Control design is especially extended to the modelling of the driver behaviour and tries to use nonlinear approaches, e.g. (18). In this area the connection of software developed for other areas of application e.g. for hydraulics and actuators starts to be used.

The integration of the vehicle in the traffic and automated driving to enlarge the efficiency of transport and road capacity are strong incentives. So for Intelligent Vehicle Highway Systems special issues of the VSD-Journal are published. The car including driver and the driver assistance systems for controlled interaction with surrounding and traffic partners seems a very comprehensive and complex area but especially interesting for future car and traffic planning, (19).

The vehicle and its modelling being part of a more global system does not put the emphasis so much on more and more details to be included in the modelling but needs more teamwork between scientists of different areas and a very balanced tuning of the integrated components and their modelling.

7. REFERENCES

(1) **Rickert P., Schunck T.E.:** *Zur Fahrmechanik des gummibereiften Kraftfahrzeugs.* Ingenieur Archiv, Band 9, 1940.

(2) **Sharp R.S., Crolla D.A.:** *Controlled rear steering for cars – a review.* Proceeding of the ImechE, International Conference “Advanced Suspensions”, C437/88, 1988.

(3) **Ellis J.R.:** *Vehicle Handling Dynamics.* MEP, London, 1994.

(4) **Kortüm W., Lugner P.:** *Systemdynamik und Regelung von Fahrzeugen.* Springer Verlag, 1993.

(5) **Lugner P.:** *Theoretische Grenzen und Möglichkeiten der Fahrzeugdynamik von PKW mit Allradantrieb.* VDI-Berichte Reihe 12, Verkehrstechnik/Fahrzeugtechnik Nr. 81, 1986.

(6) **Gillespie Th.D.:** *Fundamentals of Vehicle Dynamics.* Society of Automotive Engineers, Inc. (SAE), 1992.

(7) **Mitschke M.:** *Dynamik der Kraftfahrzeuge.* Band B: Schwingungen, Springer Verlag, 3. Auflage, 1997.

(8) **Cebon D.:** *Handbook of Vehicle-Road Interaction.* Swets & Zeitlinger 1999.

(9] **Lugner P., Lorenz R., Schindler E.:** *The Connextion of Theoretical Simulation and Experiments in Passenger Car Dynamics.* Proceedings of the 8th IAVSD-Symposium, Swets & Zeitlinger, 1983.

(10) **Pacejka H.B. (editor):** *Tyre Models for Vehicle Dynamics Analysis.* Swets & Zeitlinger, 1993.

(11) **Rahnejat H.:** Multi-Body Dynamics; Vehicles, Machines and Mechanisms. Professional Engineering Publishing Limited, UK, 1998.

(12) **Kortüm W., Sharp R.S.:** *Multibody Computer Codes in Vehicle System Dynamics.* Swets & Zeitlinger, 1993.

(13) *Getting Started Using ADAMS/Car.* Mechanical Dynamics, Inc., 1999.

(14) **Plöchl M., Lugner P.:** *Braking Behaviour of a 4-wheel-steered Automobile with an Antilock Braking System.* Proceeding of the 14th IAVSD-Symposium on "The Dynamics of Vehicles on Road and on Tracks". Swets & Zeitlinger, 1996.

(15) *SIMPACK User Manual.* INTEC GmbH, Wessling, 1997.

(16) **Rulka W.:** *Effiziente Simulation der Dynamik mechatronischer Systeme für Industrielle Anwendungen.* Diss., TU Wien, 1998.

(17) *SWIFT-Tyre*: Delft-Tyre,TNO Delft, Newletter 2000.

(18) **Plöchl M., Lugner P.:** *Passenger car and passenger car - trailer – different tasks for the driver.* Society of Automotive Engineers of Japan, Inc. and Elsevier Science BV, 1999.

(19) **Dickmanns E.D.:** *Computer Vision and Highway Automation.* VSD Volume 31, Nr. 5/6, 1999.

Predictive control of vehicle suspensions with time delay for a quarter car model

A VAHIDI and **A ESKANDARIAN**
Center for Intelligent Systems Research, The George Washington University, Ashburn, Virginia, USA

ABSTRACT

Considerable amount of research has been carried out on active vehicle suspension control during the last few decades, basically using optimal linear quadratic regulators. Effect of time delay has been neglected most of the time or availability of preview information has been assumed. In this paper a discrete predictive method is formulated and time delay as an inherent characteristic of active systems is also taken into account. Preliminary results show that the proposed method, compared to passive suspension, is capable of reducing the RMS body acceleration and also suspension deflection for low frequency road disturbances, even in the presence of time delays. For higher frequency road inputs, however, the acceleration response degrades while the maximum suspension deflection is still controlled better than passive system.

INTRODUCTION

Ride Comfort has been one of the important issues in automotive industry and has been under research for many years to enhance vehicle suspension systems for better ride performance. Active suspension control methods, as one possible way to improve ride comfort without sacrificing handling performance, have attracted much attention. Basically in an active suspension system, based on the feedback from a number of sensors and according to a control law, control forces are applied to reduce body acceleration while maintaining an acceptable level of suspension stroke and tire deflection as packaging and handling measures.

During the last two decades the research has been pursued on development of control methods as well as control hardware suitable for practical implementation of active suspension systems. A list of relevant publications can be found in [1].

Among the control methods, optimal linear quadratic theory is the most widely used [2,3,4], in which minimization of a continuous-time performance index over an "infinite" prediction horizon results in a Riccati equation. Solving this equation a continuous-time feedback control algorithm is derived [5,6] which needs to be discretized for digital control.

An alternative way for controller design is to use a discrete-time control method. Predictive control method, introduced in the late seventies, is developed in this context. In this method, a discrete time control sequence is computed at each sampling time to minimize a performance index over a finite prediction horizon. Due to its discrete nature and because of its "finite" prediction horizon, the predictive method is much easier to formulate and the formulation does not require the solution of the Riccati equation [7]. Its discrete nature is suitable for direct use in digital control and time delays in the control loop can be more easily included in the formulations. Some other features of predictive control that make it an attractive control methodology are outlined in [8,9].

Rodellar, et al. [10] have used this method in structural vibration control for earthquake excitations, and both experimental and numerical results are promising. Gopalasani, et al. [11] have also used this method for experimental studies in preview suspension control of a two DOF quarter-car model.

While time delays are unavoidable in the control system, unanticipated time delays may cause malfunction of the control system or even instability of the system. Preview control, proposed by some researchers for active suspension control [12,13,14] is one possible way to compensate for these time delays, but this is only possible with an additional cost of preview sensors. Existence of noise in the preview information is another issue to be dealt with.

In this study the objective has been to implement discrete predictive method in suspension control while including time delays in control formulations to avoid possible malfunction in the realistic case where time delay really exists. As the first step a single degree of freedom quarter-car model was used, which authors believe can give an insight into the performance of predictive method in time-delay control of vehicle suspensions.

Preliminary results show that in low frequency road inputs predictive-controlled suspension performs better than passive system, both in absolute acceleration and suspension stroke reduction. In higher frequencies, however, the acceleration performance starts to degrade while suspension deflection is always below that of the passive one.

THE PREDICTIVE CONTROL CONCEPT

The concept of a predictive control can be summarized in the following two steps:

1) At each sampling instant, k, a prediction horizon is defined, over a finite number of time steps ahead and a discrete-time model of the system is used to predict the response in this horizon.

2) A desired control sequence is computed at each instant k, by minimizing a performance index over the following prediction horizon, to make the predicted response of the system close to a desired trajectory. At each sampling instant this procedure is repeated using the latest available information.

Figure 1. shows a schematic view of the closed loop control system.

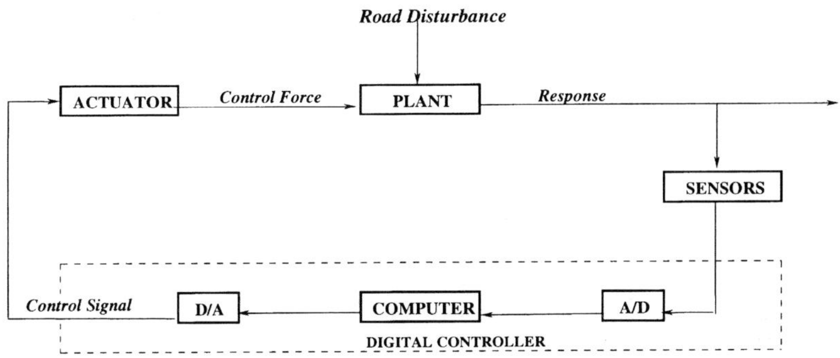

Figure 1. Schematic model of the control loop

THE DISCRETE-TIME MODEL FOR THE PLANT

A single degree of freedom quarter-car model was used in this study to evaluate the performance of predictive method in active suspension control. Dynamics of the actuator was neglected and the actuator has been modeled as an ideal force generator. Time delay in the control loop has been modeled by time-shifting the computed control force. With the coordinates shown in Fig. 2 the governing dynamic equation of the system is written as:

$$m\ddot{z}_c + c(\dot{z}_c(t) - \dot{w}_c(t)) + k(z_c(t) - w_c(t)) + u_c(t-\tau) = 0 \qquad (1)$$

where:
m: sprung mass
c: damping of the suspension
k: suspension stiffness
u: control force
τ: time delay in the control loop
z: absolute sprung mass displacement
w: road displacement input
and subscript c denotes continuous-time variables.

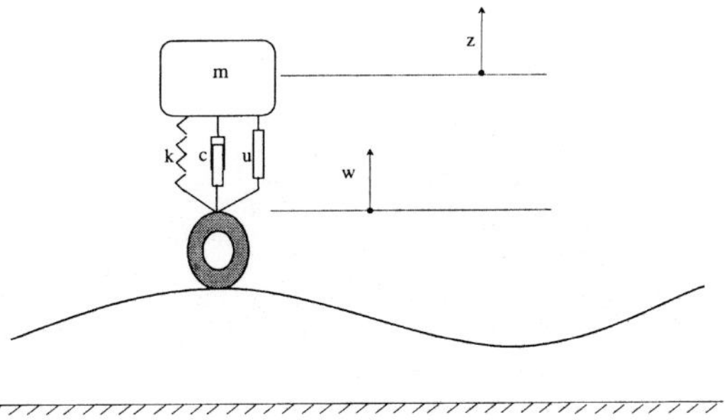

Figure 2. The single degree of freedom quarter car model

By defining the state variable as:

$$x_c = z_c - w_c \tag{2}$$

Equation (1) can be rewritten as:

$$m\ddot{x}_c(t) + c\dot{x}_c(t) + kx_c(t) + u_c(t-\tau) = -m\ddot{w}_c(t) \tag{3}$$

or in the state form:

$$\dot{\boldsymbol{X}}_c = \mathbf{F}\boldsymbol{X}_c(t) + \boldsymbol{G}u_c(t-\tau) + \boldsymbol{H}\ddot{w}_c(t) \tag{4}$$

where:

$$\boldsymbol{X}_c = \begin{bmatrix} x_c \\ \dot{x}_c \end{bmatrix} \tag{5}$$

and:

$$\mathbf{F} = \begin{bmatrix} 0 & 1 \\ -\omega_n^2 & -2\xi\omega_n \end{bmatrix} \quad \boldsymbol{G} = \begin{bmatrix} 0 \\ -\dfrac{1}{m} \end{bmatrix} \quad \boldsymbol{H} = \begin{bmatrix} 0 \\ -1 \end{bmatrix} \tag{6}$$

where

ω_n = Natural frequency of suspension system

ξ = Damping ratio of the damper

Equation (4) has the following analytical solution [15]:

$$\boldsymbol{X}_c(t) = \exp[(t-t_0)\mathbf{F}]\boldsymbol{X}_c(t_0) + \int_{t_0}^{t} \exp[(t-r)\mathbf{F}][\boldsymbol{G}u_c(t-\tau) + \boldsymbol{H}\ddot{w}_c(r)]dr \tag{7}$$

Rewriting this equation between two consecutive intervals, $t_0 = kT$ and $t = (k+1)T$ where T is the sampling period, and replacing continuous-time variables by their discrete equivalents we obtain:

$$\boldsymbol{X}(k+1) = \exp(\mathbf{F}T)\boldsymbol{X}(k) + \int_{kT}^{kT+T} \exp[(kT+T-r)\mathbf{F}][\boldsymbol{G}u_c(r-\tau)]dr + \int_{kT}^{kT+T} \exp[(kT+T-r)\mathbf{F}]\boldsymbol{H}\ddot{w}_c(r)dr \tag{8}$$

In the discretization process, the control sequence is held constant between each two sampling intervals:

$$u_c(t) = u(k) \qquad kT \le t < (k+1)T \tag{9}$$

Interpolating $w_c(r)$ linearly between the sampling intervals, kT and $(k+1)T$, equation (8) reduces to:

$$\boldsymbol{X}(k+1) = \mathbf{A}\boldsymbol{X}(k) + \boldsymbol{B}u(k-d) + \mathbf{P_1}\boldsymbol{H}\ddot{w}(k+1) + \mathbf{P_2}\boldsymbol{H}(\ddot{w}(k+1) - \ddot{w}(k)) \tag{10}$$

where $d = \frac{\tau}{T}$ that implies the time delay, τ, is considered to be an integer multiple of the sampling period. And:

$$\begin{aligned} \mathbf{A} &= \exp(T\mathbf{F}) \\ \boldsymbol{B} &= \mathbf{P_1}\boldsymbol{G} \\ \mathbf{P_1} &= \int_0^T \exp(s\mathbf{F})ds \\ \mathbf{P_2} &= -\frac{1}{T}\int_0^T \exp(s\mathbf{F})sds \end{aligned} \tag{11}$$

Now the exact solution of the system is available in discrete form. The model can then be formed as follows, using this discrete solution and eliminating ground inputs that are unknowns at each sampling time:

$$\hat{\boldsymbol{X}}(k+1|k) = \mathbf{A}\boldsymbol{X}(k) + \boldsymbol{B}u(k-d) \tag{12}$$

in which $\hat{\boldsymbol{X}}(k+1|k)$ denotes the state vector predicted at instant k for $k+1$.

If λ is the length of the prediction horizon, the response in the interval $[k, k+\lambda+d]$ can be predicted by the following state model, using equation (12):

$$\hat{\boldsymbol{X}}(k+j|k) = \mathbf{A}\hat{\boldsymbol{X}}(k+j-1|k) + \boldsymbol{B}\hat{u}(k+j-1-d|k) \quad (j=1,..,\lambda) \tag{13}$$

The model is redefined at each sampling instant, k, having the current state and previous control forces:

$$\hat{\boldsymbol{X}}(k|k) = \boldsymbol{X}(k) \tag{14}$$

$$\hat{u}(k-j|k) = u(k-j) \qquad (j=1,...,d) \tag{15}$$

Now based on this model, the predictive control formulation can be derived.

DERIVING THE PREDICTIVE CONTROL LAW

In the predictive method, first a performance index is defined which penalizes predicted response values as well as the control force along or at the end of a prediction horizon. The control sequence is the one that minimizes this performance index.

Sprung mass acceleration, representing ride comfort, and suspension deflection as packaging requirement criterion are penalized. These variables form the penalized vector, $\boldsymbol{Y}$, defined in terms of the state vector as follows:

$$\boldsymbol{Y}_c = \begin{bmatrix} z_c - w_c \\ \ddot{z}_c \end{bmatrix} = \begin{bmatrix} x_c \\ \ddot{x}_c + \ddot{w}_c \end{bmatrix} \tag{16}$$

Using (3):

$$\boldsymbol{Y}_c(t) = \mathbf{M}\boldsymbol{X}_c(t) + \boldsymbol{N}u(t-\tau) \tag{17}$$

where:

$$\mathbf{M} = \begin{bmatrix} 1 & 0 \\ -\omega_n^2 & -2\xi\omega_n \end{bmatrix} \quad N = \begin{bmatrix} 0 \\ -\frac{1}{m} \end{bmatrix} \tag{18}$$

Equation (17) can be discretized as:

$$\boldsymbol{Y}(k+j) = \mathbf{M}\boldsymbol{X}(k+j) + \boldsymbol{N}u(k+j-d) \tag{19}$$

Then the predicted penalized vector at sampling time k will be:

$$\hat{\boldsymbol{Y}}(k+j|k) = \mathbf{M}\hat{\boldsymbol{X}}(k+j|k) + \boldsymbol{N}u(k+j-d|k) \tag{20}$$

Here instead of a weighted sum of responses along the prediction horizon only the response at $\lambda + d$ time steps ahead:

$$\hat{\boldsymbol{Y}}(k+\lambda+d|k) = \mathbf{M}\hat{\boldsymbol{X}}(k+\lambda+d|k) + \boldsymbol{N}\hat{u}(k+\lambda|k) \tag{21}$$

is penalized. Assuming a constant control force along the control horizon the performance index can then be defined:

$$J = \frac{1}{2}\hat{\boldsymbol{Y}}(k+\lambda+d|k)^T \mathbf{Q}\hat{\boldsymbol{Y}}(k+\lambda+d|k) + \frac{1}{2}\hat{u}(k|k)^T R\hat{u}(k|k) \tag{22}$$

with the conditions:

$$u(k|k) = u(k+1|k) = \ldots = u(k+\lambda|k) = u(k) \tag{23}$$

$$\hat{\boldsymbol{X}}(k|k) = \boldsymbol{X}(k) \tag{24}$$

where $\mathbf{Q}$ is a positive semi-definite symmetric weighting matrix and R is a positive scalar weight.

On the other hand by successive use of Eqn. 13, Eqn. 21 may be rearranged as:

$$\hat{\boldsymbol{Y}}(k+\lambda+d|k) = \mathbf{S}\boldsymbol{X}(k) + \boldsymbol{Z_0}u(k) + \sum_{i=1}^{d} \boldsymbol{Z_i}u(k-i) \tag{25}$$

where:

$$\mathbf{S} = \mathbf{M}\mathbf{A}^{\lambda+d} \tag{26}$$

$$\boldsymbol{Z_0} = \mathbf{M}(\mathbf{A}^{\lambda-1} + \mathbf{A}^{\lambda-2} + \ldots + \mathbf{A}^2 + \mathbf{A} \quad + \mathbf{I})\boldsymbol{B} + \boldsymbol{N} \tag{27}$$

$$\boldsymbol{Z_i} = \mathbf{M}\mathbf{A}^{\lambda+i-1}\boldsymbol{B} \tag{28}$$

By substituting Eqn. 25 into the performance index and differentiating with respect to u:

$$\frac{\partial J}{\partial u(k)} = 0 \tag{29}$$

and with some more manipulations, the control law can be derived:

$$u(k) = \mathbf{D}\boldsymbol{X}(k) + \sum_{i=1}^{d} \boldsymbol{H}_i u(k-i) \tag{30}$$

where $\mathbf{D}$ is a gain:

$$\mathbf{D} = -(Z_0^T \mathbf{Q} Z_0 + R)^{-1} Z_0^T \mathbf{QS} \tag{31}$$

and:

$$\boldsymbol{H}_i = -(Z_0^T \mathbf{Q} Z_0 + R)^{-1} Z_0^T \mathbf{Q} Z_i \qquad (i=1,\ldots d) \tag{32}$$

NUMERICAL SIMULATION RESULTS

The single DOF quarter car model, shown in Fig. 2., was subjected to a sinusoidal road input and performance of active and passive methods in vibration isolation were compared.

The sprung mass is 350 kg. Both for passive and active system the suspension stiffness is 15000 N/m and damping is 20%. Tire stiffness was assumed high to simplify the model to a single degree of freedom model.

A 1 Hz sinusoidal road profile was used to evaluate and compare the performance of passive and active suspensions with different values of time delay for low frequency disturbances. A sampling time of 0.02 seconds was used in all simulations. For predictive controller, a prediction horizon of eight sampling time intervals was chosen. The first diagonal member of weighting matrix, Q, is unity while the three other members are zeros.

Figures (3) show the passive and active control results for $R=1\times10^{-5}$ and a single-interval time delay (time delay=0.02s). It is observed that the predictive control method has performed well, both in body acceleration control and also in suspension deflection reduction. With 79% lower suspension deflection the active system has reduced the RMS and maximum sprung mass acceleration by about 48% compared to the case of passive suspension.

Effect of weighting factor, R, has been shown in figure (4) for time delays of 0.02, 0.06 and 0.08 seconds. As expected the lower the value of R, less penalization is on control force and with higher control forces, body acceleration and suspension deflection are controlled more. However it can be observed that with increase of time delay, less control can be achieved on body acceleration and suspension deflection.

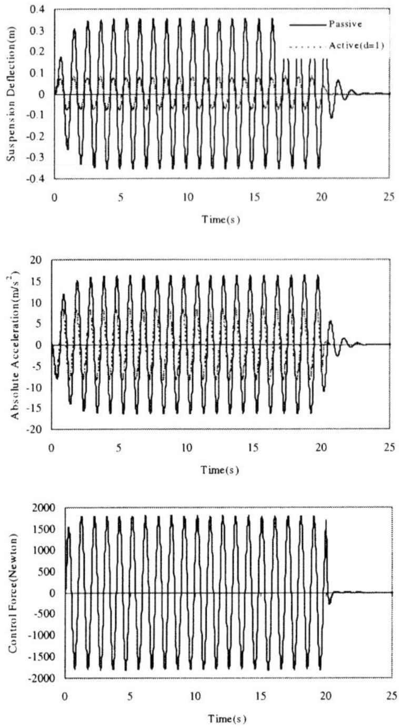

Figure 3. Passive and active control performance with 0.02 seconds time delay in the active system

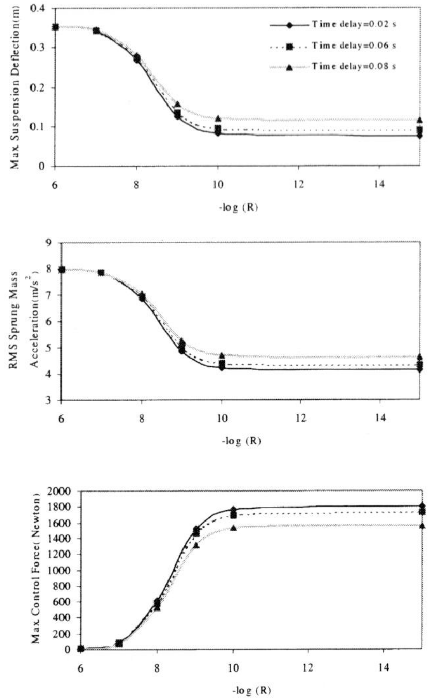

Figure 4. Effect of weighting factor, R, on control

CONCLUSION

A predictive control methodology was proposed and formulated for active suspension control of vehicles. Results indicate that predictive method has performed better than the passive system in controlling body acceleration and suspension deflection of a quarter-car model, even when there is a reasonable time delay in the control loop that happens in real situations. This has been for low frequency road inputs. Preliminary study shows that in high frequency vibrations, this method is able to reduce suspension deflection while it does not perform well in acceleration control. A different selection of penalized variables to form the performance index is currently under research, and first outcomes show that this approach can improve the active method performance in high frequency inputs.

REFERENCES

1. Elbeheiry, E.M., Karnopp, D.C., Elaraby, E.M., and Abdelraaouf, A.M., (1995) " Advanced ground vehicle suspension systems – A classified bibliography." *Vehicle System Dynamics*, **24**, 231-258.

2. Chalasani, R.M., (1986) "Ride performance potential of active suspension systems- Part I: Simplified analysis based on a quarter-car model", *ASME. AMD*, **80**, 187-202.

3. Chalasani, R.M., (1986) "Ride performance potential of active suspension systems- Part II: Comprehensive analysis based on a full-car model", *ASME. AMD*, **80**, 205-234.

4. Hrovat, D., (1993) " Application of optimal control to advanced automotive suspension design." *ASME Journal of Dynamic Systems, Measurement, and Control*, **115**, 328-342.

5. Esmailzadeh, E., and Fahimi, F., (1997) " Optimal adaptive active suspensions for a full car model." *Vehicle System Dynamics*, **27**, 89-107.

6. Gordon, T.J., Palkovics, L., Pilbeam, C., and Sharp, R, (1993) "Second generation approaches to active and semi-active suspension control system design." *Vehicle System Dynamics proceedings of the 13th IVASD Symposium on the Dynamics of Vehicles on Roads and on Tracks,* 158-171.

7. Mosca, E., (1995) *Optimal, predictive, and adaptive control.*, Prentice Hall, NJ.

8. Camacho, E.F., and Bordons, C., (1995) *Model predictive control in process industry.*, Springer-Verlag, London; UK.

9. Soeterboek, R., (1992) *Predictive control: A unified approach.*, Prentice Hall International., UK.

10. Rodellar, J., Chung L.L., Soong, T.T., and Reinhorn A.M., (1989) " Experimental digital control of structures." *ASCE J. Engrg. Mech.*, **115**, 1245-1261.

11. Gopalasani, S., Osorio, C., Hedrick, K., and Rajamani, R., (1997) "Model predictive control for active suspensions-controller design and experimental study." *Proceedings of the ASME Dynamic Systems and Control Division*, **61**, 725-733.

12. Tomizuka, M., (1976) " Optimum linear preview control with application to vehicle suspension-Revisited." *Journal of Dynamic Systems, Measurement, and Control*, **98**, 309-315.

13. Hac, A., (1992) "Optimal linear preview control of active vehicle suspension." *Vehicle System Dynamics*, **21**, 167-195.

14. Hac, A., and Youn, I., (1993) "Optimal design of active and semi-active suspensions including time delays and preview." *ASME Journal of Vibration and Acoustics,* **115**, 498-508.

15. Franklin, G.F., Powell, J.D., and Emami-Naeini, A., (1994*) Feedback control of dynamic systems:* Section 3.6.2, Addison-Wesley.

Optimization of ride comfort

O FRIBERG
Department of Structural Mechanics, Chalmers University of Technology, Göteborg, Sweden
P ERIKSSON
Structural Dynamics and Durability, SCANIA CV AB Buses and Coaches, Katrineholm, Sweden

ABSTRACT

Ride comfort optimisation, with the objective to minimise passenger and driver discomfort, of buses is studied. Transient acceleration response, due to road irregularities, at various locations in the bus is computed using a commercial FEM-program. The FEM-program input files are parameterised and linked to design variables influencing spring stiffness, beam cross-sectional properties, etc. Frequency weighting according to the comfort standard ISO 2631:1997 is applied and root mean square (RMS) values or vibration dose values (VDV) are included in the objective function to be minimised. One aim of the study is to obtain a basic understanding, without lengthy computations, of properties of such objective functions. Previous studies, using a large FE-model, have indicated presence of local minima and it is therefore of interest to gain more insight into fundamental properties of objective functions used for ride optimisation problems. A future aim is to evaluate the number of objective function evaluations using different optimisation procedures. Such information, evaluated from a small model, is of interest before optimisation of full-scale models. Needs for global optimisation algorithms are identified.

1 INTRODUCTION

Vehicle comfort is a complex subject. The human perception of comfort depends on external factors such as sound and vibration levels, exposure time, temperature and humidity as well as on intra-subject factors (body position, orientation, etc.) and inter-subject variability (age, experience, expectation, etc.) (Griffin, 1990). In this work, structural vibration in a lower frequency region, up to 80 Hz, is considered. Passengers and drivers in bus vehicles are exposed to such vibrations, especially in cities due to road obstacles and frequent start/stop manoeuvres. Designs of city buses may require low floors without steps, large interior space for rapid passenger flow and large openings for entrances/exits, see Figure 1. Design changes due to such requirements may have a serious impact on the comfort characteristics of a bus.

Computer simulations at an early design state are desirable. In order to make detailed evaluations of vibrations, fairly large computer models are required (van Asperen and Voets, 1986). If structural optimisation is added, resulting in substantially increased needs for computer resources, it is of interest to evaluate different mathematical algorithms and computational strategies for ride comfort optimisation.

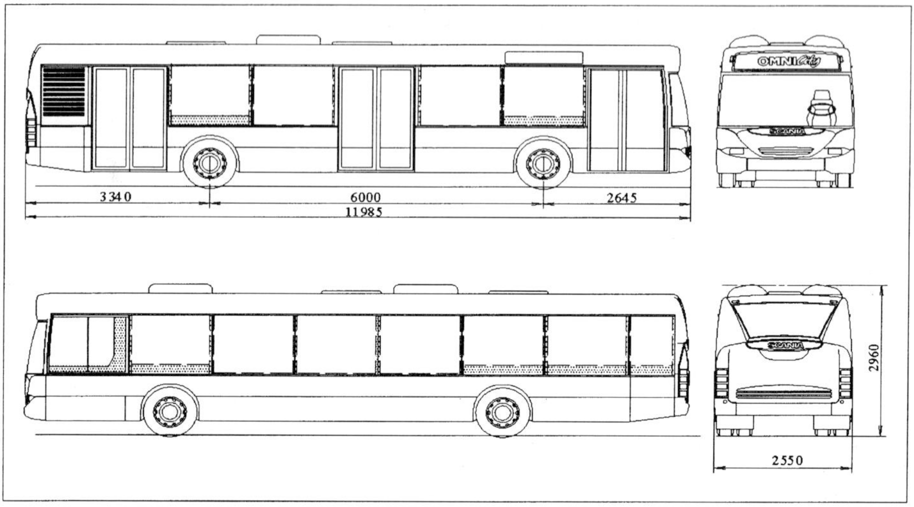

Figure 1. Main dimensions (mm) of the SCANIA OmniCity bus

2 BUS EXCITATION AND RIDE COMFORT EVALUATION

Structural vibrations are generated by road irregularities, tyre/wheel imperfections, driveline forces and driver steering wheel or gas/brake input. Herein, road irregularities are considered only. One may then study the response driving on measured road profiles (Anderson, 1998), (Eriksson and Friberg, 2000) or when passing a single obstacle (Eriksson, 2001). The latter case has the merits of better experimental repeatability and more convenient ways of comparing bus designs at different geographic locations. A driving case according to Figure 2 is considered here where an obstacle on the road is passed at 20 km/h. A rear axle passage of the obstacle is considered only. The accelerations are evaluated at locations A1, A2 and A3, corresponding to the driver seat, a standing position and the back sofa, respectively.

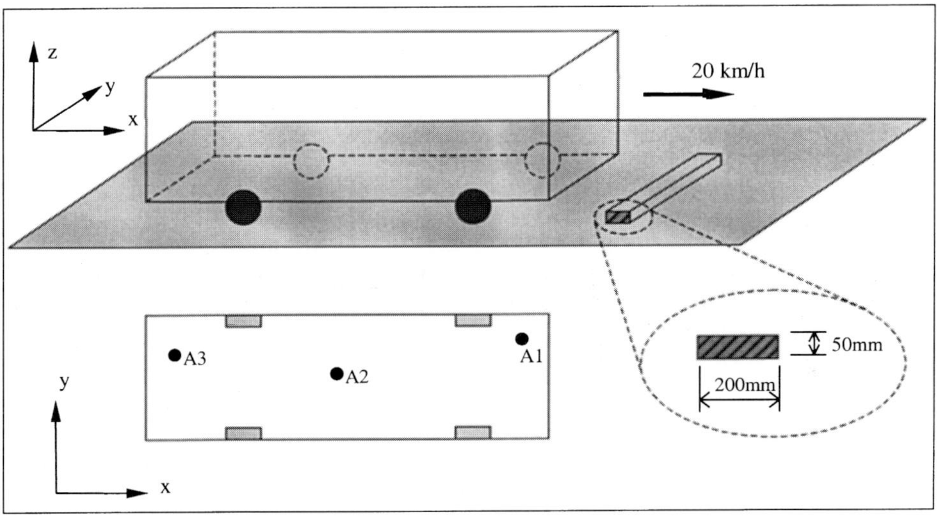

Figure 2. Obstacle passing. Locations A1, A2 and A3 for ride comfort evaluation

In order to mathematically quantify comfort, the ISO 2631:1997 (ISO, 1997) standard is applied. Based on human sensitivity to vibrations, this standard assigns weighting filters, in the frequency range 0.5-80 Hz, to acceleration response. Different filters are defined depending on directions (vertical, horizontal, etc.) and positions (seat-back, feet, etc.). In this study, only vertical accelerations are considered and the "W_k"-filter is applicable. The standard defines two methods for comfort evaluations, a so-called basic evaluation method using RMS values and an additional vibration dose value method, giving the following expressions

$$f_{rms} = \sqrt{\frac{1}{T}\int_{t_1}^{t_2} [a_w(t)]^2 dt} \tag{1a}$$

$$f_{vdv} = \sqrt[4]{\int_{t_1}^{t_2} [a_w(t)]^4 dt} \tag{1b}$$

where $T = t_2 - t_1$ is the time interval of interest and $a_w(t)$ is the frequency weighted acceleration, i.e. after the filter has been applied to the original acceleration. The basic method is intended for continuously running signals whereas the dose value method is intended for shocks and transients.

In order to define a global measure, i.e. an objective function for optimisation purposes, we evaluate f_{vdv} or f_{rms} values at various locations within the bus and perform a simple summation (leaving an open question how to give priority to drivers, back-sofa passengers et al). The following vibration dose value based objective function f_{vdv} (**x**) is thus defined to quantify such a global discomfort value of the bus

$$f_{vdv}(\mathbf{x}) = \frac{f_{vdv}^{A1}(\mathbf{x}) + f_{vdv}^{A2}(\mathbf{x}) + f_{vdv}^{A3}(\mathbf{x})}{f_{vdv}^{reference}} \qquad (2)$$

where f_{vdv}^{A1}, f_{vdv}^{A2} and f_{vdv}^{A3} are the vibration dose values, according to (1b), at locations A1, A2 and A3, respectively. The vector **x** contains normalised design variables. The quantity $f_{vdv}^{reference}$ is a normalising value defining $f_{vdv}(\mathbf{x}^{nom}) = 1$, where $\mathbf{x}^{nom}$ is the nominal design vector (all elements in **x** being equal to unity). By changing the subscript *vdv* to *rms* in equation (2) and applying equation (1a), we define a corresponding RMS-value based objective function f_{rms} (**x**).

3 COMPUTATIONAL TOOLS

The FE program ABAQUS was used (Hibbit, Karlsson & Sorensson, Inc., 1998) to compute response in the time domain of the bus passing the obstacle. After a CAD-program model creation, the ABAQUS input files were parameterised using an "in-house" technique, where real value fields were replaced by special text-combinations and linked to user-written FORTRAN-routines defining the design variables. Also, in-house routines were used for the frequency weighting. The software IDESIGN (Arora, 1989a,b) was the selected optimisation tool when minimising discomfort objective functions.

4 PREVIOUS STUDIES USING A LARGER FE-MODEL

In previous studies, (Eriksson and Friberg, 2000) and (Eriksson, 2001), a FE model according to Figure 3 was used for ride comfort evaluation and optimisation. The model contains 10694 finite elements and 7861 nodes resulting in approximately 50000 dofs. By using substructuring of the chassis frame and the body, the number of dofs was decreased to approximately 900. The implicit Newmark-scheme, with HHT-modifications, was applied for time integration.

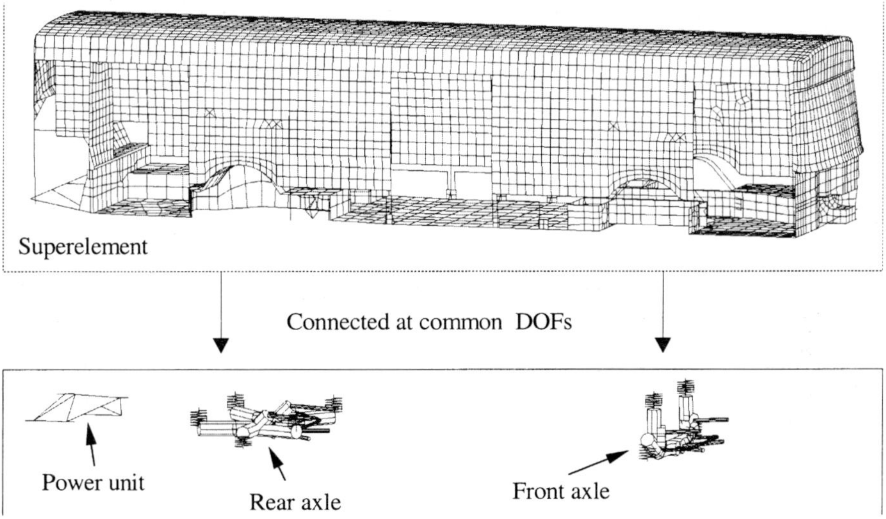

Figure 3. FE model of bus structure, power unit and axles

In Figure 4 the obstacle modelling is shown. Also, the simplified tyre model, neglecting longitudinal forces and damping, is indicated. The tyre obstacle passage takes 0.07 s, the considered time interval T for acceleration evaluation is 1.28 s and a time increment $\Delta t = 2.5 \cdot 10^{-3}$ s was used.

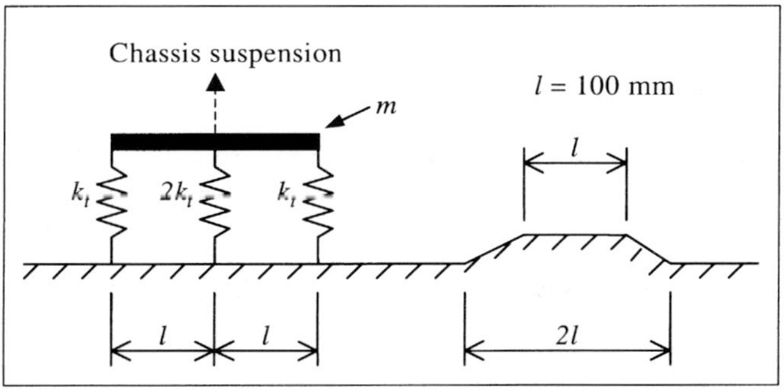

Figure 4. Rear axle/tyre and obstacle modelling. Obstacle height is 50 mm

Initially 11 design variables, influencing physical dimensions of beams and aluminium sheets within the bus body structure, were selected. Based on sensitivities at the nominal design, 6 of these design variables were selected as active and an optimisation was carried out by IDESIGN with Equation (2) as the objective function. Altogether 11 iterations, requiring 77 objective function evaluations (approximately 20 minutes of CPU-time per evaluation), were needed to achieve an optimum value reducing the objective function by 24%. However, this optimum was found to be a local one and an increase in CPU-time, choosing alternative starting designs, will result if the global optimum within the design space is sought for.

5 CURRENT STUDY USING A BEAM MODEL

Due to the presence of local minima, it was decided to create a smaller FE model for tests of optimisation algorithms. Since the substructure condensation of the bus body and chassis takes the major computational time, these parts were replaced with a continuous beam built-up from 49 beam elements. The tyre, axle, wheel, suspension and engine models were kept accounting for symmetry where appropriate. Boundary conditions corresponding to plane motion (the *xz*-plane in Figure 1) of the beam were applied. The resulting model has 97 elements, 71 nodes and approximately 400 dofs. One objective function evaluation took 50 seconds of CPU-time.

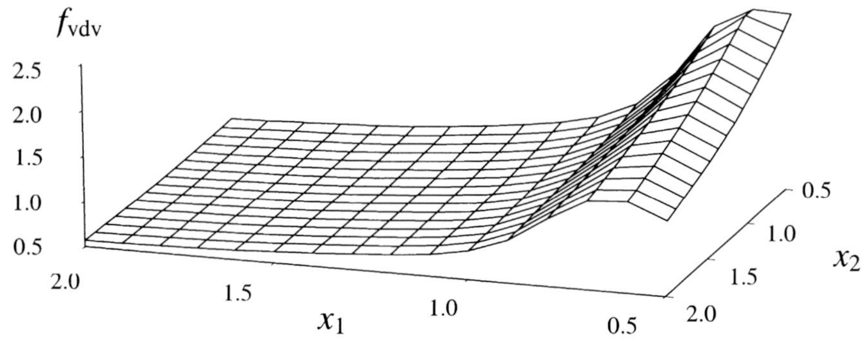

Figure 5. Objective function f_{vdv} as a function of design variables x_1 and x_2

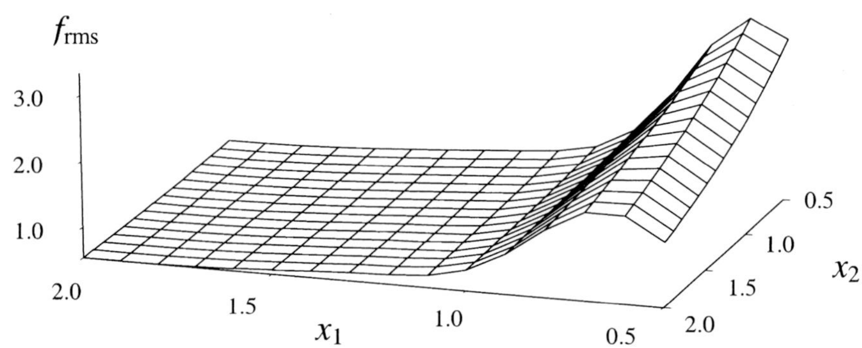

Figure 6. Objective function f_{rms} as a function of design variables x_1 and x_2

In order to visualise discomfort objective function dependencies on design variables, only 2 normalised design variables were defined. The first, x_1, is a common multiplier for stiffness and damping values of the four engine rubber insulators. The second design variable, x_2, affects the beam bending stiffness linearly and also the mass per unit beam length. Non-structural mass, constant and distributed evenly along the beam, is present.

In Figures 5 and 6, the objective function surfaces corresponding to Equation (2) for both vibration dose values and RMS-values are plotted. Each plot shows the result of 256 (16x16) objective function evaluations using 0.5 as lower and 2.0 as upper bounds for both design variables. It is concluded that an optimum searching algorithm based on gradients will give different results depending on the starting design.

An increased number of design variables will probably increase the number of local minima and the need of global search algorithms is thus identified.

6 DISCUSSIONS AND CONCLUSIONS

Comparisons between the large FE model and the beam model show that the accelerations computed in the latter case have substantially lower magnitudes then in the former (i.e. a factor 2-4 depending on location A1-A3). It must be immediately stated that the beam used had constant properties along the whole length, i.e. account for large door openings, etc. was not taken. This is not, however, the only main source of discrepancy. Design sensitivities, at the nominal design using the large FE model, are high for stiffness design variables influencing the cross-sectional "distortion" not accounted for by the beam model. Further investigations are needed to evaluate these mutual effects. However, the objective of this study was mainly to create a computationally fast model, of reasonable but limited accuracy, that can be used for test purposes of optimisation algorithms.

7 ACKNOWLEDGEMENT

The support from SCANIA CV AB Buses & Coaches and The Swedish National Council for Technical Research and Vehicle Engineering is gratefully acknowledged.

8 REFERENCES

Anderson R.J. (1998), "Multi-body dynamics of the ride quality of an off-road articulated truck", C553/026, *IMechE Conference Transactions 1998-13,* International Conference on Multi-Body Dynamics, New Techniques and Applications, December 10-11, London, UK

Arora J.S. (1989a), *IDESIGN User's Manual, Version 3.5.2, Technical Report ODL-89.7,* College of Engineering, The University of Iowa, Iowa City, USA

Arora J.S. (1989b), *Introduction to Optimum Design,* McGraw-Hill, New York, USA

Eriksson P. and Friberg O. (2000), "Ride comfort optimization of a city bus", *Structural Optimization*, Springer-Verlag (to appear)

Eriksson P. (2001), "Optimization of a bus body structure", *Heavy Vehicle Systems*, Special Series, *International Journal of Vehicle Design*, Inderscience Enterprises Ltd (to appear)

Griffin M.J. (1990), *Handbook of Human Vibration*, Academic Press Limited, London, UK

Hibbit, Karlsson & Sorenson, Inc. (1998), *ABAQUS/Standard User's Manual, Version 5.8,* Hibbit, Karlsson & Sorenson, Inc., Pawtucket Rhode Island, USA

ISO (1997), "Mechanical vibration and shock – Evaluation of human exposure to whole-body vibration – Part 1: General requirements", *ISO 2631-1:1997(E),* ISO, Geneva, Switzerland

van Asperen F., Voets H. (1986), "Optimization of the dynamic behavior of a city bus structure", *IMechE Conference Publications 1986 MEP-257,* International Conference on the Bus '86, London, UK

Passenger and carbody interaction in rail vehicle dynamics

P CARLBOM
Division of Railway Technology, Department of Vehicle Engineering, KTH, Stockholm, Sweden

SYNOPSIS

The passenger and carbody interaction has an influence on ride comfort in rail vehicles. It is therefore of interest to include simple models of this interaction in rail vehicle dynamics simulation models. This paper proposes a basic two-degree-of-freedom model of passengers and carbody. A modal representation of the carbody is assumed. The model suggests that a parameter taking into account passengers' weight and location is a suitable measure of the interaction. Experimental modal analysis of a rail vehicle carbody with passengers is carried out to test the proposed model, and it seems to be appropriate.

1 INTRODUCTION

Ride comfort is of major concern in the development of competitive rail vehicles. In the design phase of a vehicle, numerical simulation offers a powerful tool to predict vibration levels and ride comfort. In order to cut computational cost a small model size is preferable. The rail vehicle is often modelled as a multi-body system with a flexible carbody, since the structural vibrations of the carbody account for a large part of the vibrations. The carbody is usually modelled by the finite element method, and the model is then reduced by selecting low-frequency eigenmodes of the free carbody. This method is appropriate since present comfort standard and praxis (1) emphasize low frequencies up to 20 Hz. However, since there is a trend towards considering higher frequencies and towards focusing closer onto the passengers, the comfort should be evaluated at the interface between passenger and carbody, i.e. seat pan, back- and foot rests.

Simulation of rail vehicle-track dynamics is nowadays a well established discipline, and there are many software packages that meet the needs, see for instance (2). The framework of flexible multi-body dynamics for vehicle dynamics is also well developed, see for instance (3).

Human vibration is truly multi-disciplinary and "involves physics, psychology, mathematics, physiology, engineering, medicine and statistics" (4). The human body models are mainly based on impedance measurements, mostly in the vertical direction. One of the first published measurements is found in (5). Quite extensive measurements have been made recently (6). Simple few-degree-of-freedom models are defined in the standard (7).

There are, however, few published studies focusing on the interface and interaction between passengers and the rail vehicle. One study found in the literature is (8). But, in rail vehicle industry the interest is awaken and strong.

The model proposed in this paper combines human body modelling with vehicle system dynamics. The work is based on a case study of two common, almost identical, Swedish rail vehicles, the SJ-S4M (9)(10) and the SJ-B7, see Fig. 1. Experimental modal analysis of the latter is carried out with a carbody, empty as well as filled with passengers, where different "passenger distributions", i.e. passengers sit at different locations, are tested.

In the next Section a passenger-carbody model is proposed and discussed, and a "passenger load parameter" is defined. Then, in the following Section the experimental modal analysis of the SJ-B7 vehicle is described. Modal analysis results are given, and the "passenger load parameter" is tested. In the final Section the validity of the proposed model is discussed and further work is outlined.

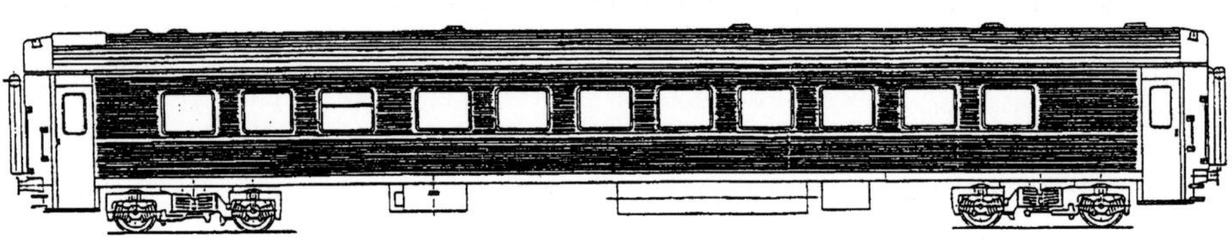

Fig. 1 The SJ-B7 (SJ-S4M) vehicle. (26.4 m long)

2 MODELLING

2.1 Human body modelling

For the purposes of the present study, models that provide a simple mathematical summary of the human body dynamics are needed. The present study is limited to vertical vibration, since, on the one hand, vertical structural vibrations proved to be the most important in the case study (9), and on the other hand, most of the studies on human whole-body dynamics found in literature treat vertical vibrations. Few-degree-of-freedom models of this type are given in references (5)(6)(7). It must be stressed that these models do not explain the motion of different body parts, but merely describe the seated human body exposed to vertical vibration by mathematical models. The models are based on impedance measurements: the subject sits on a stiff support which is excited, the exciting force and the resulting acceleration are measured and the impedance, i.e. the ratio between the force and the resulting velocity, is calculated.

Up to now some hundred persons have been measured by this method, and the investigations show similar results. The most important effect is a peak in the impedance function at between

4 and 6 Hz. The results depend to a large extent on the posture, sitting erect or slouched makes a difference. Statistically, fitted model parameters for a group of men and a group of women differ only marginally (6). Also, the excitation amplitude seems to be less important than the posture.

It is worthwhile to point out the difference between perceived ride comfort and actual body vibration, there is no proven correlation between these. The perception of vibration is complex and there is not a well-defined "centre" of perception. Therefore the present standard (1) focuses on the vibrations at the interface, that is seat pan, back- and foot rest, and relate these vibrations to discomfort. According to the standard vertical vibrations between 5 and 15 Hz are perceived as more annoying than vibrations at other frequencies.

Here a one-degree-of-freedom human body model of the type proposed in (6) is used.

2.2 Carbody modelling

For rail vehicle dynamics analysis an eigenmode representation of the free carbody is often chosen. The carbody mode shapes and their eigenfrequencies are obtained from finite element calculations. Relative damping values are either obtained from measurements, or guessed. The modal model of the carbody is inserted into a multi-body model of the vehicle consisting of carbody, bogies, wheel sets and track. The track excites the carbody via the bogies. The contact between wheel and rail is non-linear and makes numerical simulation necessary.

A reduced modal model is chosen because it is preferable to have a small number of degrees-of-freedom in numerical simulation to cut computation cost. This is feasible since most vibration comfort studies focus on lower frequencies, up to 20 Hz, making it possible to neglect higher frequency modes.

The carbody mode shapes are typical for an oblong box: lateral and vertical bending modes, shear modes, torsional modes and breathing modes, where walls, floor and roof move in and out. The lowest eigenfrequency typically lies between 8 and 12 Hz.

Here a modal model of the carbody is assumed, with some ten global deformation modes with eigenfrequencies below 20 Hz, but also rigid body modes may be considered. Each carbody mode is here dealt with separately, however.

2.3 Proposed passenger-carbody model

2.3.1 Aim and scope

The proposed model is intentionally simple, although it is not the simplest one; it is also possible to model the passenger by lumping all the passenger mass to the carbody structure. The aim of the present model is to capture and explain the main features of the passenger-carbody dynamics and it is intended to be included in a flexible multi-body model of the rail vehicle.

The proposed model has few parameters. Three of them are related to dynamical properties of the human body, and values may be taken from literature. The key parameter is a so-called "passenger load parameter" defined below, which includes the passenger mass and as well as "passenger distribution", i.e. where passengers sit.

2.3.2 The model

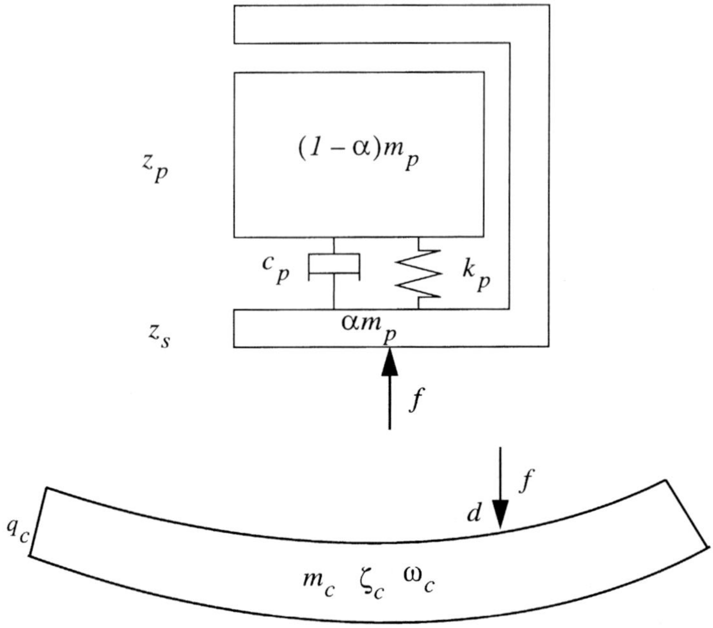

Fig. 2 Human body and carbody model

To begin with a single passenger is considered, the model is then generalized to include several passengers in Section 2.3.3. The *passenger* is represented by a human body model with two masses, see top of Fig. 2, corresponding to model "1b" in (6). The total body mass m_p is partitioned in a sprung and an unsprung mass. A vertical force f from the carbody acts on the unsprung part αm_p. Between the two masses in the model there is a linear spring and a linear viscous damper. The values of the spring constant k_p and the damper constant c_p may be obtained from impedance measurements. Most of the published measurements indicate that k_p and c_p are proportional to m_p. Here this is assumed to be the case, giving

$$k_p = \omega_p^2 m_p \qquad [1]$$

$$c_p = 2\zeta_p \omega_p m_p \qquad [2]$$

introducing the "undamped circular eigenfrequency" ω_p and the "relative damping" ζ_p of the human body model. The three parameters ω_p, ζ_p and α describe the dynamical properties of the human body, while the body mass m_p is used to scale the model. Most of the values of the relative damping found in literature are about 50% and the undamped eigenfrequency varies between 4 and 6 Hz.The fraction of unsprung mass α may be between 5% and 20%.

After attributing a vertical degree of freedom z_p to the sprung mass (p as in passenger) and a vertical degree of freedom z_s (s as in seat) to the unsprung mass, the equations of motion for the human body model may be set up

$$(1-\alpha)m_p\ddot{z}_p + c_p(\dot{z}_p - \dot{z}_s) + k_p(z_p - z_s) = 0 \qquad [3]$$

$$\alpha m_p\ddot{z}_s + c_p(\dot{z}_s - \dot{z}_p) + k_p(z_s - z_p) = -f \qquad [4]$$

The *carbody* is represented by a set of global eigenmodes. In principle the passenger introduces a coupling between all the carbody modes. However, it can be shown that the coupling is weak and negligible, due to the fact that the passenger mainly behaves as a sprung mass. As a consequence, each carbody mode can be dealt with separately. Consider one of them, the vertical bending mode for instance, cf. bottom of Fig. 2. The eigenfrequency of the vertical bending mode often lies between 8 Hz and 12 Hz. The carbody mode shape is here assumed to be mass-normalized, so that its modal mass equals 1 kg. The response to a point force f acting on the mode shape depends on the *vertical* displacement value d of the mode at this particular point. If the undamped circular eigenfrequency is denoted ω_c and the relative damping ζ_c, then the equation of motion of the carbody mode is

$$\ddot{q}_c + 2\zeta_c\omega_c\dot{q}_c + \omega_c^2 q_c = fd \qquad [5]$$

where q_c is the generalized degree-of-freedom of the carbody mode.

Assume that the unsprung mass in the human body model stays in permanent contact with the motion of the carbody at the point where the passenger sits, giving a direct relation between the carbody generalized degree of freedom q_c and the vertical degree of freedom of the unsprung mass z_s, namely

$$z_s = q_c d \qquad [6]$$

The resulting mechanical system has two degrees-of-freedom. If these are chosen as the vertical degree-of-freedom of the unsprung mass z_p and the carbody generalized degree-of-freedom q_c then, using equations [1] to [6], the equations of motion can be written as

$$\boldsymbol{M}\ddot{\boldsymbol{q}} + \boldsymbol{C}\dot{\boldsymbol{q}} + \boldsymbol{K}\boldsymbol{q} = \boldsymbol{0} \qquad [7]$$

with

$$\boldsymbol{q} = \begin{bmatrix} z_p \\ q_c \end{bmatrix} \qquad \boldsymbol{M} = \begin{bmatrix} (1-\alpha)m_p & 0 \\ 0 & \alpha m_p d^2 + 1 \end{bmatrix}$$

$$\boldsymbol{C} = 2\begin{bmatrix} m_p\zeta_p\omega_p & -m_p d\zeta_p\omega_p \\ -m_p d\zeta_p\omega_p & m_p d^2\zeta_p\omega_p + \zeta_c\omega_c \end{bmatrix} \quad \boldsymbol{K} = \begin{bmatrix} m_p\omega_p^2 & -m_p d\omega_p^2 \\ -m_p d\omega_p^2 & m_p d^2\omega_p^2 + \omega_c^2 \end{bmatrix} \qquad [8]$$

The roots of this coupled system are found by solving the characteristic equation in λ

$$\begin{aligned}&[(\omega_p^2 + 2\zeta_p\omega_p\lambda + (1-\alpha)\lambda^2)(\omega_c^2 + 2\zeta_c\omega_c\lambda + \lambda^2)\\&+ P\omega^2(\omega_p^2 + 2\zeta_p\omega_p\lambda + \alpha(1-\alpha)\lambda^2)]m_p = 0\end{aligned} \qquad [9]$$

where the "passenger load parameter"

$$P = m_p d^2 \qquad [10]$$

has been introduced. The coefficients in the polynomial in [9] are real, and therefore any complex roots are complex-conjugated pairs. The four roots of the coupled system are a function of P, as expressed by equation [9]. In this sense the parameter P measures the interaction between the passenger and the carbody. As P approaches zero, two of the roots, i.e. the "carbody mode roots", approach the roots of the empty carbody $\lambda_{1,2} = -\omega_c(\zeta_c \pm i\sqrt{1-\zeta_c^2})$. The other two roots approach the roots of the isolated human body model. The model thus predicts how the "carbody mode roots", and thereby the carbody modal eigenfrequency and damping, change as a function of the passenger load parameter P. The parameter depends on the mass as well as the location of the passenger. A passenger sitting in a node of a mode shape does not interact with the mode. In this case P evaluates to zero since then d equals zero.

2.3.3 Generalisation to several passengers

The model is here generalized to the case with several passengers. One may well assume that the passengers are alike from a mechanical modelling point of view, i.e. values for "human body relative damping" etc. do not vary too much within a population of passengers, so that mean values are meaningful.

Suppose there are N passengers sitting at different locations and having different masses m_{pi}. The index i refers to the passengers and runs from 1 to N. Denote the vertical displacement value of the mode where each passenger sits by d_i. Attribute two degrees-of-freedom z_{pi} and z_{si} to each passenger. Each passenger is described by a system corresponding to equations [3] and [4]. The force on the carbody from passenger i is now to be multiplied with d_i, and the total force is obtained by summing over all passengers, so that [5] turns into

$$\ddot{q}_c + 2\zeta_c\omega_c\dot{q}_c + \omega_c^2 q_c = \sum_i f_i d_i \qquad [11]$$

In this general case [6] is replaced by

$$z_{si} = q_c d_i \qquad [12]$$

The equations of motion for this general case may be written as [7], but now with

$$
\boldsymbol{q} = \begin{bmatrix} z_{p1} \\ \dots \\ z_{pN} \\ q_c \end{bmatrix} \quad \boldsymbol{M} = \begin{bmatrix} (1-\alpha)m_{p1} & 0 & 0 \\ & \dots & \\ 0 & (1-\alpha)m_{pN} & 0 \\ 0 & 0 & \sum_i \alpha m_{pi} d_i^2 + 1 \end{bmatrix}
$$

$$
\boldsymbol{C} = 2\begin{bmatrix} m_{p1}\zeta_p\omega_p & 0 & -m_{p1}d_1\zeta_p\omega_p \\ & \dots & \\ 0 & m_{pN}\zeta_p\omega_p & -m_{pN}d_N\zeta_p\omega_p \\ -m_{p1}d_1\zeta_p\omega_p & -m_{pN}d_N\zeta_p\omega_p & \sum_i m_{pi}d_i^2\zeta_p\omega_p + \zeta_c\omega_c \end{bmatrix} \qquad [13]
$$

$$
\boldsymbol{K} = \begin{bmatrix} m_{p1}\omega_p^2 & 0 & -m_{p1}d_1\omega_p^2 \\ & \dots & \\ 0 & m_{pN}\omega_p^2 & -m_{pN}d_N\omega_p^2 \\ -m_{p1}d_1\omega_p^2 & -m_{pN}d_N\omega_p^2 & \sum_i m_{pi}d_i^2{\omega_p}^2 + \omega_c^2 \end{bmatrix}
$$

The characteristic equation of this system has $2N+2$ roots. As in the previous case, the "carbody mode roots" approach $\lambda_{1,2} = -\omega_c(\zeta_c \pm i\sqrt{1-\zeta_c^2})$ as P approaches zero, where now

$$
P = \sum_{i=1}^{N} m_{pi} d_i^2 \qquad [14]
$$

Also as above, two of the roots approach the roots of the isolated human body mode. The remaining $2N$-2 roots correspond to modes that are not excited by the bogies, making it possible to reduce the system to a two degrees-of-freedom of the form [7][8], where now [3][4] may be regarded as describing a generalized passenger that represents the lot of the passengers, with m_p representing the total mass of the passengers and d representing a "mean deformation". The displacement z_{pi} of each passenger may be retrieved from z_p using [6] and [12]:

$$
z_{pi} = (d_i/d)z_p \qquad [15]
$$

As a measure of interaction one may study how the carbody mode eigenfrequency and relative damping change as a function of P as defined in [14]. Fig. 3 shows an example of what the proposed model predicts; relative damping increases almost linearly with P for small values of P. The parameter values are here $\omega_p = 2\pi \cdot 5$ rad/s, $\zeta_p = 0.5$, $\alpha = 0.1$, $\omega_c = 2\pi \cdot 9.2$ rad/s and $\zeta_c = 0.023$. The eigenfrequency may increase or decrease for small values of P depending on the values taken by α and ζ_p. In the next Section, in Fig. 6, corresponding measurement results are shown.

The passenger parameter P defined in equation [14] ranges typically from 0 to maybe 0.5 kgm^2 for a fully occupied vehicle, the maximum value depending on the mode shape. The maximum value of P for a particular mode shape may be estimated by multiplying the total weight of the passengers with a squared "mean deformation of the mode shape over the parts where the passengers sit". Such a mean was introduced by the author in (10), where some typical values for the mean deformation are found.

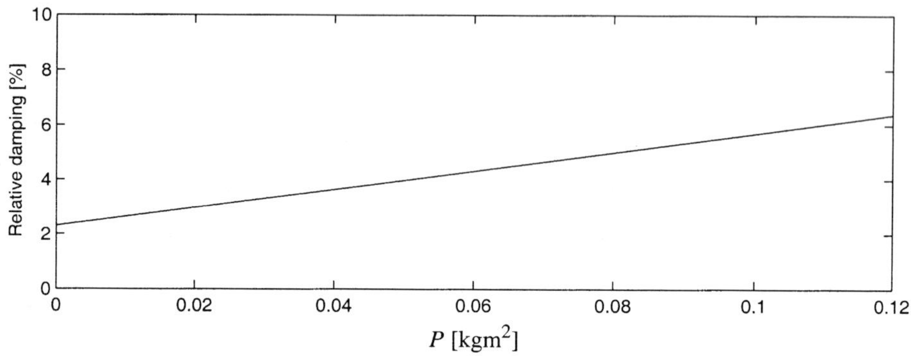

Fig. 3 Relative damping of a carbody mode as a function of passenger load parameter *P*

3 MEASUREMENTS

3.1 Experimental modal analysis of a carbody with passengers

3.1.1 Setup and excitation

A SJ-B7 vehicle was set up for experimental modal analysis, and measurements were carried out during two weeks in the spring of 1999. The attention was focused on the carbody modes that had proven to be important in previous on-track measurements (9), namely the carbody first vertical bending mode and torsion modes. An hydraulic exciter was placed to excite these modes, cf. Fig. 4, and fastened by bolts to ground and to one of the carbody "side-sills", i.e. the two beams along the junction of the floor and the two side walls. The carbody structure was excited with an almost-white-noise force spectrum from 0.5 Hz to 39 Hz. Four different excitation levels were used, but the results presented in this paper correspond to typical acceleration levels of a vehicle running on main-line tracks.

3.1.2 Response points and passengers

Vertical acceleration was measured at 24 points of the carbody as shown in Fig. 4, ten points out-doors on the side-sills, and fourteen on the inner floor at the feet of seats. The ten out-door accelerometers are intended to make it possible to identify the first vertical bending mode and torsion primarily, but also the second vertical bending mode and rigid body modes, such as roll and pitch. The fourteen accelerometers in-doors at the inner floor, close to the feet of seats, are used in order to obtain estimates of the values d_i in [14].

35 persons, mainly students, came to act *passengers* during one afternoon, and two sets of measurements were performed. The participants were asked about their weight, and the average was 66.6 kg with values ranging from 52 to 80 kg. The number of persons was chosen to make it possible to try different "passenger distributions". During half of the measurements the

passengers sat in the middle of the carbody and during the rest they sat at the ends of the carbody, see bottom of Fig. 4. In the first case there were only 33 passengers, unfortunately. In order to calculate the sum in equation [14], each occupied seat must be assigned a mass, and a vertical displacement value for each carbody mode. For each occupied seat the closest accelerometer at the inner floor is chosen. In this way the masses of the passengers in the seats belonging to an accelerometer position may be lumped, and it is these lumped values that are given, in kg, in Fig. 4.

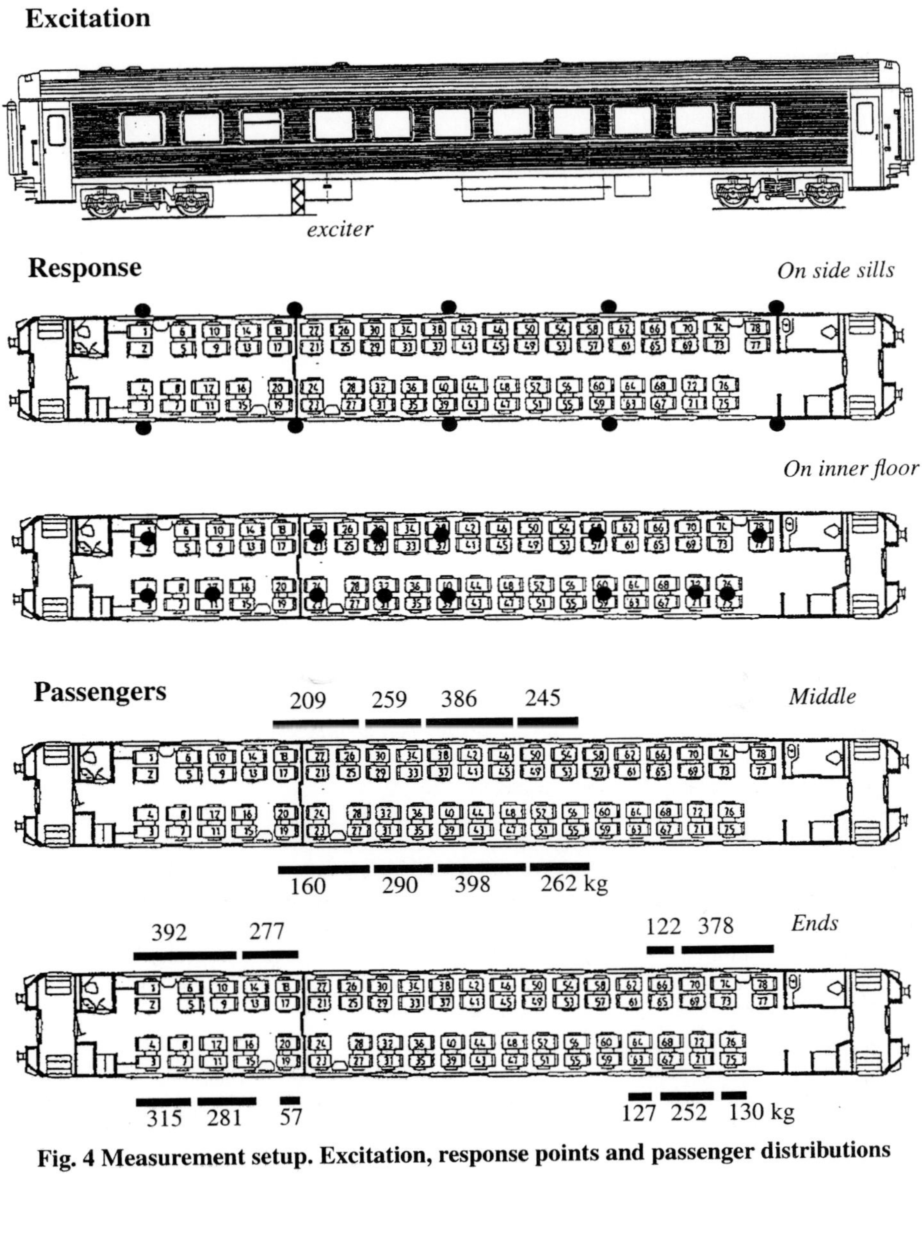

Fig. 4 Measurement setup. Excitation, response points and passenger distributions

3.2 Results

3.2.1 Mode shapes and modal parameters

Four global modes that have been identified are considered here: shear, first vertical bending and two torsion modes. The mode shapes are shown in Fig. 5.

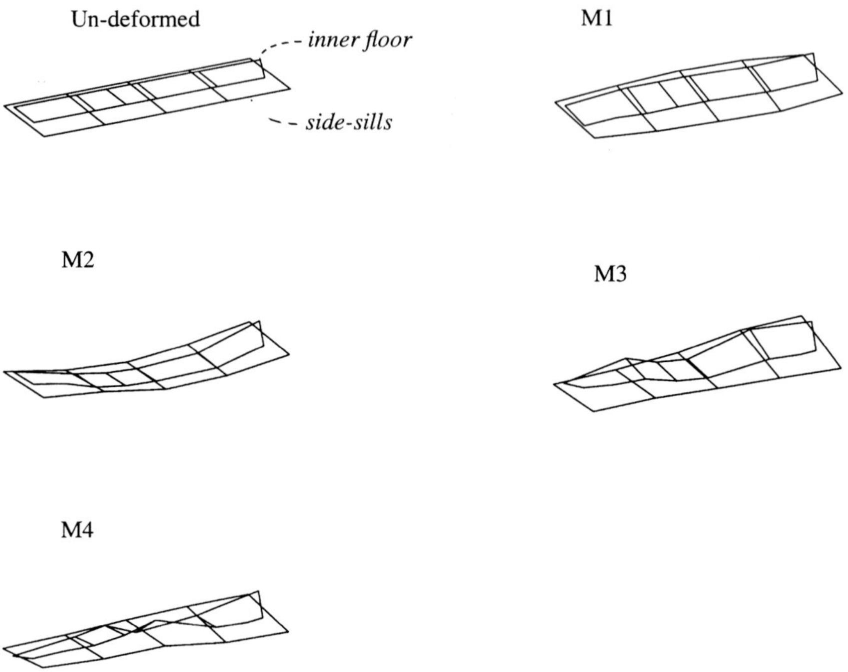

Fig. 5 Identified global mode shapes

In the figure the deformation of the inner floor as well as the deformation of the side-sills is shown. The numbers refer to Table 1, where the eigenfrequency and relative damping of the modes are given. The shear mode is difficult to identify due to the choice of response points. Looking at the shapes in Fig. 5 one notes that the inner floor follows the side-sills for shear and vertical bending. This is not the case for the torsion modes however.

Table 1 Modal parameters of identified modes

No	Shape	Frequency [Hz]	Relative damping[%]
M1	Shear	8.2	3.5
M2	Vertical bending	9.2	2.3
M3	Torsion	12.5	1.4
M4	Torsion	13.8	4.2

3.2.2 Passenger-carbody interaction

The eigenfrequency and relative damping of a carbody mode shape is a function of the passenger load parameter *P* according to equation [9]. The value of *P* is here calculated for each passenger distribution and each of the mode shapes, applying definition [14] and using the measured mode shapes and passenger weight data. Fig. 6 shows the measured eigenfrequency and relative damping as a function of *P* for these cases.

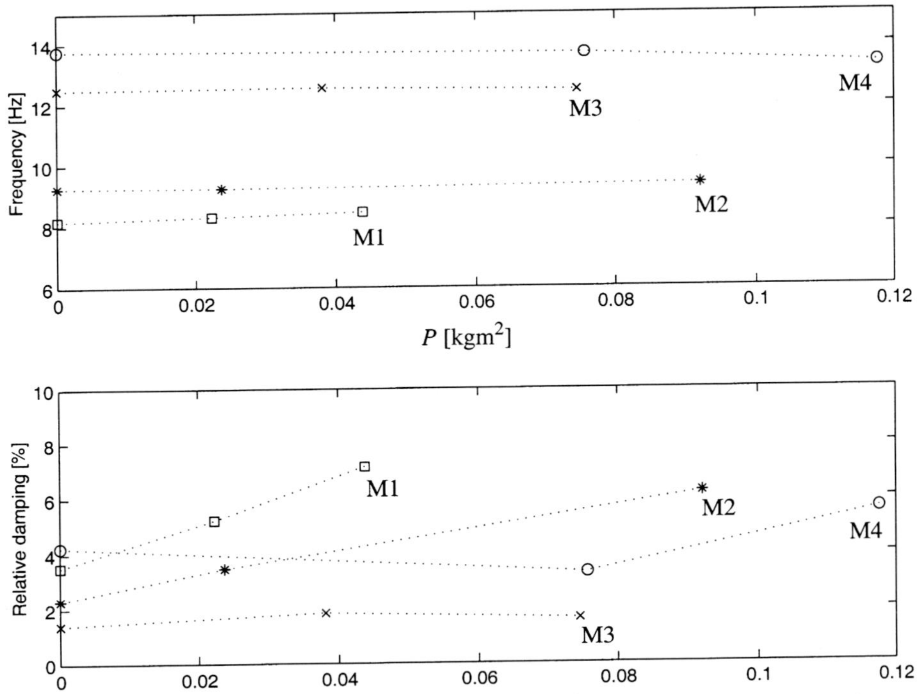

Fig. 6 Eigenfrequency and relative damping of carbody modes M1 to M4, cf. Table 1, as a function of the passenger load parameter *P*

4 DISCUSSION AND CONCLUSIONS

4.1 Conclusions from the measurements

The measurements presented here show clearly that the *passengers and vehicle interact* for this vehicle, already with only 35 passengers. Change in relative damping and eigenfrequency is here used to measure the interaction. Both eigenfrequency and relative damping are estimated from measured frequency response functions, and sometimes the interpretation is not straight-forward making this measure somewhat uncertain.

The relative damping increases with increasing values of *P*, although the trend is not clear. For the first two modes the dependency seems to be linear. Carbody mode eigenfrequency tends to increase with increasing values of *P*.

4.2 Validity of the proposed model

The main trend found in the measurements, that relative damping increases with increase in passenger load, is predicted by the proposed model. The model predicts that the change is linear in P for small values of P, that is below 0.5 kgm^2. The measurement results show that the trends are stronger for the first two modes, i.e. shear and vertical bending. This might be explained by the fact that their eigenfrequencies lie closer to the "human body eigenfrequency", and therefore interact more strongly. However, the quantitative change in carbody relative damping and eigenfrequency might not be correctly predicted by the model.

An important conclusion is however that *passengers should not be modelled by lumping masses to the carbody structure*. Rather, if passenger load is to be taken into account then the proposed model is the simplest one that may be used.

In the proposed model the seat dynamics, e.g. the seat stiffness, has not been taken into account. Also, non-linear behaviour is not considered here. Maybe the behaviour of the torsion modes, e.g. the decrease in damping, can be explained by generalizing the proposed model by including such effects. But, it is important to remember the aim of the model, namely to be included in a flexible multi-body model of the whole vehicle and the track, calling for a model with few degrees-of-freedom. Ultimately, it is the vibration comfort perceived by the passengers that matter.

ACKNOWLEDGEMENTS

The author wishes to thank Dr. Mats Berg for his support and valuable advice. The support, not only financial, from Adtranz is gratefully acknowledged.

REFERENCES

(1) ORE *Question 153B Application of the ISO 2631 standard to railway vehicles* (1989)

(2) S. Iwnicki *The Manchester Benchmarks for Rail Vehicle Simulation*, Supplement to Vehicle System Dynamics, Vol 31 (1999)

(3) O. Wallrapp *Entwicklung rechnergestützter Methoden der Mehrkörperdynamik in der Fahrzeugtechnik*, DFVLR-FB 89-17 (1989)

(4) M. Griffin *Handbook of Human Vibration*, Academic Press, ISBN 0-12-303040-4 (1990)

(5) R. Coermann *The Mechanical Impedance of the Human Body in Sitting and Standing Position at Low Frequencies*, Human Factors 4, pp 225-253 (1962)

(6) L. Wei and M. Griffin *Mathematical models for the apparent mass of the seated human body exposed to vertical vibration*, JSV 212(5), pp 855-874 (1998)

(7) International Organization for Standardization *Vibration and shock - mechanical driving point impedance of the human body* International Standard, ISO 5982 (1981)

(8) K. Andereg and P. Weichelt *Der Einfluß von Ausrüstung und Zuladung auf die vertikale Biegeschwingung eines Eisenbahn-Personenwagens,* ZEV-Glasers Annalen 114 (1990)

(9) P. Carlbom, *Structural flexibility - simulation and on-track measurements of rail vehicle car body dynamics*, paper presented at VSDIA, Budapest (1998)

(10) P. Carlbom, *Combining MBS with FEM for rail vehicle dynamics analysis*, to appear in Journal of Multibody System Dynamics (2000)

Human body modelling techniques for use with dynamic simulations

N LeGLATIN, M V BLUNDELL, and **S W THORPE**
School of Engineering, Coventry University, UK

ABSTRACT

This paper describes the initial work from a programme of research investigating the use of multibody simulations of real world pedestrian impacts. This work is underway and aims to extend simulations beyond the procedures developed to meet proposed legislation. An initial literature study is provided reviewing current test procedures and mathematical analysis techniques used to model and simulate the dynamic motion of the human body. Previous work using finite element and multibody systems analysis methods for both pedestrian and occupant impact studies has been reviewed.

The paper continues to describe how programs such as MADYMO are able to combine the multibody and finite element approach and are now moving forward from well established modelling techniques for crash dummies to full scale representations of the pedestrian. The paper concludes by discussing an analysis simulation methodology based on MADYMO that extends the pedestrian impact event beyond current test procedures. The work proposed is intended to lead to models of a moving pedestrian combined with the dynamics of a vehicle model for realistic 'real world' type simulations.

Keywords: Multibody Systems Analysis, Pedestrian Impact, Occupant Modelling, ADAMS, MADYMO, Real World

1 INTRODUCTION

During the last twenty years the focus within the automotive industry has been on the protection of the driver and occupants. This has been driven not only by legislation but also by a competitive market where the consumer has become more aware through the media of vehicle safety. As we move forward in the 21st century the automotive industry must now turn its attention to the design of 'pedestrian friendly' vehicles in order to respond to proposed developments in legislation.

In the European Union around 10,000 pedestrians are killed each year and 90,000 seriously injured [1]. An example of this is the year 1995 when there were 3621 fatalities, 45,523 seriously and 261,362 slightly injured, making a total of 310,506 casualties on the roads of Great Britain. Over the last 10 years over 46,000 people have been killed on English roads, with another 580,000 seriously injured. This does not including unreported accidents, which some estimates put at half as many again. On this basis, there have been an estimated 900,000 people killed and seriously injured on English roads in the past ten years. In most of Europe the number of pedestrian casualties has been dropping since the 1970`s [3] but the situation is still not satisfactory.

In most European countries pedestrians represent a significant proportion of the road accident casualties. As a result the European Experimental Vehicles Committee (EEVC) set up a working group to assess and develop test methods for evaluating pedestrian protection for passenger cars at the beginning of the 1990`s.

The current proposed pedestrian protection legislation based on the findings of the EEVC WG17 [4] proposes different test methods and criteria that should enable a pedestrian to survive a frontal impact without death or serious injury, with a car travelling at 40 km/h.

In 1996 the European Commission put forward a draft proposal III/5021/96EN with the intention to introduce this as a draft directive in 2000 for new vehicles and in 2003 for vehicles already in use. Since, the automotive companies realised that they would not be able to respond to the different requirements in time, the introduction of the full directive has been deferred. More realistic dates may be 2007 for existing vehicles and 2002-2004 for new ones.
In order to evaluate the pedestrian friendliness of a vehicle the current legislation proposes three different tests [5,6]:

- Headform to bonnet Tests
- Upper leg to bonnet leading edge (BLE) tests
- Lower leg to bonnet tests

For the headform to bonnet tests child and adult impacts are carried out using two different sizes of impactor. They are fired at the same speed, 40 km/h, but at different angles to simulate the difference in height between an adult and a child.
Using recorded acceleration, the Head Impact Criteria (HIC) [5] can be calculated. The maximum acceptable HIC value is 1000.

For the upper leg to bonnet leading edge tests a 350 mm long instrumented tube is fired at the Bonnet Leading Edge (BLE) and the loads, at the top and bottom and bending moments along the central area of the tube are measured. The mass, velocity and impact angle of the upper leg are set according to the geometry of the front of the vehicle. Impact parameters for this test are provided in the draft proposal III/5021/96EN).There are two criteria to meet in order to pass the test although it should be noted that the loads quoted below are currently under revision.

(i) The total instantaneous load can not exceed 4 kN
(ii) The bending moment must be less than 220 Nm

The lower leg to bumper test uses an impactor made up of a simulated upper femur and lower tibia connected by replaceable steel elements that represent a knee joint. Instrumentation around the knee joint records the acceleration, bending angle and the longitudinal shear displacement. There are three criteria to meet to pass the test:

(i) Tibia acceleration of less than 150g
(ii) Peak bending angle of less than 15°
(iii) Maximum displacement of 6 mm

In the current situation [7] all production vehicles are shown to exceed the different criteria of the first two tests. The third one, the lower leg to bumper test creates even more problems. In order to overcome those problems two basic requirements need to be fulfilled before compliance can be achieved [8]:

(i) The space between the underside of the bonnet and the rigid components
(ii) The need to establish energy absorption material around the front of the vehicle

2 ANALYSIS METHODS

As mentioned earlier, in most of Europe pedestrians account for a significant proportion of the road accident casualties. The tests that have been developed to address this problem can also be simulated using advanced numerical analysis. These methods use multibody, finite element techniques or a combination of the two. Finite Element programs such as DYNA3D have been used in this area. Multibody programs such as ADAMS also include features such as ADAMS/Android that can be used to represent the human body. The MADYMO software, which is based on a multibody approach, has become well established for occupant studies and there are MADYMO models for all EEVC WG-17 sub-systems impactors.

The MADYMO headform impactors represent a child or an adult headform. Both have a spherical shape and are made of a semi-rigid material, covered by rubber skin. The inner ball is modelled by a rigid body while the rubber skin is modelled by finite elements. The friction between the rubber skin and the inner ball, and between the skin and the bonnet is described by the standard contact algorithms of MADYMO. The models are validated by drop and impactor experiments.

The upper legform impactor represents an adult femur. The impactor consists of a foam-covered tube mount at either end through load cells to a support frame, which is in turn mounted through a torque limiting joint to a propulsion system. Supplementary weights can be attached to the support frame to meet the impact conditions. Most parts are modelled by multibodies except for the tube and the foam which are modelled by finite elements. The foam is modelled by solid elements and the tube by shell elements. The model is validated by impactor experiments.

The legform impactor consists of two foam-covered rigid segments representing the lower leg and upper leg of an adult, connected by a simulated knee joint that will translate and rotate laterally. This lateral bending and shearing at the knee joint of the model is resisted by deformable elements, which are replaced after each test. The model consist of two rigid bodies representing the lower and upper leg. The foam is modelled by finite elements and the knee is modelled by a combination of revolute joints and a translational joint. The model is validated by impactor experiments. These three mathematical models illustrate the powerful hybrid approach of MADYMO, i.e the combination of multibody analysis and finite element analysis.

In order to develop and evaluate pedestrian safety technologies under more realistic conditions than the proposed European legislation, computer models of Vehicle-Pedestrian impacts have been developed. Padgaonkar [9] carried out a 3D mathematical simulation of pedestrian-vehicle impacts by using the Calspan Gross Motion Simulator. Wijk [10] and Jansen [11] developed pedestrian MADYMO models where the model parameters (geometry, mass, moments of inertia and joint characteristics) were derived from the mechanical Hybrid III dummy. The pedestrian mathematical models enhance understanding of the impact responses of pedestrians by facilitating the analysis of gross-motion.

Mathematical models also allow the assessment of potential injuries reduction due to improved car front contours [12]. However, in the above mentioned computer simulations, the impact response of the pedestrian models seemed excessively stiff, compared with impact response of a human subject. Consequently, such models are limited when it comes to simulating dynamic response of human body segments.

Gibson et al [13] developed a pedestrian 2D mathematical model to investigate head impact responses in car-pedestrian accidents. However, the 2D model was not suitable for simulating dynamic responses of pedestrians in crash environments. A complicated spatial motion of an impacted pedestrian may result from: (1) different initial postures of the pedestrian; (2) successive impacts to the body segments in a large relative movement between pedestrian and a moving car; (3) a 3D distribution of the centre of gravity of body segments. Some of the injury parameters related 3D motion can not be analysed by a 2D mathematical model [14]. Ishikawa et al [15} developed a pedestrian 3D mathematical model to analyse the influence of car front parameters on pedestrian injuries. Biomechanical data were used to describe this model, which shows more realistic response in modeling car-pedestrian collisions than can do a pedestrian mathematical model based on dummy data. One of the disadvantages of this model was a lack of sufficient detail in the description of joints and leg segments, so that it cannot give a detailed simulation of the response of the knee and leg.

Yang [16] developed and validated a whole human-like pedestrian model to study the dynamic response and injury mechanisms of pedestrians in a car-front impact, with focussing on the lower extremity impacts. This study focused on developing a human-like knee joint and breakable leg to study kinematics, injury measures and predict the risk of leg injury. Yang also developed and validated a FEM model of the lower extremity to better understand its dynamic responses and injury mechanisms by means of stress analysis.

A significant advantage of humanoid models is the ability to predict injury mechanisms during pedestrian/vehicle impacts, and to evaluate different design solutions to overcome these injury mechanisms. The humanoid modelling approach can also allow the study of active safety systems that deploy during an impact. [17]. Compared with dummies [18], a humanoid model is advantageous since it can reproduce real world collisions from accident data, and it does not have the inherent stiffness which a dummy has.

The latest research in this area has been carried out at the Cranfield Impact Centre and has been used by the Ford Forschungszentrum Aachen (FFA) to simulate real world pedestrian accidents [19]. Using a complete Finite Element humanoid model engineers are able to reproduce any pedestrian accident, knowing the impact speed and the pedestrian position at impact. This work led to a new approach to pedestrian safety research developed by FFA several years ago. The aim of that programme of research is to study pedestrian accidents under more realistic real world conditions than the proposed European legislation by developing, firstly more accurate vehicle models and secondly more accurate pedestrian humanoid models.

3 CONCLUSIONS

The literature reviewed here indicates that multibody systems analysis tools play a significant part, and will continue to do so, in the development of vehicles that offer improved protection to the pedestrian on impact. Computer modelling of the tests outlined in proposed legislation is well established with specialised software modules available to model the various impactors. Research is also underway in several areas looking at more detailed modelling of the human body and simulations that are more representative of the real world rather than recreating test procedures.

In line with this approach the author has undertaken a programme of study aimed at combining realistic vehicle dynamics with pedestrian motion, i.e. a walking or running pedestrian on impact. The vehicle dynamics may be established using software such as ADAMS and importing time histories to a MADYMO body that has a realistic representation of vehicle structural stiffness. It is intended that the simulations will extend the impact event beyond the proposed test procedures and will focus on producing realistic dynamic motion of the overall event. Correlation of such work presents obvious problems compared with traditional engineering analysis simulation of prototype testing procedures. One method could be to use an investigative approach where the simulation recreates a known accident. The accident report can then be used as a basis for establishing whether the simulation outputs are realistic.

Work on realistic real world modelling at this stage will help to develop the simulation tools that will be needed by automotive designers in the future. It is likely that during the next ten to twenty years that in addition to improvements in vehicle shape and material at the front end external devices similar in concept to occupant airbags will start to be considered as part of the solution to pedestrian protection. Work with such novel devices will present problems in areas such as accurate sensing. Realistic multibody systems based models that can accurately simulate the initial conditions of a pedestrian on impact will be needed to solve these problems and improve vehicle protection for pedestrians.

REFERENCES

[1] : Lawrence G.J.L., Hardy B.J, " Costs and Benefits of the EEVC Pedestrian Impact Requirements." Project Report 19, TRL, UK, (1993)

[2] : Lines C.J., " Safety and the road, " Automotive Environmental Impact and Safety, Autotech`97, ImechE.

[3] : Vallee H., Thomas C. and Terriere C., " Pedestrian Casualties: The Decreasing Statistical trend", Proceedings of the 12th International Conference on Experimental Safety Vehicles, Goteborg, May 1989.

[4] : European Commission, " Draft proposal for a European Parliament and Council Directive relating to the protection of pedestrians and other road users in the event of a collision with a motor vehicle and amending Directive 70/156/EEC" III/5021/96 EN, Brussels, 7 February 1996.

[5] : Brown G., " Techniques for the development pedestrian friendly vehicles", Automotive Environmental Impact and Safety, Autotech` 97, ImechE.

[6] : Harris J., " A Study of Test Methods to Evaluate Pedestrian Protection for Cars" , EEVC WG10, Proceedings of the 12th International Conference on Experimental Safety Vehicles, Goteborg, May 1989.

[7] : Clemo K.C., Davies R.G., " The Practicalities of Engineering Cars for Pedestrian Protection",
Motor Industry research Association; UK, Paper No 98-S10-P-16.

[8] : Dickison and Davies, " The Pedestrian Friendly Car ", 62nd Road Safety Congress, ROSPA, 24-26 February 1994

[9] : Padgaonkar, A.J., "A3D Mathematical simulation of Pedestrian-Vehicle Impact with experimental Verification." Transactions of the ASME, Journal of Biomechanical Engineering, (1977)

[10] : Wijk J.J, Wismans J, "MADYMO Pedestrian Simulations. SAE Paper 830060, P-121, Pedestrian Impact & Assessment, Int. Congress and exposition, Detroit, USA. Pp. 109-117, (1983)

[11] : Jansen E.G, Wismans, J., (1986): "Experimental and Mathematical Simulation of Pedestrian-Vehicle and Cyclist-Vehicle Accidents. Proc of the 10th Int. Tech.Conf on Experimental Safety Vehicles, Oxford England, July 1-4, 1985. US Department of Transportation, NHTSA, USA, pp.977-988.

[12] : Wismans, J., Van Wijk, J.(1982): " Mathematical Model for the Assessment of Pedestrian Protection Provided by a Car Contour. Proc. of the 9th Int. Tech. Conf on ESV, Kyoto, Japan, Nov 1-4.

[13] : Gibson T.J, Hinrichs, R.W, (1986): "Pedestrian Head Impacts: Development and Validation of a Mathematical Model." Proc. of the IRCOBI Conference on Biomechanics of Impacts. Sept 2-4, Zurich.

[14] : Schlumpf, M.R, Niederer P.F, (1987): " Motion Patern of Pedestrian Surrogates in Simulated Vehicle-Pedestrian Collisions. Journal of Biomechanics, 20(:4). Pp.371-384

[15] : Ishikawa H, Kajzer J., (1993): " Computer Simulation of Impact Response of the Human Body in Car –Pedestrian Accidents." Proc. of the 37th Stapp Car Crash Conference, Nov 8-10, San Antonio, Texas. SAE, Warrendale, PA, USA. pp .235-248

[16] : Yang J.K, Lovsund P., (1997): " Development and Validation of a Human –Body Mathematical Model for Simulation of Car-Pedestrian Impacts." Proc. of the IRCOBI Conference on Biomechanics of Impacts. Hannover, Germany, IRCOBI Secretariat, Bron, France.

[17] : Song E., Lizee E., " Development of a 3D Finite Element Model of the Human Body ", 42nd Annual Stapp Car Crash conference, Tempe, Arizona, USA, 1998.11.02 - 1998.11.04, SAE Paper No. 983152.

[18] :Risa D., Scherer, S.L., " SID-IIs Beta - Prototype Dummy Biomechanical Responses" , Task Group Of the Occupant Safety research Partnership. SAE Paper No: 983151.

[19] : Howard M., Watson J., Hardy R., " The simulation of real world car to pedestrian accidents using a pedestrian humanoid finite Element Model " International Journal of Crashworthiness.

Dynamic tyre testing for vehicle handling studies

S HEGAZY, H RAHNEJAT, and **K HUSSAIN**
Department of Mechanical and Medical Engineering, University of Bradford, UK

Abstract

This paper describes a number of tyre models available and widely used for vehicle handling studies. The performance of these models has been compared against the results obtained experimentally, using a tyre-testing rig. The paper describes the experimental set up. The procedure used for the purpose of comparison has involved the creation of a multi-body model of the test rig, employing the various tyre models. The results indicate good agreement between the experimental findings and some of the output from the evaluated tyre models.

1- Introduction

Tyres rank among the most important vehicle components as they represent the interaction points between the vehicle and the road. The tyre-road contact forces are mainly dependent on the mechanical properties of the tyre (i.e. stiffness and damping), the road condition (i.e. friction coefficient between tyre and road, and the road profile), and the motion of the tyre relative to the road.

Modelling of tyre forces, acting at the contact patch between the tyre and road surface can be considered to be one of the most important aspects in vehicle ride and handling studies. Tyre models are generally developed according to the type of analysis in mind. As an example, for vehicle handling studies the tyre model is often required to analyse the lateral force and the side slip angle. Therefore, the relationship between the lateral tyre force and the side-slip angle must be determined carefully. On the other hand, for vehicle braking studies tyre models require to analyse the tyre longitudinal force and the longitudinal slip ratio. For ride quality and vibration studies, tyre models are often required to transmit the effect of road surface undulations to the vehicle body. Therefore, the vertical force must be carefully determined. This paper focuses on tyre theory, describing the different tyre models used for vehicle handling simulation studies. These include such models as the Fiala (1), Pacejka's Magic Formula (2), the Smithers tyre model (3), the Arizona tyre model (4) and the System

Technology's STI (5). In addition, a tyre testing rig and experimental results obtained are described, as well as simulation studies of the same conditions with the aforementioned tyre models within a multi-body dynamic model of the same test rig.

2- Experimental Tyre Testing

The dynamic tyre test rig consists of the main components; a single tyre traction tester, measuring equipment and data acquisition system, test track and the control room. It is designed to investigate the dynamic properties of tyres at a maximum vertical load of 5 kN, which depends upon the tyre size. Basically, the test rig consists of the tested tyre, powered by an electric AC motor via a transmission system. The test rig guides the tyre to travel on different surfaces such as on soil, concrete or asphalt. The characteristic values such as the camber angle, wheel load and tyre inflation pressure can be adjusted.

The length of the test track is 25 m, comprising a sample of an actual road. The tester is equipped with different transducers to measure the driving torque, number of tyre revolutions and the actual travelled distance in order to obtain the slip ratio and the instantaneous tyre deflection on a solid surface or through tyre penetration. The output voltage from these transducers are recorded and processed using a computer. Figure 1 shows a view of the single tyre traction tester.

The rig frame is a load carrying beam structure, consisting of two longitudinal beams with a U cross-section and several cross members. The frame forms a base to which all the main parts and units are attached. The main dimensions of the frame are 2.45 m in length and 0.52 m in width. It also include the following:

i. Two rigid wheels to guide the vehicle on the rail. Each wheel is assembled on a shaft with 2 bearings and mounted onto the transverse beam.

ii. Two side caster wheels on either sides guide the test tyre inside the testing track.

The driving system consists of an electric motor, a four speed mechanical gearbox, a propeller shaft, the final drive and the wheel hub. These parts of the driving system are chosen to withstand the forces under the testing conditions.

The tested tyre is mounted on a drum, in a manner that the line of action of the vertical load is passed through the tyre centre. The frame is designed for testing different tyres up to 1 m in diameter, requiring a special interface for each tyre size.

The vertical load is applied using different weights of predetermined values, according to the tyre type load carrying capacity and tyre inflation pressure.

The measuring equipment include a torque transducer to measure the tyre driving torque, a displacement transducer to measure the instantaneous tyre deflection or tyre penetration, a load cell sensor to monitor the net traction force and a revolution counter to compute the number of revolutions of the tyre. The data acquisition system is equipped to record the output data from the transducers and transfer it to the microcomputer.

3- Modelling of Tyre Test Rig in ADAMS

In order to validate the various numerical tyre models against experimental measurements a multi-body model of the tyre testing rig has been developed in ADAMS. The interface between the multi-body model and ADAMS TIRE is through a TIRE statement (6). In order to formulate the mathematical tyre model, it is necessary to incorporate a Fortran subroutine, linking this with ADAMS to provide an executable solver. The interface between ADAMS and a Fortran program is through the TIRSUB subroutine. TIRSUB is a user-written tyre-terrain force evaluation routine. It defines a set of three forces and three moments, acting at the tyre contact patch in the SAE co-ordinate system for the tyre (7) (see figure 2). Equations used to formulate these forces and moments, have been programmed into the subroutine to represent the previously mentioned tyre models.

3.1 The multi-body tyre test rig model

The tyre test rig has been modelled in ADAMS to compare the tyre lateral force and the self-aligning moment for different mathematical tyre models at different values of vertical load, slip and camber angles. The tyre rig contains a tyre, which rolls forward on a flat road surface similar to the actual testing surface. Table 1 lists the inertial components, while table 2 lists the constraints in the tyre test-rig multi-body model.

Table 1 The parts and inertial properties

Part No.	Part name	Mass, Kg	Moment of inertia, Kg-mm^2		
			I_{xx}	I_{yy}	I_{zz}
1	Ground	-	-	-	-
2	Tyre	5	25.09E+4	33.56E+4	25.09E+4
3	Carrier	10	1	1	1
4	Frame	350	1	1	1
5	Sliding carrier	1	1	1	1

Table 2 The constraints in the tyre test rig

No.	Constraint type	Part I	Part J	No. of constraint
1	Revolute joint	Tyre	Carrier	5
2	Revolute joint	Carrier	Frame	5
3	Cylindrical joint	Frame	Sliding carrier	4
4	Translational joint	Slider carrier	Ground	5
5	Motion	Joint 2		1
6	Motion	Joint 3		1
7	Motion	Joint 4		1

In figure 3 the tyre part 2 is connected to the carrier part 3 by a revolute joint, aligned with the spin axis of the tyre. The carrier part 3 is connected to the frame part 4 by a revolute joint, which is aligned with the direction of travel. A rotational motion is applied to this joint to set the required camber angle during the simulation of the test tyre. The frame part 4 is connected to a sliding carrier part 5 by a cylindrical joint, which is aligned in the vertical direction. This joint allows the frame part 4 to slide up or down relative to part 5 and, therefore, to transmit the applied vertical force on the frame 4 to the tyre. A rotational motion is applied to the cylindrical joint to set the required slip angle of the tyre during simulation studies. The sliding carrier wheel part 5 is connected to the ground part 1 by a translational joint aligned with the direction of travel. A specified motion input is applied at the translational joint to control the forward velocity of the tested tyre.

According to the Gruebler – Kutzbach expression, the single tyre traction tester has two degrees of freedom. One degree of freedom is associated with the spin motion of the tyre, while the other degree of freedom is the height change of wheel centre above the road. A static equilibrium analysis, followed by dynamic simulation has been carried out. The tyre model must receive all information about the position, orientation and velocity at each wheel centre and also the topography of the road surface at each time step in order to calculate the forces and moment at the contact patch.

3.2 A review of theoretical tyre models

The theoretical tyre models include; Fiala, Smither, Pacejka, Arizona and STI tyre models.

The Fiala tyre model has the advantage of simplicity, and requires a smaller number of input parameters than other tyre models. The parameters R_1, R_2, C_Z *and* ξ are used to formulate the vertical load in the tyre and are required for all the tyre models. In the case of fiala tyre model, the tyre forces include: the vertical, the lateral and the longitudinal forces, whilst the moments include the rolling resistance and the self-aligning moment.

The vertical tyre force is dependent on the tyre deflection and its rate of change; both measured along the tyre vertical directional vector. The deflection is obtained by an instantaneous evaluation of the distance between the position of the wheel centre marker and the road surface plane. The time rate of change of deflection is obtained by the vector scalar product of the instantaneous tyre radius vector and the wheel centre global velocity. The quantities, thus obtained, are employed to find the stiffness and damping contributions to the tyre vertical force, as indicated below.

$$F_Z = F_{ZK} + F_{ZC} \qquad [1]$$

$$F_{ZK} = -C_Z\ \delta \qquad [2]$$

$$F_{ZC} = -2\sqrt{m_t\left|C_Z\right|}\ \xi\ \dot{\delta} \qquad [3]$$

The longitudinal tyre force is dependent on the longitudinal tyre stiffness and the slip ratio. The utilised adhesion coefficient is dependent on the slip ratio and the tyre-road material combination.

The instantaneous value of the tyre road coefficient, ignoring the effect of camber change is:

$$S_{S\alpha} = \sqrt{S_s^2 + \tan^2\alpha} \qquad [4]$$

where, the coefficient of friction is given by linear interpolation as:

$$\mu\left(S_{S\alpha}\right) = \mu_0 + \left(\mu_1 - \mu_0\right)S_{S\alpha} \qquad [5]$$

A critical value for longitudinal slip ratio is calculated using:

$$S^* = \left|\frac{\mu F_Z}{2\,C_S}\right| \qquad [6]$$

and:

$$F_X = \begin{cases} -C_s\,S_s & \left|S\right| \le \left|S^*\right| \\ -\left(F_{x1} - F_{x2}\right) * Sign\left(S_s\right) & \left|S\right| > \left|S^*\right| \end{cases} \qquad [7]$$

Where:

$$F_{x1} = \mu * |F_Z| \quad [8]$$

$$F_{x2} = (\mu * F_Z)^2 / (4 * |S_s| * C_s) \quad [9]$$

The lateral tyre force depends on the ratio between the slip angle and the critical slip angle. A critical value for the slip angle is calculated using:

$$\alpha^* = \left| \frac{3\mu F_Z}{C_\alpha} \right| \quad [10]$$

Then:

$$F_Y = \begin{cases} -\mu|F_Z|(1 - H^3)Sign(\alpha) & |\alpha| \le |\alpha^*| \\ -\mu|F_Z|Sign(\alpha) & |\alpha| > |\alpha^*| \end{cases} \quad [11]$$

Where:

$$H = 1 - \frac{C_\alpha|\tan(\alpha)|}{3\mu|F_Z|} \quad [12]$$

The rolling resistance is negative, when the tyre is rolling forward and positive, when the tyre is rolling backward. It is computed as:

$$M_r = \mp C_r F_Z \quad [13]$$

The self-aligning moment is similar to the lateral force calculation and depends on the ratio between the slip angle and the critical slip angle as follows:

$$M_Z = \begin{cases} 2\mu|F_Z|\ R_2\,(1-H)\,H^3 Sign(\alpha) & \textit{Elastic deformation} & |\alpha| \le |\alpha^*| \\ 0.0 & \textit{For complete sliding} & |\alpha| > |\alpha^*| \end{cases} \quad [14]$$

The Arizona tyre model (4) is based upon the assumption of a rectangular contact patch, and a parabolic pressure distribution. It is, therefore, modelled as a beam on an elastic foundation. The evaluation of the longitudinal and lateral tyre forces and the self-aligning moment are based on the precise operating conditions that needs some slip parameters and the modified lateral friction coefficient.

The critical slip ratio due to camber angle, is given by $S_{\gamma c}$, where:

$$S_{\gamma c} = \mu \frac{F_Z}{C_\gamma} \quad [15]$$

The critical slip ratio due to longitudinal slip is given by S_{sc}, where:

$$S_{sc} = 3\mu \frac{F_Z}{C_s} \quad [16]$$

The critical slip ratio due to the slip angle is given by $S_{\alpha c}$, where:

$$S_{\alpha c} = \frac{C_s}{C_\alpha}\sqrt{S_{sc}^2 + S_s^2} - 3C_\gamma \frac{S_\gamma}{C_\alpha} \quad [17]$$

The term critical stands for the maximum value, which allows an elastic deformation of the tyre. The total slip ratio, $S_{s\alpha}$ is determined as:

$$S_{s\alpha} = \frac{B_2 + \sqrt{B_2^2 - B_1 B_3}}{B_1} \quad [18]$$

Where:

$$B_1 = (3\mu F_Z)^2 - (3C_\gamma S\gamma)^2$$

$$B_2 = 3C_\alpha S_\alpha C_\gamma S_\gamma$$

$$B_3 = -\left[(C_S S_S)^2 - (C_\alpha S_\alpha)^2\right]$$

The contact patch length, l is determined as:

$$l = 1 - S_{s\alpha} \quad [19]$$

The modified lateral coefficient, μ_{ym} is evaluated as:

$$\mu_{ym} = \mu \sin\beta - \left(\frac{C_\gamma S_\gamma}{F_Z}\right) \quad [20]$$

The longitudinal tyre force depends upon the adhesion limit, the rolling elastic deformation of the tyre or the complete sliding motion.

a- For elastic deformation condition only:

$$F_X = C_s S_s l^2 + \mu \cos\beta\, F_Z \left(1 - 3l^2 + 2l^3\right) \quad [21]$$

b- For complete sliding condition only:

$$F_x = \mu \cos\beta\, F_Z \quad [22]$$

The lateral tyre force is also dependent upon the adhesion limit, the elastic deformation or the complete sliding motion.

a- For elastic deformation condition only

$$F_Y = C_\alpha S_\alpha l^2 + \mu_{ym} F_Z \left(1 - 3l_2 + 2l^3\right) + C_\gamma S_\gamma \quad [23]$$

b- For complete sliding condition only:

$$F_Y = \mu \sin\beta \; F_Z \quad [24]$$

The self-aligning moment, M_Z is the sum of components of the moment generated by the slip angle, $M_{z\alpha}$ in addition to the two components, $M_{zs\alpha}$ *and* $M_{zs\gamma}$ produced by the longitudinal force, which has an offset between the wheel centre plane and the tyre tread base due to slip and camber angles. Therefore:

$$M_Z = M_{z\alpha} + M_{zs\alpha} + M_{zs\gamma} \quad [25]$$

a- For elastic deformation condition only

$$M_{z\alpha} = \left[C_\alpha S_\alpha \left(-\frac{1}{2} + \frac{2}{3} l \right) + \frac{3}{2} \mu_{ym} F_Z S_{s\alpha}^2 \right] \sqrt{8R_1\delta}\,\left(l^2\right) \quad [26]$$

$$M_{zs\alpha} = \frac{2}{3} C_s S_s S_\alpha l^2 + \frac{3\mu^2 \sin\beta \cos\beta \; F_Z^2 \sqrt{8R_1\delta}}{5C_\alpha} \left(1 - 10l^3 + 15l^4 - 6l^5\right) \quad [27]$$

$$M_{zs\gamma} = S_\gamma F_X \sqrt{\frac{R_1^2 - 8R_1\delta}{4}} \quad [28]$$

b- For complete sliding condition only:

$$M_{z\alpha} = 0 \quad [29]$$

$$M_{z\alpha s} = \frac{3\mu^2 \sin\beta \cos\beta \; F_Z^2 \sqrt{8R_1\delta}}{5C_\alpha} \quad [30]$$

$$M_{zs\gamma} = S_\gamma F_X \sqrt{\frac{R_1^2 - 8R_1\delta}{4}} \quad [31]$$

Where: the term $\left(R_1^2 - 8R_1\delta\right)$ is equal to zero, if it has a negative value.

The Magic formula tyre model is mainly of an empirical nature and contains a set of mathematical formulae, which are partly based on a physical background. The first Magic Formula version was presented by Bakker et al (8). The basic idea for using the sine and arcsine functions was described for mainly pure slip conditions. The second version of the Magic formula was presented by Pacejka et al (2) for combined braking and cornering conditions. Tyre relaxation lengths were introduced in order to have a first order approach for the transient tyre behaviour. This model was improved later for combined slip calculations (9). Bayle (10) proposed to have a more empirical approach, reducing the complexity of the force calculations under combined slip conditions and yielding a considerably higher calculation speed. The latest of Magic Formulae was developed by Pacejka (11) and is known as the Delft-Tyre. It combines the advantages of the previous versions.

The mathematical equations introduced in this section for the tyre model are special versions of the Magic Formula, which was developed by Pacejka (2). This formulae are used to calculate the longitudinal and the lateral tyre forces and the self-aligning moment as functions of the vertical force, the slip ratio and of the side slip and camber angles.

The equation yielding the longitudinal force as a function of the slip ratio is:

$$F_X = D \sin\left(C \arctan(B\phi)\right) + S_v \qquad [32]$$

Where:

$$\phi = (1-E)(S+S_h) + (E/B)\arctan\left(B(S+S_h)\right) \qquad [33]$$

and:

$$C = a_0$$
$$D = a_1\, F_Z^2 + a_2\, F_Z$$
$$B = \left(a_3\, F_Z^2 + a_4\, F_Z\right) * e^{-a_5 F_Z} / CD$$
$$E = a_6\, F_Z^2 + a_7\, F_Z + a_8$$
$$S_h = a_9\, F_Z + a_{10}$$
$$S_v = 0$$

The equation yielding the lateral force as a function of the slip angle is:

$$F_Y = D \sin\left(C \arctan(B\phi)\right) + S_v \qquad [34]$$

where:

$$\phi = (1-E)(\alpha + S_h) + (E/B)\arctan\left(B(\alpha + S_h)\right) \qquad [35]$$

and:

$$C = b_0$$
$$D = b_1\, F_Z^2 + b_2\, F_Z$$
$$B = b_3 \sin\left(2 * \arctan(F_Z / b_4)\right)\left(1 - b_5\, |\gamma|\right) / CD$$
$$E = b_6\, F_Z + b_7$$
$$S_h = b_8\, \gamma + b_9 F_Z + a_{10}$$
$$S_v = b_{11}\, \gamma\, F_Z + b_{12}\, F_Z + b_{13}$$

The equation yielding the self-aligning moment as a function of the slip angle is:

$$M_Z = D \sin\left(C \arctan(B\phi)\right) + S_v \qquad [36]$$

where:

$$\phi = (1 - E)(\alpha + S_h) + (E / B)\arctan\left(B(\alpha + S_h)\right) \qquad [37]$$

and:

$$C = c_0$$

$$D = c_1\ F_Z^2 + c_2\ F_Z$$

$$B = (c_3\ F_Z^2 + c_4\ F_Z)(1 - (c_6\ |\gamma|))e^{-c_5\ F_Z}\ / CD$$

$$E = (c_7\ F_Z^2 + c_8\ F_Z + c_9) * (1 - (c_{10}\ |\gamma|))$$

$$S_h = c_{11}\ \gamma + c_{12} F_Z + c_{13}$$

$$S_v = (c_{14}\ F_Z^2 + c_{15}\ F_Z)\gamma + c_{16} F_Z + c_{17}$$

The values of B, C, D, E, S_v *and* S_h are expressed as functions of a number of coefficients, a_i, b_i *and* c_i which can be considered as characteristic of any specific tyre, but it depends also on road conditions and speed. The coefficient values for brake force, lateral force and self aligning moment are given in Tables 3-5 (12) in the appendix.

The Smithers tyre model is a particular version of the *Magic Formula*. The only differences between the two models are the techniques used for the determination of the lateral force and the self-aligning moment coefficients. For Smithers, the lateral tyre force and the self-aligning moment are dependent on six parameters, which represent the cornering stiffness, the stiffness factor, the peak factor and the curvature factor coefficients in addition to the vertical and the horizontal offsets, as:

$$F_Y = S_v + D * \sin[C_\alpha / (B * D) * \arctan\{B * (1 - E) * shift + E * \arctan(B * shift)\}] \qquad [38]$$

where:

$$Shift = \alpha - S_h$$

F_Y can be replaced by M_Z in equation [38], together with different parameters.

The STI (System Technology Inc.) tyre model was developed by Allen et al (5) for vehicle simulations for full range of operating conditions (slip, camber, normal load) on both paved and off-road surfaces. The tyre model simulation is based on the composite slip formulation, σ which is the heart of the model and is a function of the lateral and the longitudinal slip:

$$\sigma = \frac{\pi l^2}{8\mu_0 F_Z}\sqrt{C_\alpha^2 \tan^2\alpha + C_s^2 (S_s / 1 - S_s)^2} \qquad [39]$$

where:

$$l = l_0\left(1 - C_a \frac{F_{Xest}}{F_Z}\right) \qquad [40]$$

$$l_0 = \frac{0.076\sqrt{F_Z F_{ZT}}}{2R_2 P} \qquad [41]$$

$$F_{Xest} = \left(\frac{C_S}{F_Z} \right) * S_s * F_Z \qquad [42]$$

$$f(\sigma) = \frac{C_1\sigma^3 + C_2\sigma^2 + (4/\pi)\sigma}{C_1\sigma^3 + C_3\sigma^2 + C_4\sigma + 1} \qquad [43]$$

The longitudinal force, F_X is expressed as:

$$F_X = -\frac{\mu F_Z f(\sigma) C_s' S_s}{\sqrt{C_\alpha^2 \tan^2\alpha + C_s'^2 S_s^2}} \qquad [44]$$

The lateral tyre force, F_Y is evaluated as:

$$F_Y = \frac{\mu F_Z f(\sigma) C_s \tan\alpha}{\sqrt{C_\alpha^2 \tan^2\alpha + C_s'^2 S_s^2}} + C_Y \gamma \qquad [45]$$

Where:

$$C_s' = C_s + (C_\alpha - C_s)\sqrt{\sin^2\alpha + S_s^2 \cos^2\alpha} \qquad [46]$$

The self-aligning moment, M_Z is evaluated as follows:

$$M_Z = \frac{C_m l^2 \tan(\alpha)}{1 + G_1\sigma^2}\left[\frac{C_\alpha}{2} - G_2 C_s \frac{S_s}{1 - S_s}(2 + \sigma^2)\right] \qquad [47]$$

The lateral and longitudinal stiffness coefficients, camber thrust stiffness, tyre contact patch length, and the tyre road coefficient are variable and depend on the vertical load.

4-Results and discussion

Using the measured parameters, mentioned in section 2, the dynamic tyre-road characteristics are evaluated. The relationship between the brake force coefficient and the tyre slip at a load of 2 kN is shown in figure 4. The values obtained using the Magic Formula are also shown in the figure, facilitating a comparison to be made. In both this figure and figure 5, illustrating the coefficient for the lateral force, the value of unity has been exceeded. These cannot be true for the given conditions, because the maximum adhesion coefficient is around 0.8 for a tyre against road contact with no stiction. For pure longitudinal or pure cornering, the coefficient should be 0.8 as indicated by the experimental results in figure 4. The maximum theoretical adhesion coefficient is dependent upon the peak factor D in equations [32] and [34]. This factor is evaluated as a function of the tyre vertical load, as previously indicated and can be modified as shown below, when an appropriate power index is selected.

$$D = a_1 F_z^n + a_2 F_z \qquad [48]$$

Typical plots for the longitudinal force coefficient for different power indices in equation [48] are given in figure 6. The instantaneous lateral tyre force at a vertical load of 3.5 kN, and slip angle change $-35:35^{O}$ for Fiala (1), Pacejka (2), Smithers (3), Arizona (4), and the experimental tyre force measurements are shown in figure 7. In this figure the change between the lateral tyre forces for the Pacejka (2), Smithers (3) and the experimental values (13) show small differences, because of their coefficients. The lateral tyre force for the Fiala (1) and the Arizona (4) tyre models are coincident and are similar in shape, but different in their values. Furthermore, the self-aligning moment for the above models are similar in their shapes, but different in their values as shown in figure 8. The Arizona tyre model gives the lowest values for the self-aligning moment, because the kinematics of tyre remains distinct.

In conclusion better qualitative conformance has been obtained between the Pacejka (2) and the experimental measurements (13).

5-Acknowledgements

The authors wish to express their gratitude to the Egyptian Military Attache's Office for the financial support extended to this research project.

Nomenclature

a_i, b_i, c_i	: Brake and lateral forces, and self-aligning moment coefficients
B	: Stiffness factor
$C_1 - C_4$	: Constants for STI tyre model
C_a	: Tyre contact patch elongation coefficient
C_m	: Aligning moment stiffness
C_s	: Longitudinal tyre stiffness
C_Z	: Vertical tyre stiffness
C_α	: Cornering stiffness due to slip angle
C_γ	: Cornering stiffness due to camber angle
D	: Maximum amplitude factor
E	: Shape factor
F_X	: Longitudinal tyre force
F_Y	: Lateral tyre force
F_Z	: Vertical tyre force
F_{ZT}	: Rated design load
G_1, G_2	: Shape factors for self-aligning moment
l	: Contact patch length
M_r	: Rolling resistance moment
M_Z	: Self-aligning moment
m	: Mass of tyre
P	: Tyre inflation pressure
R_1	: Unloaded tyre radius

R_2 : Carcass radius
S_s : Longitudinal slip ratio
S_α : Longitudinal slip ratio due to slip angle
S_γ : Longitudinal slip ratio due to camber angle
S_{sc} : Critical slip ratio due to longitudinal slip ratio
$S_{\gamma c}$: Critical slip ratio due to camber angle
$S_{s\alpha}$: Total tyre slip ratio
S_h : Horizontal shift
S_v : Vertical shift
α : Slip angle
α^* : Critical slip angle
γ : Camber angle
σ : Composite slip
δ : Tyre deflection
μ : Friction coefficient
μ_{ym} : Modified lateral coefficient

6-References

[1] Fiala, E., "Seitenkrafte am rollenden luftreifen", VDI-Zeitschrift 96, 1964, 973.

[2] Gim, Gwanghun, " Vehicle dynamic simulation with a comprehensive model for pneumatic tires", Ph.D. Thesis, the university of Arizona, 1988.

[3] Pacejka H.B., Bakker.E., and Lidner L., " A new tire model with an application in vehicle dynamic study", SAE paper 890087, 1989 pp.101-113.

[4] Schuring.D.J, Pelz.W, and Pottinger.M.G, " The BNPS model – an automated implementation of the Magic Formula concept", SAE paper 931909, Vol.102, pp. 120-130, 1993.

[5] Allen, R.W. and Rosenthal, T.J., "A vehicle dynamics tire model for both pavement and off-road conditions", SAE paper 970559, 1997, pp. 27-38.

[6] ADAMS/Subroutine users' Guide – Version 8, Mechanical Dynamic Inc., November 1994.

[7] "Vehicle dynamic terminology", SAE J670d, 1975.

[8] Pacejka, H.B., Bakker,.E., and Nyborg, " Tyre modelling for use in vehicle dynamics dtudies", SAE paper 870421, 1987, pp. 2.190-2.204.

[9] Pacejka, H.B., and Bakker, E., " The Magic Formula tyre model", Vehicle System Dynamics, Vol.21, 1991, pp. 1-18.

[10] Bayle, P., Forissier, J.F. and Lafon, S., " A new tyre model for vehicle dynamic simulations combined cornering and braking-driving manoeuvre", Michelin France, 1990.

[11] Pacejka, H.B., " The tyre as a vehicle component", In Proceedings of 26th FISITA Congress, Prague, 1996, pp. 1-19.

[12] Genta, G., Motor Vehicle Dynamics : Modeling and Simulation, Series on Advances in Mathematics for Applied Sciences, Vol. 43", SAE, Warrendale, 1997.

[13] Hegazy, S., Multi-body dynamic analysis for assessment of vehicle handling under transient maneouvres, Ph.D. thesis, University of Bradford, 2000.

Appendix:

Table 3 The coefficient values for brake force

a_0	a_1	a_2	a_3	a_4	a_5
1.65	-7.6118	1122.6	-0.00736	144.82	-0.076614
a_6	a_7	a_8	a_9	a_{10}	
-0.00386	0.085055	0.075719	0.023655	0.023655	

Table 4 The coefficient values for lateral force

b_0	b_1	b_2	b_3	b_4
1.6929	-55.2084	1271.28	1601.8	6.4946
b_5	b_6	b_7	b_8	b_9
0.0047966	-0.3875	1	-0.045399	0.0042832
b_{10}	b_{111}	b_{112}	b_{12}	b_{13}
0.086536	-7.973	-0.2231	7.668	45.8764

Table 5 The coefficient values for self-aligning moment

c_0	c_1	c_2	c_3	c_4	c_5
2.2264	-3.0428	-9.2284	0.500088	-5.56696	-0.25964
c_6	c_7	c_8	c_9	c_{10}	c_{11}
-0.00129724	-0.358348	3.74476	-15.1566	0.0021156	0.00346
c_{12}	c_{13}	c_{14}	c_{15}	c_{16}	c_{17}
0.00913952	-0.244556	0.100695	-1.398	0.44441	-0.998344

Figure 1 Single tyre traction tester

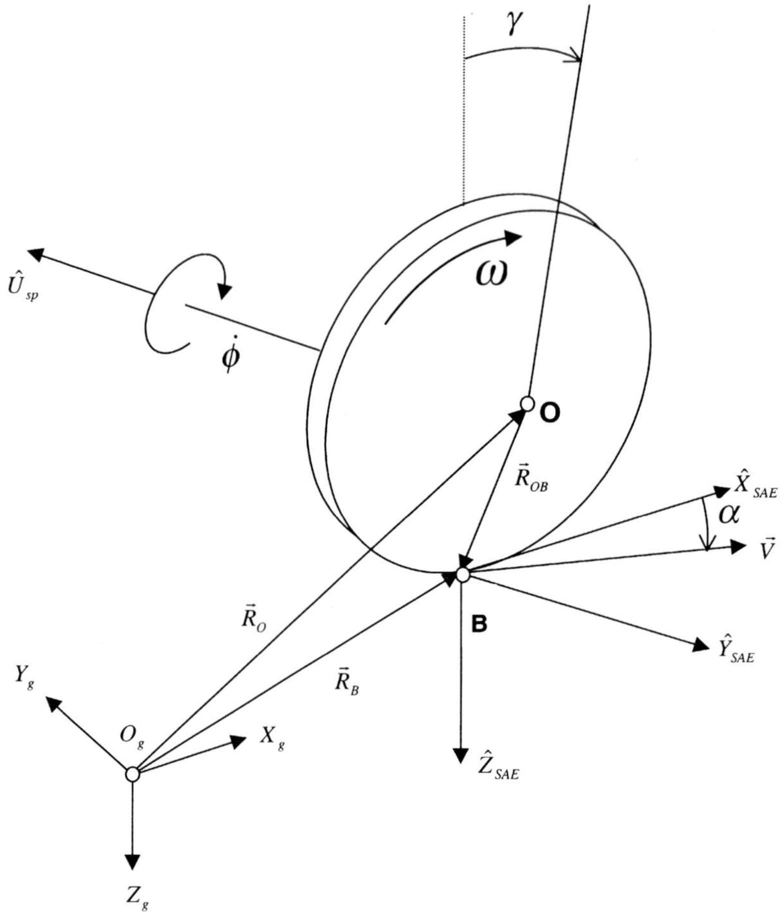

Figure2 The SAE co-ordinate system

Parts

No.	Part Name
1	Ground
2	Tyre
3	Carrier
4	Frame
5	Slider

Constraints

No.	Constraint Type
1	Revolute joint
2	Revolute joint
3	Cylindrical joint
4	Translational joint

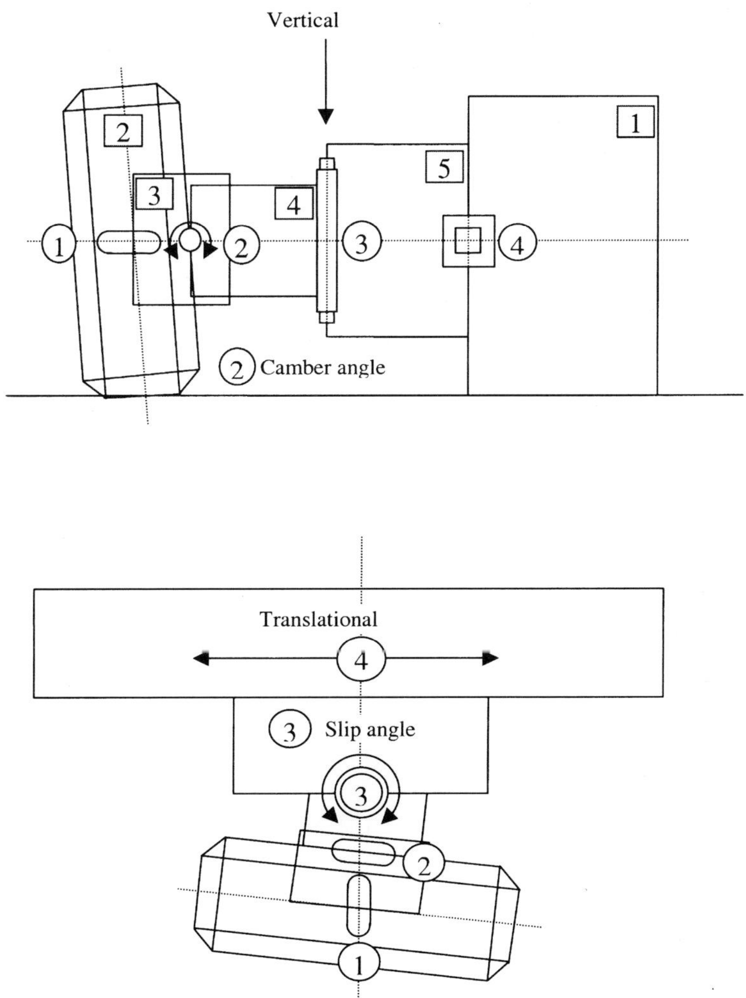

Figure 3 Schematic views of the single tyre tester multi-body model

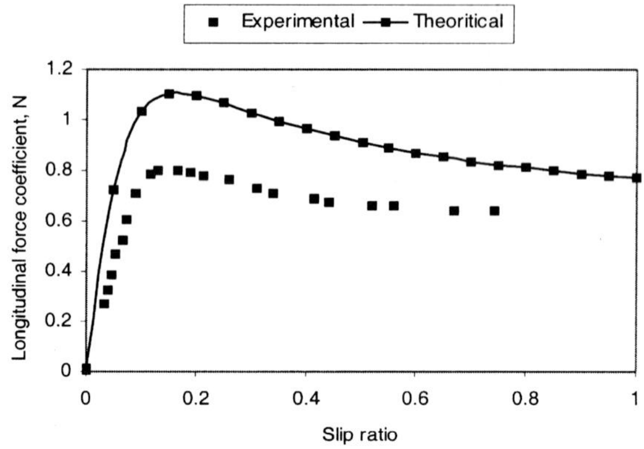

Figure 4 Experimental and theoretical longitudinal tyre force coefficient

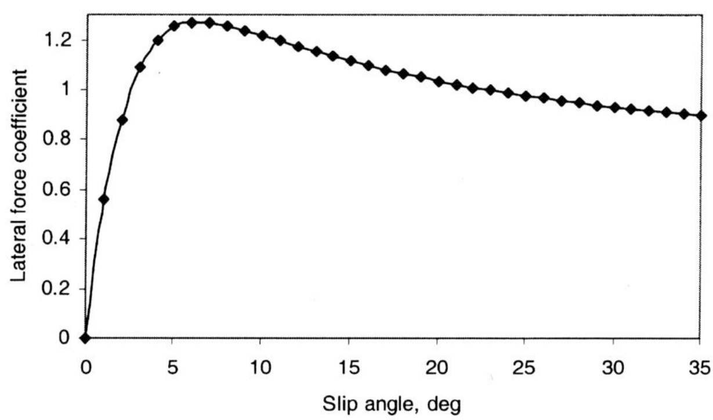

Figure 5 Variation of lateral force coefficient with slip angle

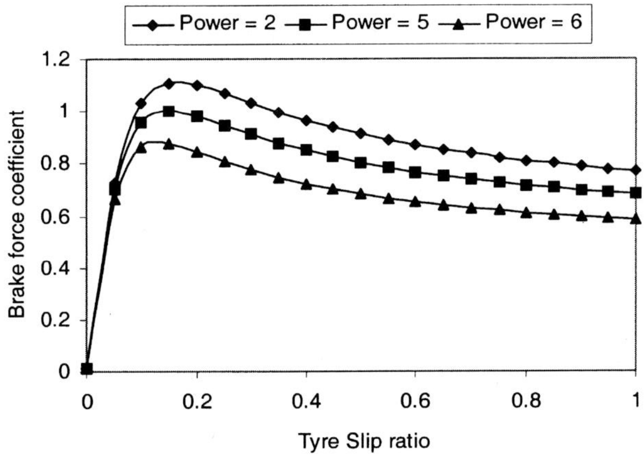

Figure 6 Variation of brake force coefficient with slip ratio

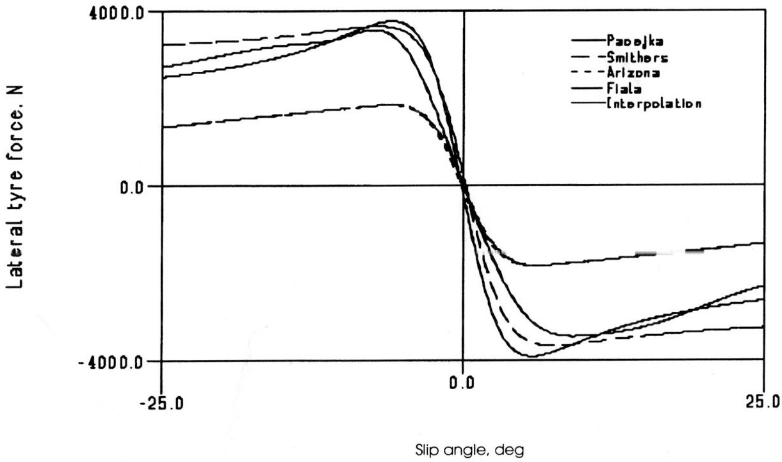

Figure 7 Lateral tyre force change with slip angle

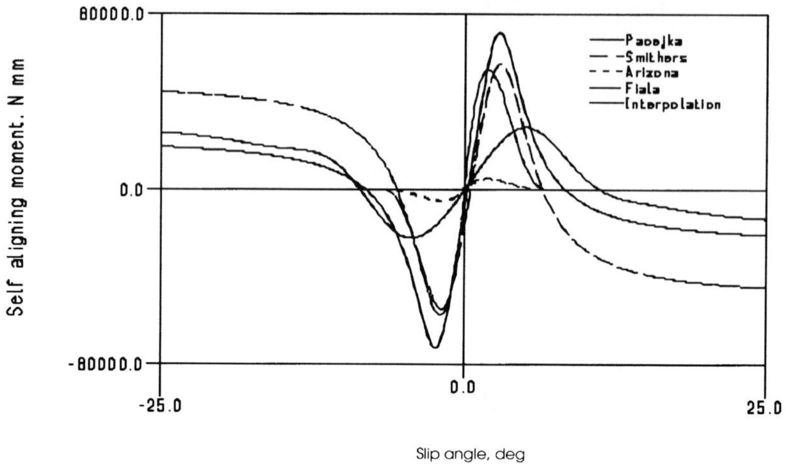

Figure 8 Self-aligning moment change with slip angle

Engine Dynamics

Analysis of crankshaft and cylinder block vibration in operation, coupling by means of non-linear oil film characteristics and dynamic stiffness

N HARIU
Isuzu Motors Limited, Hokkaido, Japan
K SATOU
Isuzu Motors Limited, Fujisawa, Japan
K NISHIDA and **K SAITOH**
Muroran Institute of Technology, Muroran, Japan

SYNOPSIS

This paper describes a method for analysing the non-linear noise and vibration characteristics of a V6 diesel engine. This engine is light as the cylinder block is made from aluminium, while it has high power/torque, which would eventually end up in a high level of noise and vibration if no measures were taken. Recently, the noise and vibration of crankshafts has been analysed up to a frequency of around 1kHz for individual vibration orders, which is not enough, and should be analysed up to as high as around 4 kHz, because noise radiating from cylinder block surfaces is proportional to the squares of the vibration velocities of the surfaces, originated from a crankshaft and propagated through oil film. In addition, the dynamic stiffness of a cylinder block and the non-linear effects of oil film on journals should be taken into account, considering that the peaks of noise are found in a frequency range from 1kHz to 3 kHz. As a result, good agreement was obtained between the measurement and calculation for the vibrations of the cylinder block surfaces at frequencies up to 4kHz, as well as for the axial vibrations of the rotating crankshaft. This analysis technique has turned out useful in reducing the noise and vibration level of the multiple-cylinder diesel engine.

1. INTRODUCTION

Recently diesel engines have becoming lighter and lighter while a high power/torque has been demanded by users. Unless measures are taken, light engines end up in a low stiffness of engine structure, and high power/torque results in high combustion pressure. All these lead to a high level of noise and vibration. One of the major parts of an engine is a crankshaft, and according to some reports[(1)], crankshaft vibration accounts for about 30 % of total noise radiating from an engine. Therefore, a method for accurately estimating the levels of cylinder block surface vibrations originated from a crankshaft has been needed to solve noise and vibration problems.

A crankshaft in a running engine is rotated under the forces of combustion pressure and inertia forces. The vibration of a crankshaft, on which such parts as a damper pulley, a flywheel, a timing gear are mounted (this assembly is hereinafter referred to as 'the crankshaft system'), is coupled with that of a cylinder block, and it has been known that the coupling is greatly affected by oil film on the bearings. The oil film has a higher stiffness than that of a crankshaft at certain crank angles, while it also has a damping characteristic, and crankshaft system vibrations are damped down. The stiffness and the damping coefficient change in a non-linear manner as engine speed changes, which makes the vibration mechanism very complicated. Such non-linear vibration characteristics should be analysed to explain the behaviour of the crankshaft system. Moreover, the mechanism of crankshaft system vibrations being coupled with cylinder block surface ones should be analysed, considering the non-linear relationship, to accurately estimate and lower the level of the cylinder block surface noise and vibration transmitted from the crankshaft system. Experiments and calculations have been made to explain the mechanism of crankshaft system vibration [(2),(3),(4),(5),(6)]. Experimental methods, however, have a problem of the stiffness of a cylinder block seat of the crankshaft being harmed by providing grooves for sensors, and have restrictions in determining the locations of the sensors. With the conventional numerical analysis method considering the non-linear characteristics of oil on a rotating crankshaft, the levels of up to 10th-order vibration components can be estimated at an engine speed. At a maximum speed of 4500 rpm, the highest frequency of the 10th-order crankshaft vibration component is 750 kHz, while the measured peaks of noise and vibration of cylinder block surfaces are in a range from 1 kHz to 3 kHz. Therefore, the frequencies of predictable noise and vibration levels with the conventional calculation method are not high enough. Also, static stiffness is used in the conventional method, and the coupling of the crankshaft and cylinder block vibration is not considered, which results in discrepancy between the measurement and calculation results. Therefore, the conventional calculation method is not considered adequate to estimating the peaks of noise and vibration of cylinder block surfaces.

The objective of this study is to achieve an accurate numerical analysis method for predicting the levels of noise and vibration of cylinder block surfaces up to a frequency of 4 kHz, originated from the crankshaft system of a running engine and propagated through film-like oil and a cylinder block. The authors have improved the conventional method by considering the non-linear stiffness and damping coefficient of oil film on the journal bearings and using dynamic stiffness of the cylinder block. In this study, a V-type 6-cylinder engine was used while in the previous study, the cylinder-in-line type engine had been used.

The accuracy of the numerical analysis was verified by comparing the calculation results with measurement ones obtained by using a laser vibration meter to measure vibration levels in the crankshaft axial direction, and vibration meters on the cylinder block surfaces to measure vibration levels in three orthogonal directions simultaneously. The calculation results agree well with the measurement ones, which demonstrates the reliability of the new analysis method.

2. ANALYSIS METHOD

Figure 1 shows the V-type 6-cylinder, 4-cycle diesel engine used in this study, as well as the major specifications of the engine. Considered in this analysis are stiffness of the crankshaft system, the non-linear stiffness and damping coefficient of oil film, the non-linear characteristics of excitation by combustion pressure and reciprocal inertia forces, and the dynamic stiffness of the cylinder block to accurately estimate the vibration levels of the crankshaft system of the running engine up to 4 kHz.

Figure 2 shows the calculation model, and Figure 3 the calculation flowchart. Three-dimensional solid elements were used to build up the FEM model of the crankshaft. The natural vibration modes of the V-type engine crankshaft need to be analysed in three dimensions, while those of the conventional in-line type engine have to be analysed only in two dimensions: inside and outside the crankshaft plane. The calculated natural modes and frequencies for the crankshaft, flywheel, ring gear, and damper pulley FEM models were compared with the measured ones respectively to verify that the models were accurate enough to be used in the subsequent calculations. Then, to verify the accuracy of the crankshaft system and the cylinder block FEM models, the calculated amplitude of the accelerance transfer function was compared with the measured one, and good agreement was obtained.

The calculation was done by using equivalent oil film stiffness and damping coefficient, equivalent to those of oil film on the journal bearings, which were determined by the Sommerfeld method [7] (to be explained in detail in section 2.1). Exciting force made up of combustion pressure and inertia force was determined in the time domain, and then translated in the frequency domain. In this process, relationship between the time and order increments was determined to have consistency of the exciting force frequency with the response one. In the process of calculating the rotating crankshaft system vibration characteristics, a rotating coordinate system should be used to express exciting forces, and thus the calculated exciting forces in the static coordinate system were translated in the rotating one (to be explained in detail in section 2.2). Simulation software developed in-house was used in determining the non-linear relationship of the oil film stiffness and exciting forces, and in the coordinate transformation.

Then, the function matrix for the vibration transfer from the excitation points to the bearing contact points was determined by using versatile software, NASTRAN [8], with the equivalent oil film stiffness and damping coefficient given to the crankshaft system FEM model. The dynamic stiffness matrix for the crankshaft system considering the equivalent oil film stiffness and damping coefficient is the reverse matrix of the transfer function, and this reverse matrix was determined by the Gauss-Jordan method (to be explained in detail in section 2.3). As the equivalent oil film stiffness and damping coefficient in the dynamic stiffness matrix differ by engine speed, the dynamic stiffness matrix was determined for each engine speed.

The dynamic stiffness matrix for the cylinder block was calculated in the same manner on the cylinder block FEM model, which was also determined by the transfer function matrix derived from the experiment results.

The dynamic stiffness matrix for the crankshaft system considering the equivalent oil film stiffness and damping coefficient was composed with that for the cylinder block by transfer function synthesis method (dynamic stiffness composition). The composed matrix was calculated for each frequency to determine the characteristics of vibrations at the crankshaft end in the 'x' direction and at the bearing-block contact points. Rotational vibration orders were derived from the calculated amplitudes of the vibration velocities in the 'x' direction at the crankshaft end. As the vibration characteristics at the bearing-block contact points in the 'y' and 'z' directions were determined in the rotating coordinate system, the vibration characteristics were translated in the static coordinate system to obtain the vibration characteristics of the cylinder block surfaces (to be explained in detail in section 2.4).

The calculation results were compared with the measurement ones to verify the effectiveness of the numerical analysis method. In the experiment, velocity in the crankshaft axial direction ('x' direction shown in Figure 3) was measured at the pulley end by using a laser vibration meter.

In this study, the vibration characteristics determined by using the cylinder block dynamic stiffness were compared up to 4 kHz with the results obtained by the conventional method of using static stiffness determined by applying a unit load to each bearing-block contact point.

The new analysis method is explained in detail in the following sections:

2.1 Calculation of equivalent oil film stiffness and damping coefficient

Figure 4 shows the oil film model. The equivalent oil film stiffness and damping coefficient were calculated by using the following Sommerfeld number, S:

$$S = \frac{2\mu NLR}{P}\left(\frac{R}{C_p}\right)^2 \quad \cdots\cdots\cdots\cdots\cdots\cdots\cdots\cdots\cdots (1)$$

where R: journal radius, C_P: average clearance, μ: oil viscosity coefficient, N: engine speed, L: bearing width, P: load on bearing

Then, non-dimensional oil film stiffness and damping coefficient, so and do, were determined by using the bearing characteristics data[9] derived from the Sommerfeld number, S, by the Reynolds equation. The equivalent oil film stiffness and damping coefficient, So and Do, were determined by using the non-dimensional ones.

$$\begin{bmatrix} So_y & So_z \\ Do_y & Do_z \end{bmatrix} = P/C_p \begin{bmatrix} so_y & so_z \\ do_y & do_z \end{bmatrix} \quad \cdots\cdots\cdots (2)$$

As seen in equation (1), the Sommerfeld number, S, depends on load on bearing, P, and engine speed, N. The equivalent oil film stiffness and damping coefficient are considered to have non-linear relationships with engine speed and vibration frequency as the load on bearing, P, has a non-linear relationship with vibration frequency with combustion pressure considered, and the bearing characteristics data has a non-linear relationship. Figure 5 shows an example of calculating equivalent oil film stiffness at an engine speed of 3000 rpm by using equations (1) and (2). (The suffix of $So_{y\text{-}i}$ in the figure indicates the vibration direction, 'y', and the journal number, 'i'.) This figure indicates that the equivalent oil film stiffness has a non-linear relationship with crank angle, considering the change in a complicated manner. The broken line in the figure shows the stiffness of the crankshaft in the 'y' direction. This figure also indicates that the equivalent oil film stiffness is higher than that of the crankshaft system at certain crank angles. Therefore, taking the oil film characteristics into consideration is critical for analysing the crankshaft system vibration characteristics.

2.2 Calculation of exciting forces on crankshaft

Exciting force calculation flowchart is shown in the left-hand side of Figure 6. Exciting force made up of combustion pressure and inertia force is determined in the time domain. Combustion pressures were determined by interpolating the measurement results by the spline approximation method. Inertia forces were determined by multiplying accelerations and reciprocating plus rotating inertia masses. Conventionally, 2nd-or-higher order rotational acceleration terms have been neglected in achieving approximate inertia forces, but should have been accurately calculated to determine the characteristics of the crankshaft vibrations at the crankshaft end in the 'x' direction. Therefore, the authors determined the 2nd-or-higher order acceleration terms by using the Runge-Kutta method of obtaining the 2nd-order differential equation of the piston displacement. Then, the exciting forces in the time domain were translated in the frequency domain by fast Fourier transform (FFT). In order to have the frequencies of the transformed exciting forces consistent with those of the rotational order components which change, depending on engine speed, the following relationship between crank angle increment, engine speed and FFT Nyquist frequency was determined as follows:

$$\delta_t = \frac{\delta_{deg}}{6N} \quad \cdots\cdots (3)$$

where δ_{deg}: crank angle increment, N: engine speed, δ_t: sampling time.

According to the Nyquist sampling theory, there is the following relationship between maximum frequency, f_{max},, resolution frequency, δ_f

$$f_{max} = \frac{1}{2\delta_t}, \; \delta_f = \frac{1}{n\delta_t} \quad \cdots\cdots (4)$$

where n: the number of exciting forces in the time domain

$$O_{max} = \frac{N f_{max}}{60}, \quad \delta_f = \frac{N\delta_0}{60} \quad \cdots\cdots (5)$$

where δ_O: rotational order increment, O_{max}: highest analysable rotational order
Equation (6) is derived from equations (3), (4) and (5):

$$\delta_{deg} = \frac{360}{\delta_0 n} \quad \cdots\cdots (6)$$

With this equation, crank angle increment, δ_{deg}, is defined by the number of exciting forces, n, and order increment, δ_o. Then, the exciting forces in the static coordinate system were translated in the rotating coordinate system by a generally-used coordinate transformation method. Shown in the left-hand side of Figure 7 are vectors F_x and F_y in the static coordinate system, as well as a radial vector, F_r, and a tangential vector, F_t, in the rotating coordinate system. Shown in the right-hand side of the figure are calculated major exciting force order components. This figure indicates that the 1st- to 4th-order rotational components of the exciting force change as engine speed changes. With these results, contributions of the order components of the exciting force to the vibration peaks can be evaluated.

2.3 Calculation of dynamic stiffness matrix

The equation of motion for calculating the dynamic stiffness of the crankshaft system an the cylinder block with the equivalent oil film stiffness and damping coefficient considered is expressed as follows:

$$[M]\{\ddot{x}\} + [C]\{\dot{x}\} + [K]\{x\} = \{f\} \quad \cdots\cdots (7)$$

where [M]: mass matrix of FEM model, [C]:stiffness matrix, {x}: displacement vector
Dynamic stiffness matrix [B] is expressed by using Laplace transform, s, as follows:

$$[B]=[H]^{-1}=[[M]s+[C]s+[K]]\cdots\cdots\cdots(8)$$

where [H]: transfer function matrix which is a reverse dynamic stiffness matrix

In this analysis, the authors determined the equivalent oil film stiffness and damping coefficient at each engine speed, composed them with the crankshaft system FEM model, and then determined the individual transfer functions for the vibration transfer from the excitation points to the bearings by using NASTRAN to achieve the dynamic stiffness matrix.

The equation of motion for the crankshaft with the oil film effects considered is expressed with the dynamic stiffness B_{ij} of the matrix $[B_{crk+oilfilm}]$ and the displacement {x} as follows:

$$[B_{crk+oilfilm}]\{x_{crk+oilfilm}\}=\begin{bmatrix} B_{11} & \cdots & B_{15} & B_{16} & \cdots & B_{110} \\ & \ddots & & & & \\ & & \ddots & & & \\ & & & \ddots & & \\ & SYM. & & & \ddots & \\ & & & & & \ddots \end{bmatrix}\begin{Bmatrix} x_{p1x} \\ \vdots \\ x_{p4z} \\ x_{cyl-j1x} \\ \vdots \\ x_{cyl-j5z} \end{Bmatrix}=\begin{Bmatrix} f_{p1x} \\ \vdots \\ f_{p4z} \\ f_{cyl-j1x} \\ \vdots \\ f_{cyl-j5z} \end{Bmatrix}\cdots(9)$$

where $\{f_p\}$: exciting force, suffix $_p$: exciting force or displacement of crank pin 'p', suffix $_{cyl-ji}$: force or displacement of bearing 'i'

In the same manner, the equation of motion for the cylinder block is expressed with the dynamic stiffness B_{cylij} of the matrix $B_{cyl}]$ as follows:

$$[B_{cyl}]\{x_{cyl}\}=\begin{bmatrix} B_{cyl\,11} & \cdots & B_{cyl-15} & & \\ \ddots & & & & \\ & \ddots & & & \\ & & \ddots & & \\ & SYM. & & \ddots & \\ & & & & \ddots \end{bmatrix}\begin{Bmatrix} x_{cyl1x} \\ \vdots \\ x_{cyl5z} \\ x_{spc1x} \\ \vdots \\ x_{spc5z} \end{Bmatrix}=\begin{Bmatrix} f_{cyl1x} \\ \vdots \\ f_{cyl5z} \\ f_{spc1x} \\ \vdots \\ f_{spc5z} \end{Bmatrix}\cdots(10)$$

where suffix $_{SPC}$: points of cylinder block being constrained to engine mount
The obtained dynamic stiffness matrices were composed by the transfer function synthesis method to obtain a dynamic stiffness matrix [B].

$$[B(\omega)]\cdot[X(\omega)]=\{F(\omega)\}\cdots\cdots(11)$$

The amplitude of velocity at the crankshaft end those of accelerations at the bearing-block contact points of the cylinder block were determined by applying the iterative method to equation (11) for each frequency.

2.4 Coordinate transformation from rotating coordinate system to static one for the vibration characteristics

The 'y' and 'z' vibration components obtained in the previous section 2.3 are expressed in the crankshaft rotating coordinate system, and thus should be translated in the static coordinate system. Explained in this section is the method for transforming coordinate of the vibration components in the frequency domain. Given that the crankshaft vibration components and orders are y(t), z(t), i in the rotating coordinate system, and Y(t), Z(t), j in the static coordinate system respectively, the i-th order vibration component is expressed as:

$$\begin{Bmatrix} y(t) \\ z(t) \end{Bmatrix} = \begin{bmatrix} a_i & b_i \\ c_i & d_i \end{bmatrix} \begin{Bmatrix} cos(i\omega t) \\ sin(i\omega t) \end{Bmatrix} \quad \cdots\cdots\cdots\cdots\cdots (12)$$

where a, b, c, d: amplitudes of the i-th order vibration component

The coordinate in the rotating coordinate system are translated in the static coordinate system as shown in the following equation:

$$\begin{Bmatrix} Y(t) \\ Z(t) \end{Bmatrix} = \begin{bmatrix} cos\,\omega t & -sin\,\omega t \\ sin\,\omega t & cos\,\omega t \end{bmatrix} \begin{Bmatrix} y(t) \\ z(t) \end{Bmatrix} \quad \cdots\cdots\cdots (13)$$

Equation (14) is obtained by developing equations (12) and (13) and applying the half-angle formula.

$$\begin{Bmatrix} Y(t) \\ Z(t) \end{Bmatrix} = \begin{bmatrix} A_j & B_j \\ C_j & D_j \end{bmatrix} \begin{Bmatrix} cos(j\omega t) \\ sin(j\omega t) \end{Bmatrix} \quad \cdots\cdots\cdots\cdots (14)$$

where

$A_j=[a_{i-1}+d_{i-1}+a_{i+1}-d_{i+1}]/2$ (15)
$B_j=[b_{i-1}-c_{i-1}+b_{i+1}+c_{i+1}]/2$ (16)
$C_j=[c_{i-1}-b_{i-1}+c_{i+1}+b_{i+1}]/2$ (17)
$D_j=[d_{i-1}+a_{i-1}+d_{i+1}-a_{i+1}]/2$ (18)

These equations indicate that the i-th order vibration component in the rotating coordinate system is transmitted as the i+1 and i-1-th order vibration components for the j-th order vibration in the static coordinate system.

3. RESULTS OF VIBRATION ANALYSIS FOR THE ROTATING CRANKSHAFT SYSTEM AND CYLINDER BLOCK SURFACES

3.1 Verification of effectiveness of the numerical analysis method for the axial vibration

In the case of analysis dealing with non-linear characteristics, even the characteristics of the 0.5-th as well as high order vibration components can be calculated by using equation (6), but, in this study, only the 3rd- and 6th-order rotational vibration components are dealt with because they are so significant in the crankshaft axial direction for the V6-typ engine.

Compared in figure 7 are the results of measuring the characteristics of the 3rd-order rotational vibration component in the 'x' direction of the crankshaft with the results of the numerical analysis using the non-linear characteristics. The horizontal axis in the figure is engine speed, an the vertical one is the amplitude of vibration velocity. The plots in this figure, ▲, △ and □, are respectively the experiment results, the calculation ones in the case of using the dynamic stiffness for the cylinder block, and the calculation results in the case of using the conventional static stiffness. This figure suggest that the peaks at 2000 and 3500 rpm in the experiment are attributed to the peak of the 3rd-order component of the exciting force shown in figure 6. The peaks of vibration velocity amplitude, attributable to the exciting force, are easy to identify in the results of analysis using the dynamic stiffness, while not in the results obtained by using the conventional method. This shows that the amplitude of vibration attributed to the exciting force can be accurately estimated by using the dynamic stiffness of the cylinder block.

Figure 8 shows the amplitude of the 6th-order rotational vibration velocity in the 'x' direction at the crankshaft end. As in the case of the 3rd-order component, good agreement between the calculation and experiment results was obtained for the 6th-order component. Effects of exciting force on peaks are more noticeable when dynamic stiffness is used in calculation than when static one is used.

In summary, the calculation and experiment have turned out that the method of using the dynamic stiffness of the cylinder block considering oil film effects is more reliable in terms of the accuracy of predicted results than the conventional method.

3.2 Verification of effectiveness of the numerical analysis method for the cylinder block surface vibration acceleration

Figure 9 shows the results of 1/3-octave-band analysis for the cylinder block surface vibrations in the 'x' direction considering the dynamic stiffness. The horizontal axis is frequency (Hz), and the vertical one is acceleration amplitude (dB) with the calculation/measurement locations indicated by the arrow marks. The plots in this figure, △, ○ and ■, are respectively the experiment results, the calculation ones in the case of using the dynamic stiffness for the cylinder block, and the calculation results in the case of using the conventional static stiffness.

As seen in this figure, better agreement between calculation and experiment for the surface vibrations is obtained by the method of using the dynamic stiffness of the cylinder block than by the conventional method of using the static stiffness. Also, with the conventional method, the calculation results do not agree with the experiment ones in a frequency range from 500 Hz to 2000 Hz. This is because the frequency of the natural vibration mode for the cylinder block is 1.5 kHz, and the coupling of the crankshaft system vibrations with the cylinder block ones is not considered.

In general, knowing cylinder block surface vibration peak frequencies is essential to lower vibration level. Considering the experiment results in the figure, indicating that the peak of the cylinder block surface vibration originated from the crankshaft exists at 3 kHz, calculation should be done up to 4 kHz to determine the frequencies of the cylinder block surface vibration peaks. This suggests that dynamic stiffness should be considered in analysing the characteristics of cylinder block surface vibrations.

It has been known that exciting forces on a crankshaft lie at frequencies lower than 1 kHz, and those at 3 kHz are very small. However, our experiment and calculation results tell that the peak of the cylinder block surface vibration exists at 3 kHz. Also, the authors verified that a peak does not exist at 3 kHz when exciting forces act on the cylinder block alone, at the bearing contact points. This indicates that the vibration with the peak at 3 kHz is transmitted through the crankshaft system and the oil film. The detailed mechanism will be revealed in the future.

Figures 10 and 11 show the results of analysing the cylinder block surface vibrations in the 'y' an 'z' directions respectively. These figures show that peaks exist at 1.5 kHz and 3 kHz, which probably results from the coupling of the crankshaft system vibration with the cylinder block one. Better agreement between the calculation and experiment results is obtained up to 4 kHz with the method proposed here than with the conventional one.

4. CONCLUSION

The characteristics of the rotating crankshaft and cylinder block surface vibrations were analysed by mean of the transfer function synthesis method, considering the dynamic stiffness of oil film on the crankshaft bearings and that of the cylinder block. The following conclusions are reached:

1.The method for calculating the characteristics of the crankshaft system and cylinder block surface vibrations for the V-type engine, which requires three-dimensional analysis instead of the conventional two-dimensional one, has been developed, and the estimation with the method has turned out accurate.

2. The accuracy of calculating the 3rd- and 6th-order rotational vibration components in the 'x' crankshaft direction has been improved by applying the dynamic stiffness to the cylinder block vibration analysis from the accuracy obtained by using the conventional static stiffness. The effects of the exciting forces are found in the results of analysing the characteristics of the vibration in the 'x' direction.

3. The cylinder block surface vibration analysis has turned out that the peaks at 1.5 kHz and 3 kHz, attributable to the coupling of the crankshaft and cylinder block vibrations, can be achieved by calculation using the dynamic stiffness for the cylinder block, which cannot be achieved by using the conventional static stiffness. Good agreement between calculation and experiment is achieved up to 4 kHz with the analysis method proposed here, which indicates that the new analysis method is reliable.

5. ACKNOWLEDGEMENT

The authors would like to express great thanks to Mr. Hitoshi Yamashita and Mr. Kenichi Yamashita with Isuzu Advanced Engineering Centre, and Mr. Kazuyoshi Ikuno with Isuzu Engine Development Promotion Department for advice and support.

REFERENCES

(1)Okamura, et al, Cases of Noise Reduction, Nikkan Kogyo Shinbun, 1990,P5-1•P5-31

(2)Alan P.Drushitz at al, "Influence of Crankshaft Material and Design on the NVH Characteristic of a Modern , Aluminium Block, V-6 Engine", 99-01-1225, SAE, 1999

(3)N. Hariu, "Stress Prediction for an Operating Crankshaft System Using Nonlinear Vibration Analysis", 2333, No.95-1, JSME, 1995.
(4)N. Hariu and T. Nakada, "Nonlinear Vibration Analysis for an Operating crankshaft", 2448, No.940-10, JSME, 1994.
(5)N. Hariu and A.Okada, "A method of NVH and Stress under Operating crankshaft", 970502, SAE, 1997
(6)N. Hariu and T.Nakada, "Nonlinear vibration analysis for rotating crankshaft", C487/011/94, I Mech. E., 1994.
(7)H. Okamura, "Influence of Crankshaft-Pulley Dimensions on Crankshaft Vibrations and Engine-Structure Noise and Vibrations", 931303, SAE, 1993.
(8)NASTRAN, User's Manual, ver69, 1998
(9)Japan Machinery Association, Documents on Static/Dynamic Bearing Characteristics, 1984, Japan Industry Press

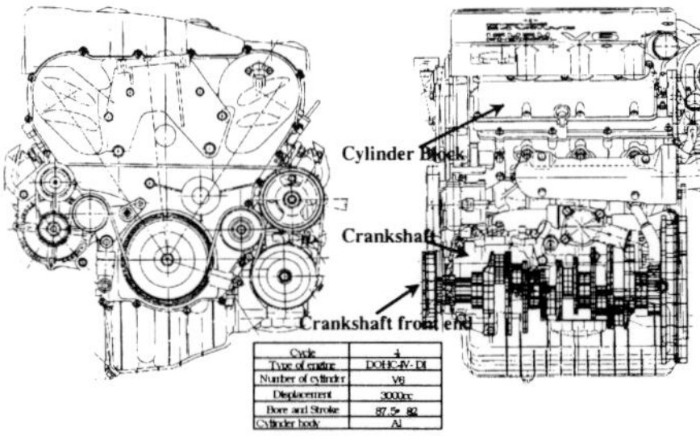

Fig. 1 The drawing of the ISUZU - V6 Diesel Engine

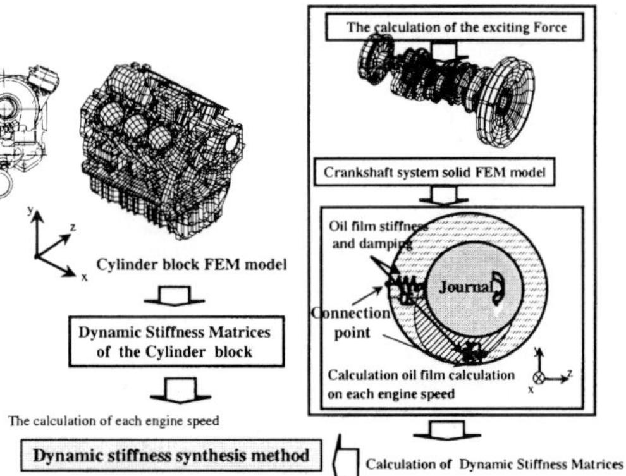

Fig.2 The nonlinear model of the dynamic stiffness synthesis method

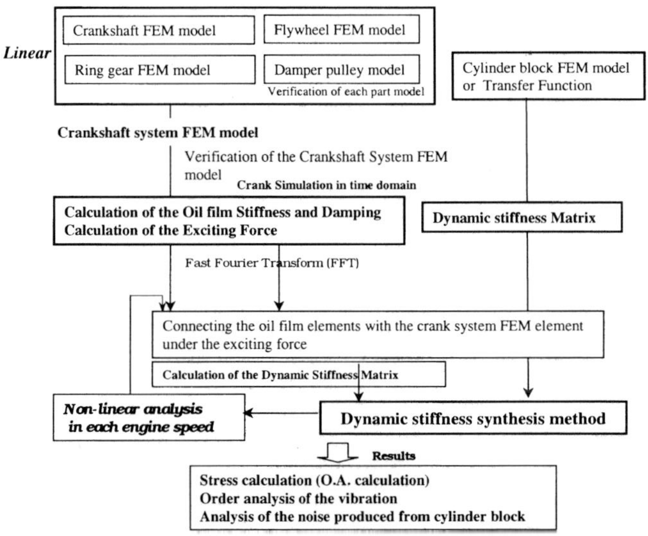

Fig. 3 The process of the non-linear analysis by the dynamic stiffness synthesis method

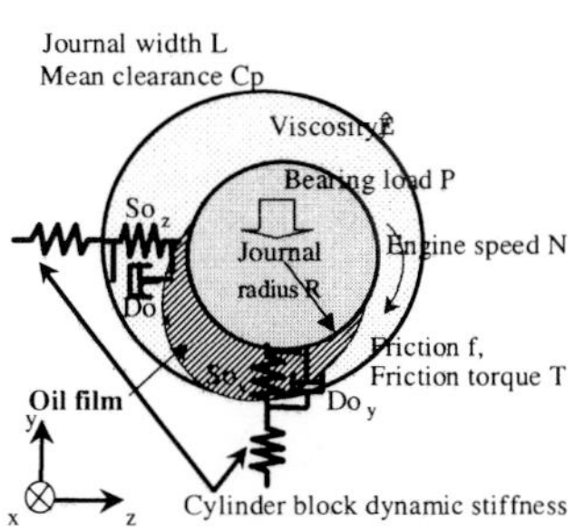

Fig.4 The detail of the nonlinear oil film stiffness So and damping Do(at 3000rpm)

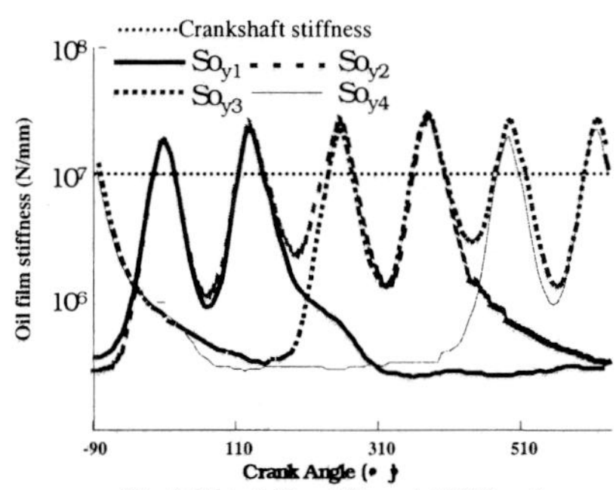

Fig. 5 The oil film stiffness (at 3000rpm)

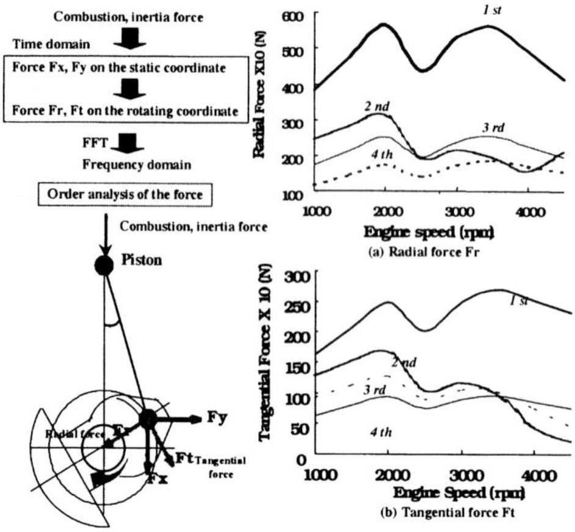

Fig.6 The boundary condition of the exciting force

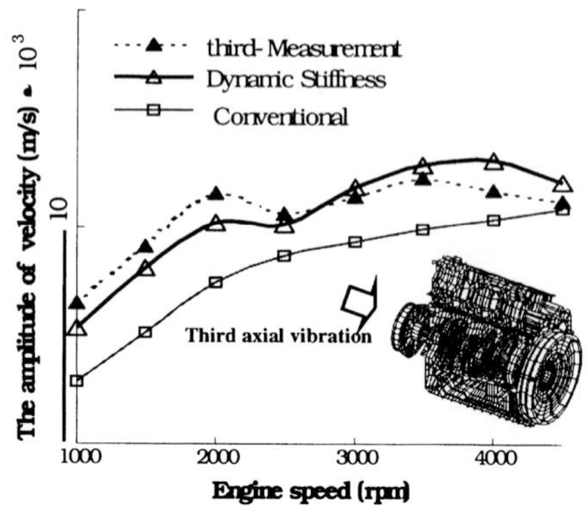

Fig.7 The third axial vibration at the crankshaft front end

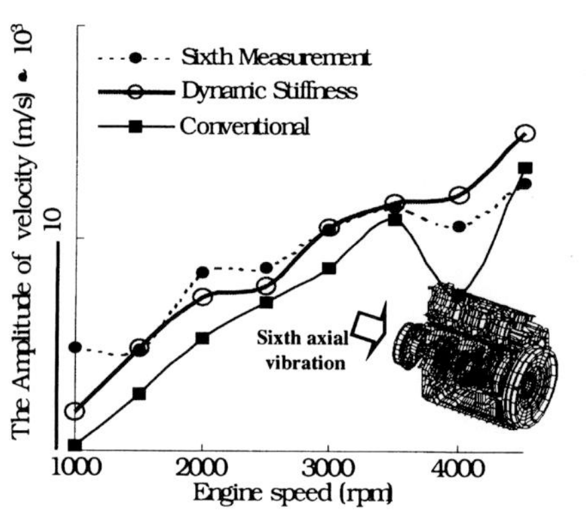

Fig.8 The Sixth axial vibration at the crankshaft front end

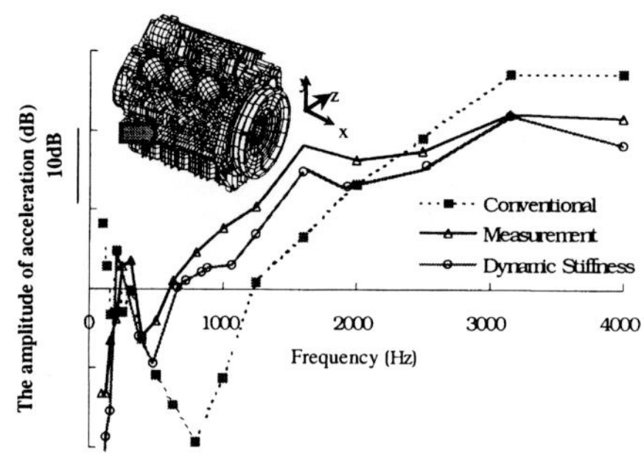

Fig.9 The amplitude of the acceleration(third octave band) on the cylinder bock surface in the x-axis direction (at 2000rpm)

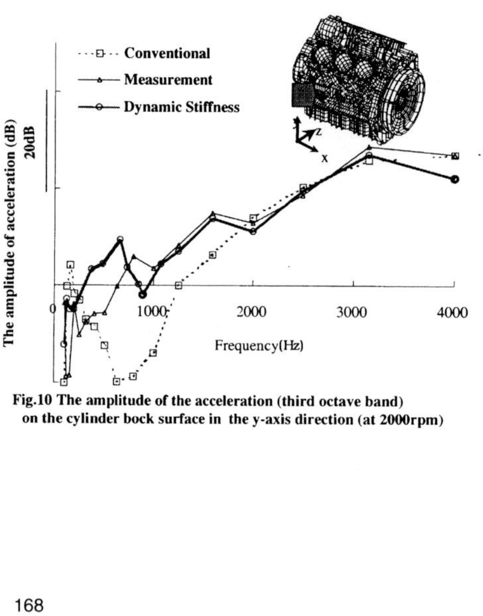

Fig.10 The amplitude of the acceleration (third octave band) on the cylinder bock surface in the y-axis direction (at 2000rpm)

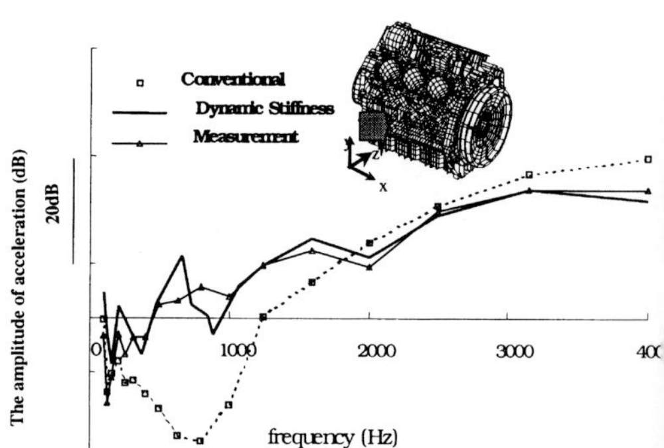

Fig.11 The amplitude of the acceleration vibration (third octave ban on the cylinder bock surface in the z-axis direction

Simulation of flexible engine block, crank, and valvetrain effects using DADS

J ZEISCHKA, D KADING, and **J CROSHECK**
Computer Aided Design Software Inc., Iowa, USA

Abstract

This paper discusses how DADS, Dynamic Analysis and Design System, is being applied to predict deformation of engine components due to structural flexibility. By combining multibody dynamics to account for large nonlinear part displacements and modal coordinates for each part from appropriate finite element analyses, a more sophisticated simulation model has been developed. The benefits of this improved simulation method are accurate prediction of loads acting between all parts in the assembly, as well as accurate prediction of local part deformations superimposed on their gross motion. These results provide insight into the interaction and coupling of multiple parts flexing under high engine loads and accelerations. Advanced flexible body animation methods allow visualization of the motion of the complete flexible system and convenient ways to focus on flexible behavior in specific locations. Position, velocity, and acceleration of all parts as well as individual finite element nodes are reported as are reaction forces in the kinematic constraints and loads in the force elements.

These simulation methods have been applied to two distinct realms within the engine. One area is the engine block and crank, while the other is the valvetrain consisting of the cam, lifters, rocker arms and valves. In both cases, the objective is the same: use computer simulation to accurately predict the dynamic response of the flexible system.

In the case of the crank and block model, an additional capability had to be incorporated to represent the oil film in the journal bearing. The hydraulic behavior of the oil keeps the crank riding in a protective film and prevents contact between it and the bearings. This complex phenomenon has been characterized and included in the simulation using both an impedance force algorithm and a quasi-static finite element method developed by Booker [11]. The force due to the oil film is calculated and applied to the flexible crank and block at each of the journal bearing locations and is the only connection between those flexible bodies. This oil film force is a function of the position and relative velocity of the crank and includes logic to account for details like the oil holes drilled in the bearing and cocking of the cylindrical surfaces. No kinematic joints are used to model the crank and block interaction. Joints, which are algebraic constraints in DADS, are used to model the connections between other parts in the engine.

The flexible body modeling approach used in DADS is well developed and documented over the last 15 years by Haug [8] and others. Flexible body models of the engine crank and block interaction have been applied to several different designs over the last 5 years by Zeischka and others [10]. The current state of the art now extends the previous work to include flexible valvetrain components as well. The basic premise of the DADS flexible body methodology is to use the multibody dynamics equations of motion for a system of rigid bodies with algebraic constraints, and include modal coordinates from suitable finite element analyses. The original set of seven generalized coordinates used in the rigid body case, three Cartesian translations and four quaternion rotations with a normalization constraint, are augmented by some number of modes from a mixture of static and vibration finite element runs. The original equations of motion have been extended to include the modal coordinates with all the resulting coupling. This approach has the advantage of efficiency over purely finite element methods for modeling the structural dynamic response of moving parts within an engine. Experience has pointed out the value of using a mixture of static and vibration modes to best represent part flexibility. The static and vibration modes are orthogonalized to ensure linear independence and the new set of modes are used for block and another set for the crank. Likewise, separate sets of modes are used for each valvetrain part.

To better represent the critical parts of the valvetrain, special attention was given to the cam and follower and to the valve springs themselves. The cam and follower are treated as a force contact relation instead of a kinematic constraint to allow separation of the cam from the follower. The valvesprings are now treated as flexible bodies, and important mass effects and coil contact events are captured during the simulation. The mass effects are associated with spring surge that occurs at high speeds. Coil contact occurs when the individual coils in the spring collide. These features are now readily simulated.

Results demonstrate the importance of including flexibility when considering simulation. Accurate prediction of loads and part displacements provide analytical tools that improve engine design. Additionally, the ability to see part flexibility and overall system motion during animation gives insight into complex mechanical behavior.

Introduction

Publications on the theory of multibody dynamics including the effects of flexible bodies based on the modal synthesis method appear in 1985 [1,2]. Several subsequent publications have been made discussing its application to industrial problems. Multibody dynamics with flexible bodies initially received more attention from engineers in aerospace [3,4,5] than from other industries. Currently, the automotive analysis and design communities have demonstrated a keen interest this technology and are very active in applying it to a variety of subsystems. This paper focuses on the engine.

Early simulation work [7] used a simple DADS model for flexible engine analysis. The crankshaft was represented by a few beam elements in this initial model. The deficiencies of this coarse representation led the authors to develop more elegant methods which can deal with refined meshed finite element (FE) models with 10,000 or more nodes that are frequently used for engine vibration and stress analysis. These more accurate models are combined with hydrodynamic journal oil film effects to provide very high fidelity simulation methods. When these advanced methods are used, and the equations of motion are solved in the time domain, the results provide the best possible representation of "dynamic stress" and true response of the coupled system.

Recent modeling work has extended to the valvetrain where there are more parts and complexity. The DADS model uses the same basic method to include flexible effects for each part using modes obtained from FE analysis. In addition, the cam contact with the lifters (or rockers depending on the design) is modeling in detail so parts can separate and impact again. The valve springs are also modeled in detail to account for "mass effects" and coil contact as the springs deform. Many important effects are predicted by the simulation including, torsional wind-up of the cam, valve float, spring coil clash, and forces throughout the system. As always, results include position, velocity, acceleration, and reaction forces for all parts in the simulation model. Since DADS is very numerically stable, these models can represent high engine speeds. Naturally, the simulation time steps are small for the high speeds, but the method is very practical.

The next section outlines the model used to create the flexible crank and block model, and finally for a flexible valvetrain model.

Model Description – Flexible Engine Block and Crank

The rigid body model can be created directly in DADS or imported through one of the interfaces to CATIA, IDEAS, Pro/E, or SolidWorks. Joints and force elements are defined between pairs of rigid or flexible bodies in the model. The kinematic joints and constraints prescribe ideal mathematical relationships between parts and are honored throughout the simulation, no matter what position the parts are in. The force elements apply forces and torques on single bodies, or between a pair bodies, based on time, position, or velocity. As such, force elements do not constrain motion, but do prescribe the acceleration of parts at an instant in time. DADS solves the equations of motion for all dependent degrees of freedom, and then numerically integrate the accelerations to determine the positions at the next time step. This process is repeated at each time step during the simulation.

The DADS model requires flexible modes for each separate part of interest. The block and crank each had their own distinct modal coordinates. Other parts were treated as rigid in this simulation. For this project, MSC NASTRAN was used to perform the FE analysis and create the flexible modes. Several DADS models were run to compare the effects of using different modes. Mode selection procedures are beyond the scope of this paper, but more detail on the effects of mode selection can be found in the literature [9,10]. It can be shown that the modal implementation found in DADS is equivalent to Craig-Bampton methods.

The crankshaft and engine block for a 1.8 Liter four-cylinder engine are depicted in Figure 1. The engine block is connected to the ground at its support points. The connecting rods and pistons are included as rigid bodies. The remainder of the system kinematics is modeled with mechanical joints defined in DADS between the different bodies of the engine.

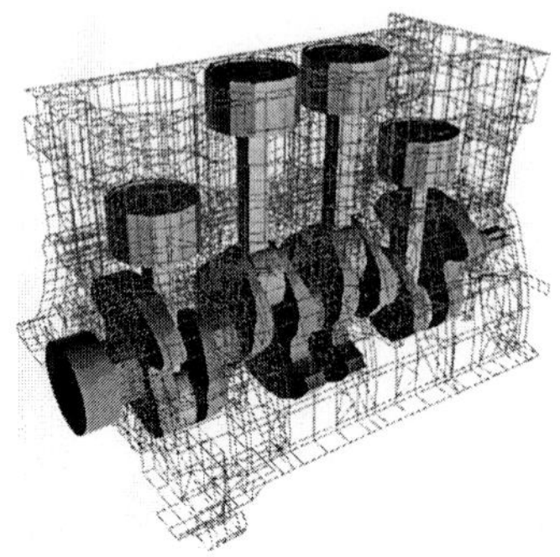

Figure 1 DADS model of engine block and crankshaft

During the simulation, the journal forces are calculated at every time step as a function of relative position and velocity of the reference nodes on the flexible engine block and crankshaft. The hydrodynamic journal effects are implemented using an impedance method [7,11]. The impedance charts provide the journal forces as a function of journal parameters such as dimension, oil viscosity, eccentricity and relative velocity. A detailed discussion of this implementation is provided in references [7,11]. The parameters for each crankshaft and block journal are provided by reference nodes in the DADS model.

Through solution of the equations of motion, the position, velocity and acceleration of each body and reaction forces for mechanical constraints are reported to the result files. For flexible bodies, DADS also reports the scale factors for the mode shapes at each time step as well as position, velocity and acceleration data for nodes from the FE models. These scale factors can then be used for efficient stress recovery back in the original finite element model.

Results of the Flexible Crank and Block Model

The following results are for a fired engine at a crankshaft rotational speed of 4000 rotations per minute (rpm). The main journals are identified by 1 to 5 the plots, where journal 5 is closest to the flywheel.

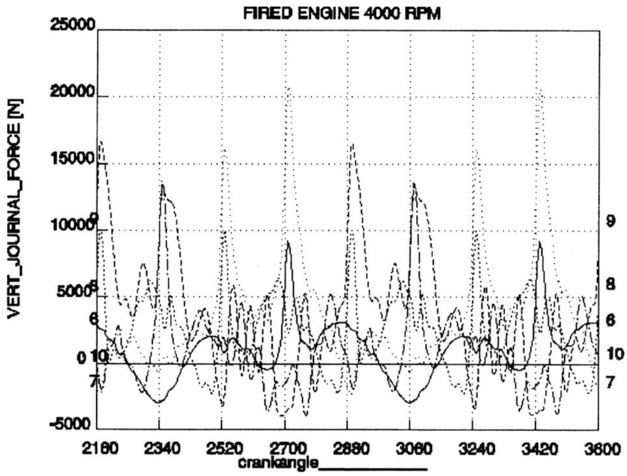

Figure 2 Vertical journal forces of fired engine model.

Figure 2 shows the vertical hydrodynamic forces (piston direction of motion) in all journals for four revolutions. The influence of the flywheel can be seen at journal 5 (curve 10) compared to the force in journal 1 (curve 6). If the flywheel is omitted, congruent loading conditions for journal 1 and 5 would be observed.

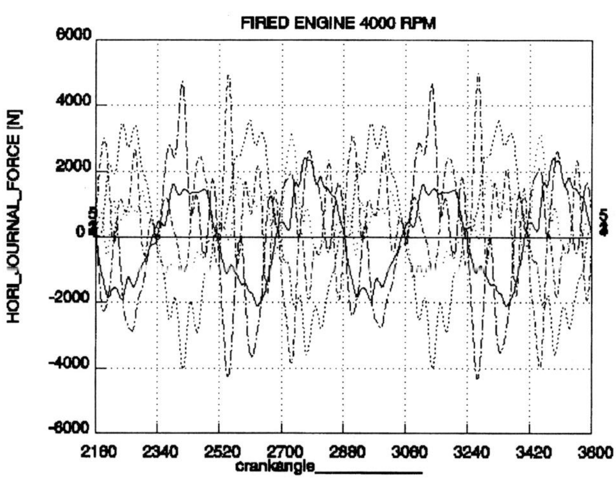

Figure 3 Horizontal journal forces of fired engine model

Figure 3 shows the horizontal hydrodynamic forces in all journals for four revolutions.

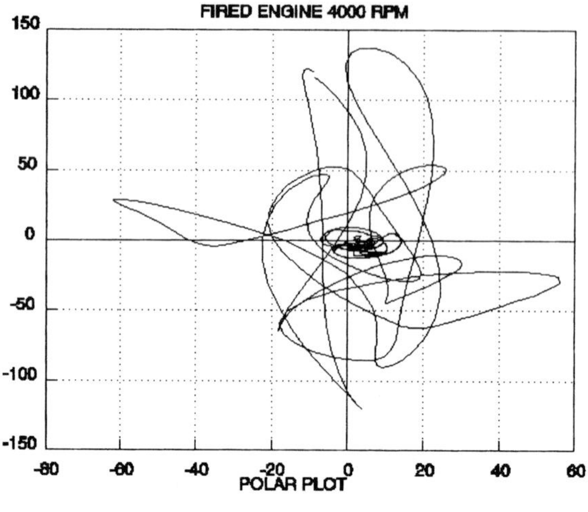

Figure 4 Polar plots of crankshaft to engine block, relative motion (micrometers)

The relative motion of pins at the crankshaft center with respective to reference points on the engine block is shown in Figure 4. The polar plots indicate a steady state response of the engine. The transfer of combustion and inertia loads from neighboring cylinders is clearly visible. Flywheel effects are also present.

Figure 5 Deformation (wireframe) of a rotating flexible crankshaft during animation.

Often, the overall motion and interaction of flexible parts is complex enough that it is hard to fully understand from plots alone. The ability to animate flexible bodies in DADS proved to be useful in visualizing the dynamics of such complex flexible systems. Figure 5 shows the scaled-up deformation of a rotating crankshaft during animation. In this figure the deformed shape is displayed as a wireframe image, while the original geometry is solid. The entire flexible model can be animated and interactively viewed in 3D. Displacements are scaled to make them more clearly visible.

Recent work has extended the FE reduction methods and hydrodynamic bearing algorithms used in the previous results. The new work was performed with General Motors and applied to a truck engine to predict the flexible dynamic response and loads in the journal bearings. Two schemes were used in the hydrodynamic bearing feature for the GM project. The algorithms from Booker [11] provided an impedance approach where force values were calculated during the run, rather than looked up from a table as in the original model outlined above. The other approach used a full FE method to model the oil film in the bearing. This method is much more detailed, and consequently runs more slowly but is important for validating the accuracy of the faster impedance method.

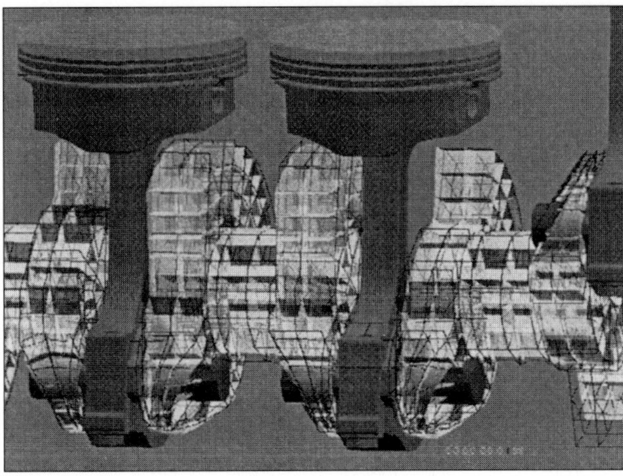

Figure 6 Flexible crankshaft
(Courtesy of General Motors Corporation)

Simulation of the flexible block and crank system can provide valuable insight in numerous ways. Orbit plots of individual journals can diagnose "knock" or "hammer" due to cross-over and impact. Thin oil film conditions can be studied. Vibration of the engine block due to firing can be used to estimate radiated noise. Finally, realistic loads lead the way to calculating stresses more accurately, and in turn, provide the necessary building blocks needed for fatigue life estimates.

Model Description – Flexible Camshaft, Valves, and Rockers

The valvetrain is one of the most complex and critical parts of the engine system, and small variations in the dynamic behavior can measurably affect performance. The valve lift characteristics contribute to engine efficiency and impact overall emissions. High-fidelity flexible multibody dynamic system models provide a powerful tool to predict valvetrain performance and gain insight into complex phenomena that are difficult to visualize or measure experimentally.

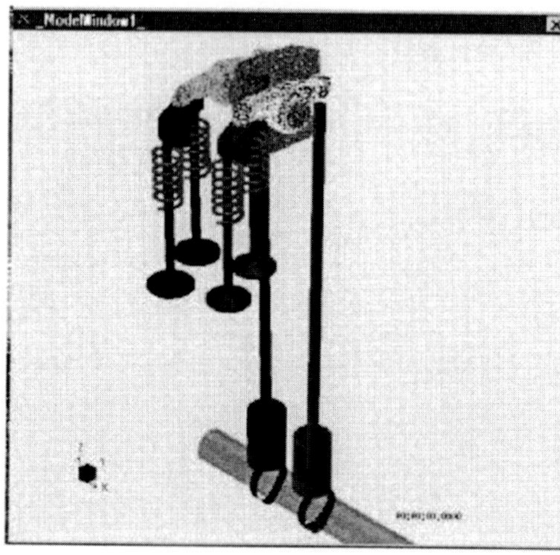

Figure 7 Flexible cam, valves, springs, support, and rocker (Courtesy of DAF Trucks)

Figure 7 shows a single valvetrain for one cylinder of an in-line six cylinder engine manufactured by DAF Trucks. The complete DADS model included all six cylinders and the associated parts. The complete model contained a total of 80 flexible parts, each having multiple modes to represent flexibility.

There are several issues considered in the flexible multibody model of the valvetrain. A contact force element is used rather than a kinematic constraint for the cam and follower interaction to allow the cam and follower to separate. The cam a spline function fit to a table of points from the lift versus cam angle description. Contact force is a function of the stiffness and damping properties of the cam and follower. In addition to the cam and follower, there are contact force elements used throughout the model, including the rocker to pushrod and valve to rocker. Since the numerical integration scheme handles intermittent contact events well, the contact feature of the model works well.

Special attention was given to the valve springs. Normally, a spring in DADS is treated like an idealized force element between two points on two parts where force is a function of displacement and velocity. In the valvetrain, springs have important dynamic properties that affect system performance, especially at high speeds. The mass of the wire coils used in the spring becomes a significant factor, and coils can contact each other in extreme cases. These effects are all accounted for in the valvetrain model.

Input to the simulation model was provided by a constant angular velocity driving constraint at one end of the cam. Different operating speeds were modeled by changing the angular velocity.

Another feature of the model was the combustion force acting on the valves. As in the previous discussion of the crank and block, the combustion force was incorporated as a function of the crank angle and shifted in phase for each cylinder.

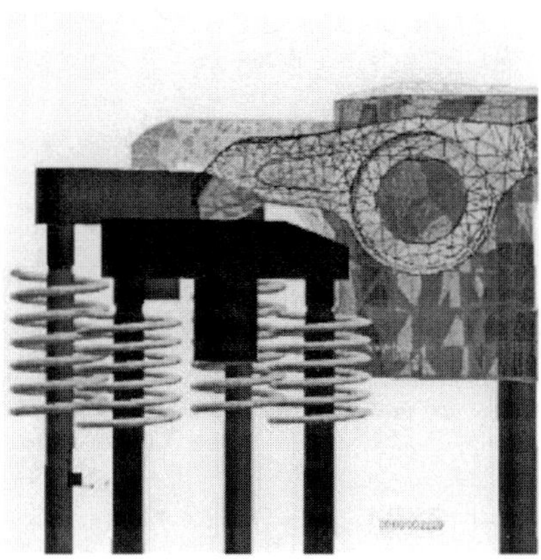

Figure 8 Flexible rocker arms, pedestal, valves, and springs where the deformation has been scaled up by a factor of 100

Figure 8 shows results of the flexible rocker arms, pedestal, and springs where the deformation has been scaled up by a factor of 100 to make it easier to see. The wireframe geometry shows the flexible mesh that originated in the FE models of each of the parts. Several flexible static modes were used for each part. These modes were selected to best represent the way a given part can deform under the constraint and loading conditions. The modal deformations included:

pushrods	axial deformation
rockers	bending
supports	deforming through loads in rocker joints
valve springs	axial deformation
valves	axial deformation
cam	bending and torsion

As with the crank and block model, results can be animated to better visualize the complex flexible multibody behavior.

Results of the Flexible Valvetrain system

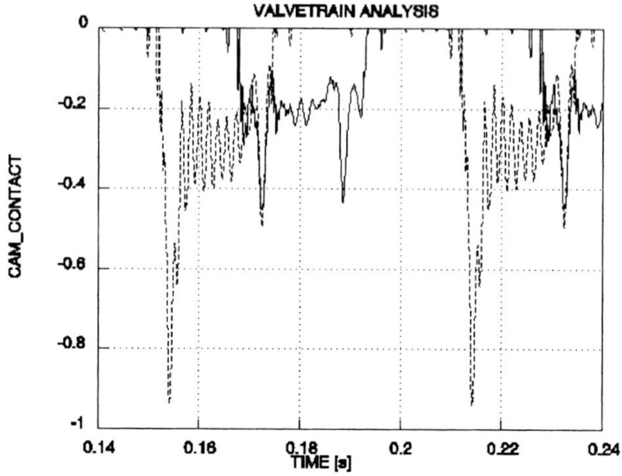

Figure 9 Cam contact force for inlet (solid) and outlet (dash) valves

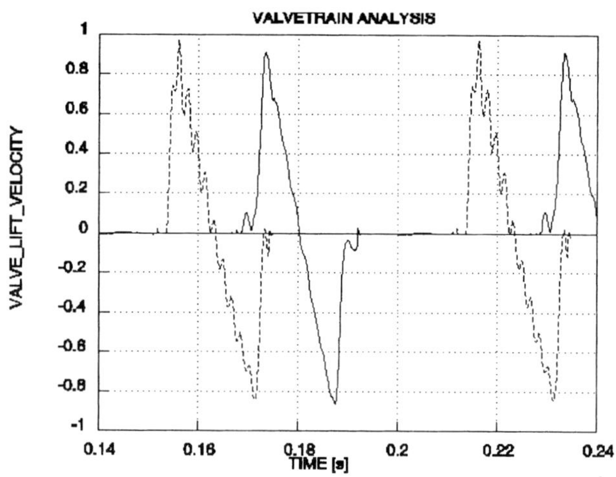

Figure 10 Lift velocity for inlet (solid) and outlet (dash) valves.

The outlet valve is opened against a high gas pressure. This force affects the dynamic response of the exhaust valve and can be seen in both figure 9 for the cam contact force and figure 10 of the valve list velocity.

Recent enhancements to valve spring modeling provide better accuracy and greater ease-of-use. Figure 11, courtesy of BMW AG, shows an example where the spring has been treated in more detail. Figure 12 shows the velocity of FE nodes of the valve spring. The surge effects are visible as the valve is opened and closed.

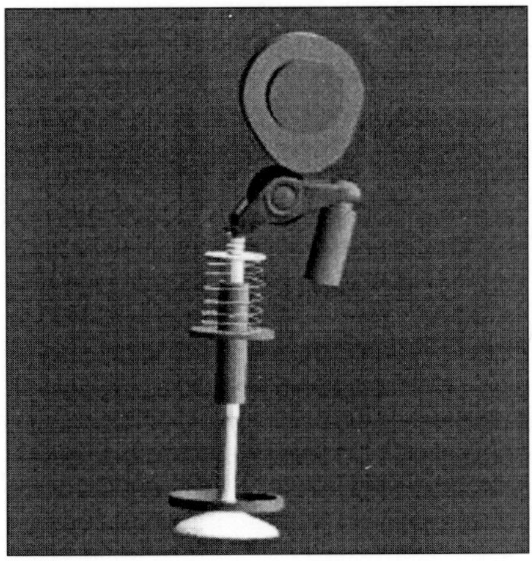

Figure 11 Enhanced valve spring model.
(Courtesy of BMW AG)

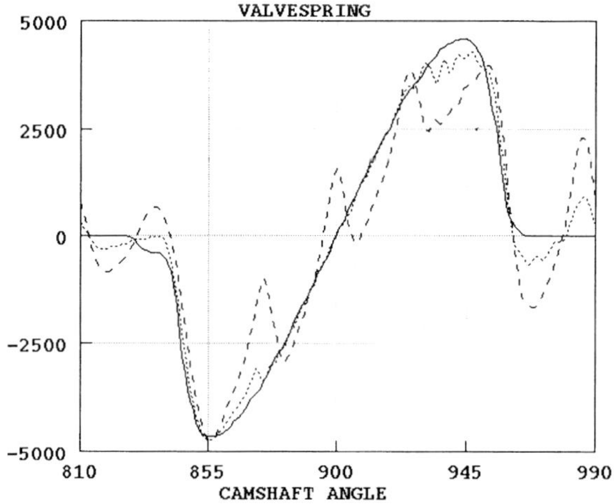

Figure 12 FE node velocity versus cam angle for
4000 (solid), 5000 (dash), and 6000 (dot) RPM

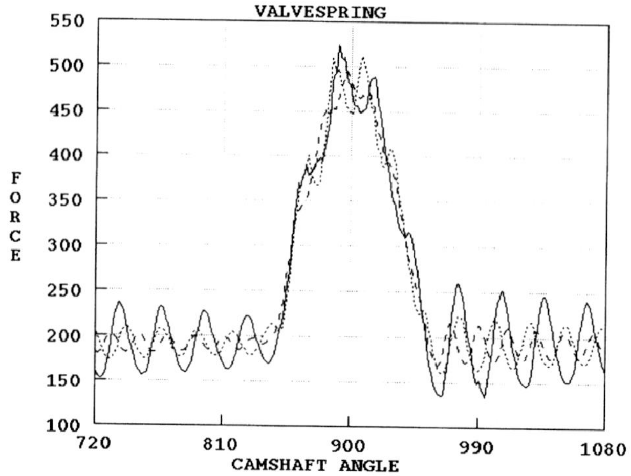

Figure 13 Valve spring force versus cam angle for 4000 (solid), 5000 (dash), and 6000 (dot) RPM

Results plotted in figures 12 and 13 provide clear evidence of the internal vibration of the valvetrain at different engine speeds. Forces in the valve spring can cause impact and acoustic excitation.

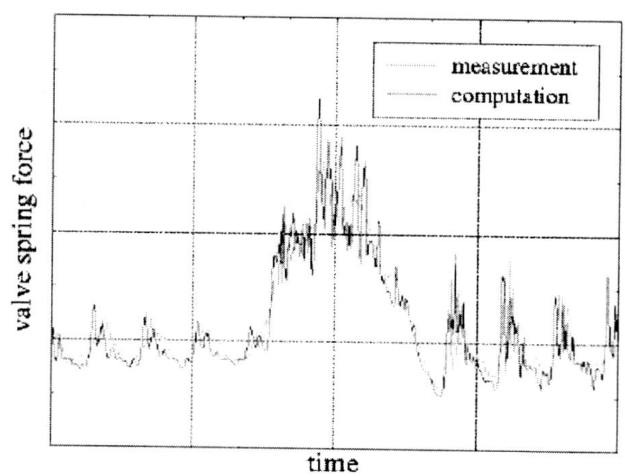

Figure 14 Experimental versus simulation results of the valve spring force

Finally, figure 14 shows a comparison of experimental versus simulation results. The correlation is quite good, and has encouraged BMW to dedicate more engineering resources to this field of simulation.

Conclusion

A description of flexible multibody dynamics using DADS has been presented. Results were presented modeling the interaction between a rotating crankshaft and engine block as well as complete valvetrain. These models show non-linear large displacements of the multibody system, with the small displacements from linear modal terms superimposed. This complete simulation approach offers high-fidelity results for displacement, velocity, acceleration, and reaction forces throughout the flexible multibody model. Resulting loads can be used in subsequent FE analysis and used to predict dynamic stress. Overall, results indicate good correlation with experimental data.

Research and development are proceeding with enhancements to the current solution method. Since many FE models that are conveniently available to analysts were built with stress analysis in mind, they are too large and detailed to be ideal for flexible multibody analysis. Reduction methods are being developed to make it easier to use these large FE models and avoid costly re-modeling to get modes for flexible multibody analysis.

New methods are also being tested to improve the efficiency of the hydrodynamic bearing algorithms used in the crank and block model. Currently two schemes can be used depending on the fidelity desired. One is faster and the other is more accurate.

DADS provides one simulation model that is general purpose enough to be used for a variety of engine, drivetrain, and full car models. The combination of user-interface, graphing, animation, and numerically robust flexible multibody algorithms provides users the ability to efficiently simulate even the most complex systems.

References

1 **W. S. Yoo, E. J. Haug**
Dynamics of Flexible Mechanical Systems using Vibration and Static Correction Modes. Technical report University of IOWA, May 1985.

2 **A. A. SHABANA**
Substructure Synthesis Methods for Dynamic Analysis of Multibody Systems. Computer & Structures, Vol 2, nr. 4, pp. 737-744, 1985.

3 **N. Vukasovic, J.T. Celigüeta, J. Garcia de Jalon, E. Bayo**
Flexible Multibody Dynamics Based on a Fully Cartesian System Support Coordinates. Proceedings of Dynamics of Flexible Structures in Space, Cranfield, U.K., 15-18 May 1990.

4 **C. Garcia Marirrodriga, J. Zeischka, E. C. Boslooper**
Numerical Simulation of Ulysses Nutation.
Fifth European Space Mechanisms and Tribology Symposium, ESTEC, Noordwijk, The Netherlands, 28-30 October 1992.

5 **O. Friberg, V. Karhu**
Simulation of deploying flexible satellite booms using DADS and ABAQUS. Dept. of Structural Mech., Chalmers Univ. of Technology, Göteborg, Sweden. C 1991.

6 **D. J. Nefske and S. H. Sung**
Coupled Vibration Response to the Engine Crank - Block System. Engineering Mechanics Department General Motors Research Laboratories, Warren, Michigan.

7 **G. Knoll, H. Peeken, Ph. Troppmann, J. Zeischka, F. Maessen**
Simulation und Dynamik elastischer Kurbelwellen unter Berücksichtigung der Grundlagersteifigkeit sowie der hydrodynamischen Schmierfilmreaktionen.
6. Internationaler Kongress :"Berechnung im Automobilbau", 21.-23. September 1992 in Würzburg.

8 **E. J. Haug**
Computer Aided Kinematics and Dynamics of Mechanical Systems
Allyn and Bacon, 1989.

9 **S.S. Kim, E. J. Haug**
Selection of Deformation Modes for Flexible Multibody Dynamics, Center for Computer Aided Engineering, College of Engineering, The University of Iowa. Technical Report 86-22. Research under U. S. Army Tank-Automotive R & D Command Grant #DAAE07-85-C-R046

10 **L. Mayer, J. Zeischka, M. Scherens, F. Maessen**
Analysis of Flexible Rotating Crankshaft with Flexible Engine Block Using MSC/NASRAN and DADS, 1995 MSC World User's Conference, Universal City, CA May 1995

11 **J.F. Booker, K.H Heubner**
Application of Finite Element Methods to Lubrication: An Engineering Approach, ASME Journal of Lubrication Technology, vol. 94, 1972

Analysis of knock intensity in spark-ignition engines

M FOOLADI MAHANI and **W J SEALE**
Department of Mechanical and Medical Engineering, University of Bradford, UK
M KARIMIFAR
Department of Mechanical Engineering, University of Kerman, Iran

ABSTRACT

An analysis of the pressure oscillations, knock intensity and knock occurrence in spark ignition engines has been performed. Successive cylinder pressure data were generated cycle-by-cycle from a Ricardo E6 laboratory variable compression ratio engine under varying knock intensity conditions. Knock intensity is calculated from the pressure data. The effects of some operating conditions on the occurrence knock and knock intensity is discussed.

A formulated dimensionless knock parameter (K_P) based on the values of maximum rate of pressure rise, maximum of cylinder pressure and ignition angle is calculated. It is shown that when knock happens in a cycle the value of (K_P) remains at a constant level. Also it is shown that the use of this knock indicator generates results that are in good agreement with experimental results.

1. INTRODUCTION

A spark ignition engine reaches knocking condition operation by increasing compression ratio or by advancing spark discharge timing or by decreasing fuel antiknock. Variation in any of these parameters may cause substantial changes in engine performances, even if knock does not occur.

Knock has two important specifications: knock occurrence cranks angle (*KOCA*) and knock intensity (K_I). Knock occurrence crank angle is defined as the crank angle at which, the engine begins to operate under knock condition. In knock condition the rate of energy release is different so one of another specification of knock is knock intensity (K_I). Leppard [1] showed that knock usually occurs earlier when the burning rate (rate of energy release) is fast in high pressure cycles in the fixed operation conditions. From this point of view an engine may operate under three conditions: without knock, weak knock and heavy knock.

One of the simplest definitions of knock intensity is how much the pressure oscillates under knock condition. For knock intensity measurement the maximum amplitude of the pressure signal commonly used. Other researchers use the difference between maximum and minimum high-pass filtered signal pressure as knock intensity. The occurrence of knock and knock intensity in a firing cycle are random and depend on the operation conditions.

Engine measurements involve the problems of knock specifications. Knock has high potential to damage the components of an engine e.g. piston or cylinder head especially in heavy level. In spite of a good progress has been done to analysis knock quantification in recent years but some problems have yet to be solved. A common difficulty of knock quantification is about analysis of signals that have information about combustion process e.g. signal pressure from pressure transducer under knock condition.

2. EXPERIMENTAL RIG

A Ricardo E6 single-cylinder four stroke spark-ignition engine with variable compression ratio was used for experiments. The engine specification and dimensions are given in table 1. Figure1 shows the experimental Rig and data acquisition system used. A Kistler- Piezo-Electric 701A with a natural frequency of 50 kHz was used to measure the pressure in the combustion chamber (the pressure transducer was cooled with a water circulated system). In this method the pressure wave acts on the diaphragm of the pressure transducer, which generates a voltage proportional to the pressure wave to charge amplifier. The data acquisition was at 8 point per crank angle or about 87 kHz.

For accurate measurements low-pass and high-pass filters are necessary. The low-pass filter eliminates high-frequency noise signals, e.g. on the closure of the valves. High-pass filter filters long-term drift resulting from changes of temperature. Selection of the correct high pass-filter guarantees that only signals from actual combustion process are evaluated.

3. ANALYSIS OF PRESSURE DATA AND KNOCK INTENSITY

Experimental data for prediction of knock and knock intensity with a Ricardo E6 research engine are used as a basis for comparison between prediction and experiment. A common problem in the investigation of pressure data, knock classification and knock intensity is related to the rate of data acquisition [8], method of determination of knock intensity, filter specification, characterisation, position of pressure transducer [9], dimensions of engine and operating conditions. For a particular engine it is possible to fix some of these parameters and get better knock specification in a meaningful way. Figures 2 shows typical measured cylinder pressure in two knocking cycles, which are selected and plotted versus crank. The experimental conditions were as follow: Fuel isooctane, stoichiometric mixture, speed 1500 RPM, intake pressure 95.5 kpa, ignition angle at 30° BTDC, Wide open throttle equivalence ratio 1 and knock intensity (K_I = 13.88 kpa). The figure 2.a, shows a steady rate of pressure increase when the spark plug is discharged and ignition take place. After ignition occurred the pressure increased suddenly as the energy of the mixture is released and reaches a maximum value after TDC and before knock, immediately knock take places and causes the pressure begins to fluctuate. Figure 2c, indicates the first fluctuations are very fast but in the end of knock process the fluctuations are not visible that is due to the frequency dependency [5]. Figure 2b and 2d shows the cylinder pressure and first fluctuations for the second measured cycle, which has a knock intensity of K_I = 12.88 kpa.

4. KNOCK INTENSITY MEASUREMENT METHODS

Many researchers have presented different classification of knock and knock intensity. Punzinauskas [2] compared some methods that were used for characterisation of knock and knock intensity. Chun et.al [3] have investigated the occurrence of the knock at different crank angle. Cowart et.al [4] have studied knock intensity measurements. In this section first different methods for knock intensity measurement are given and finally new techniques are proposed.

Maximum Amplitude of Pressure Signal: Maximum amplitude of pressure is commonly used for knock intensity as in the works of [3,6]. This method also called maximum amplitude in time domain. (Because knock happens randomly, therefore many samples are needed for evaluating knock occurrence or knock intensity).

K_I 20 Method: Konig and Shepard [7] suggested a knock intensity concept (KI20) to measure knock intensity in spark-ignition engines. In this method knock intensity is defined the maximum pressure in the 20 degree after onset of knock (KO).

Maximum Rate of Pressure Rise: A few researches [8,9] have used the maximum rate of pressure rise (1st derivative) for measure knock intensity. Knock intensity in this method is not evaluated well because normal pressure rise rate varies considerably over a range of operating conditions.

Peak Pressure Summation: A few researchers [8] have used an imperial relation for calculation of knock intensity in the base of summation of the individual cycle knock pressure as:

$$K_I = \frac{20}{N_C} \sum (P_K - 1.0) \qquad (1)$$

Where N_c is number of individual sample cycles, P_K is the peak knock pressure and 20 is a factor. In this equation the minimum sample should be about (300-10000) cycles. However, normally the maximum pressure varies even in during steady operation. So peak pressure is not a certain criteria for calculating knock intensity.

Band-Pass Filtered of the Pressure, and Third Derivative of Signal: Figures 3a and 3b show the knock intensity in the base of band-pass filtered pressure signal. As shown in these figures knock frequency occurred between 5-12 kHz. Figures 3c and 3d show the knock intensity in the base of third derivative of cylinder pressure signal. As shown in the graphs cycle-to-cycle variations observed in knock intensity. Figure 4 shows the power spectral density of the two measured cycles in figure 2 a and 2b.

5. KNOCK PREDICTIVE MODEL

Compression ratio and ignition angle are the main parameters that result in engines to operate under knocking conditions. Maximum cylinder pressure and the rate of pressure rise vary with any change in ignition angle. So it is suggested that a dimensionless knock indicator (K_P) can be defined as the ratio of the following parameters.

$$K_p = \frac{p_{max}}{\left(\frac{dp}{d\theta}\right)_{max} Iga} \qquad (2)$$

Where p_{max} is the maximum cylinder pressure in the cycle, $\left(\frac{dp}{d\theta}\right)_{max}$ is the maximum rate of pressure rise that should be calculated between maximum crank angle and 10-20° degree after inlet valve closed that the pressure increases nearly from this crank angle with piston movement. And *Ign* is the spark timing or crank angle at which, the mixture begins to burn and increases cylinder pressure. Figure 5 shows the variations of (K_P) versus knock intensity (K_I) at different engine speeds. It can be seen that increasing knock intensity results in lower K_P values.

6. CONCLUSION:

In this paper using the experimental data a knock prediction method is presented. A knock parameter predictor (K_P) can be used to check for occurrence and level of knock for any set of operating conditions. Statistical analyses, showed that the knock occurrence and knock intensity are random and only depend on the condition of each firing cycle. It was observed that the spark timing strongly affected the occurrence of knock and knock intensity.

ACKNOWLEDGMENTS

The authors would like to express their gratitude to the Iranian Ministry of Culture and Higher Education and the University of Kerman for sponsoring this research project.

REFERENCES:

1. Leppard,W.R., "Individual-Cylinder knock Occurrence and Intensity in Multicylinder Engines," SAE Paper 820074, 1982.
2. Punzinauskas, P.V., "Examination Of Methods Used To Characterize Engine Knock," SAE Technical Paper Series, SAE Paper 920808, 1992.
3. Chun, K.M., Heywood, J.B.,"Characterization of Knock in a Spark-Ignition Engine", SAE Paper 890156, 1989.
4. Cowart, J.S., Haghgooie, M., Newman, C.E., Davis, G.C., Pitz W.J. and Westbrook, C.K. "The Intensity of Knock in an Internal Combustion Engine: an Experimental and Modelling Study", SAE Paper 922327, 1992.
5. Burgdrof, K., Karlstrom, A.,"Using Multi-Rate Filter Banks to Detect Internal Combustion Engine Knock," SAE Paper 971670, 1997.
6. Gao, X., Stone,R., Hudson, C. and Bradbury, I.,"The Detection and Quantification Of Knock In Spark Ignition Engines" SAE Paper 932759, 1993.
7. Konig, G., Sheppard, C.G.W., "End-Gas Autoignition and Knock in a Spark Ignition Engine," SAE Transactions, SAE Paper 902135, 1990.

8. Brunt, F.J., Christopher R.P. and Biundo, J.,"Gasoline Engine Knock Analysis Using Cylinder Pressure Data" SAE Paper 980896, 1998.
9. Burgdorft, K., Denbrutt, I.,"Comprasition of Cylider Pressure Based Knock Detection Methods" SAE Paper 972932, 1997.

Table 1: Geometrical Dimensions of RicardoE6 Engine

Bore	76.2 (mm)
Stroke	111 (mm)
Displaced Volume	507 (cm^3)
Compression Ratio	4.5-20
Connecting Rod	242 (mm)
Clearance Volume	63 (cm^3)
Valve Timing: Inlet Valve Opens	10° Deg BTDC
Inlet Valve Closes	36° Deg ABDC
Exhaust Valve Opens	43 °Deg BTDC
Exhaust Valve Closes	8° Deg ATD
Ignition Timing,(IA)	10-55° Deg BTDC
Equivalence Ratio	1.0
Throttle	10.0 (Wide open)
Speed	2000-3000 (rpm.)
Pressure Transducer	Kistler 701A

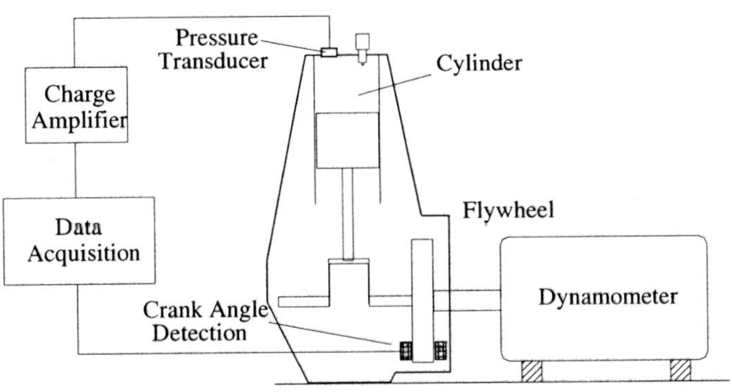

Figure 1 Schematic of test engine test rig

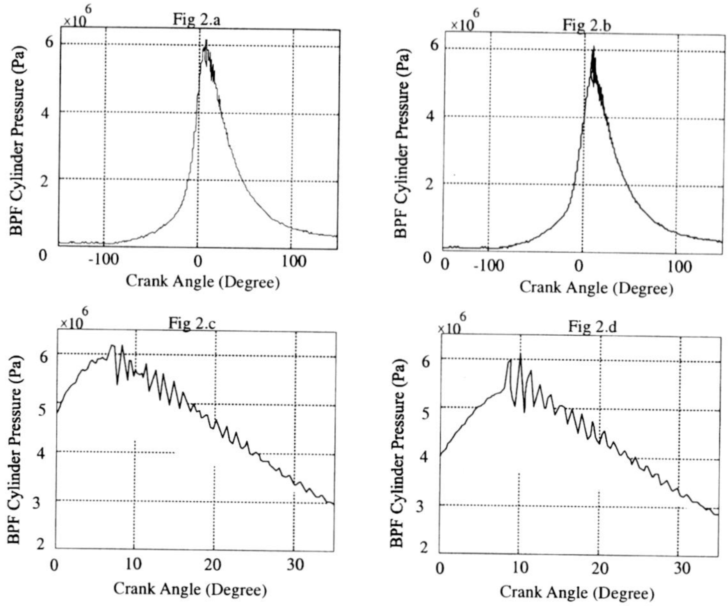

Figure 2 Diagram of pressure versus crank angle for 2 knocking cycles

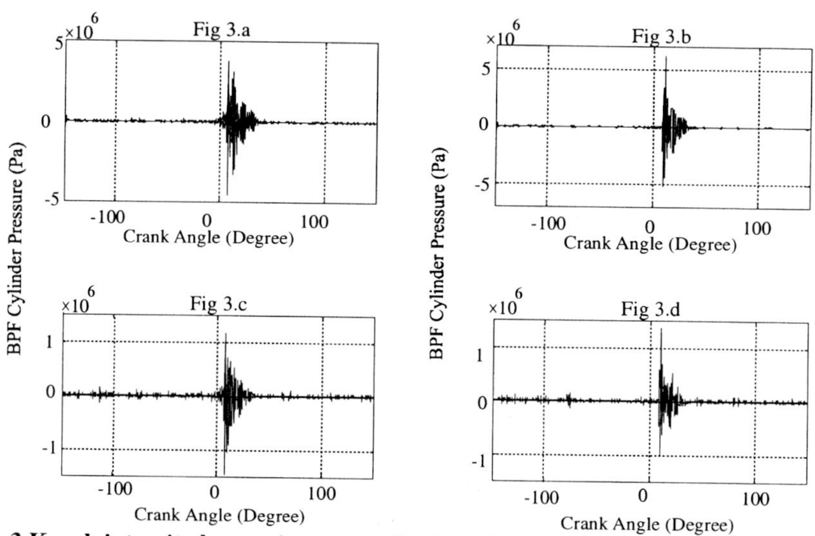

Figure 3 Knock intensity by maximum amplitude of band-pass and third derivative of cylinder pressure signal.

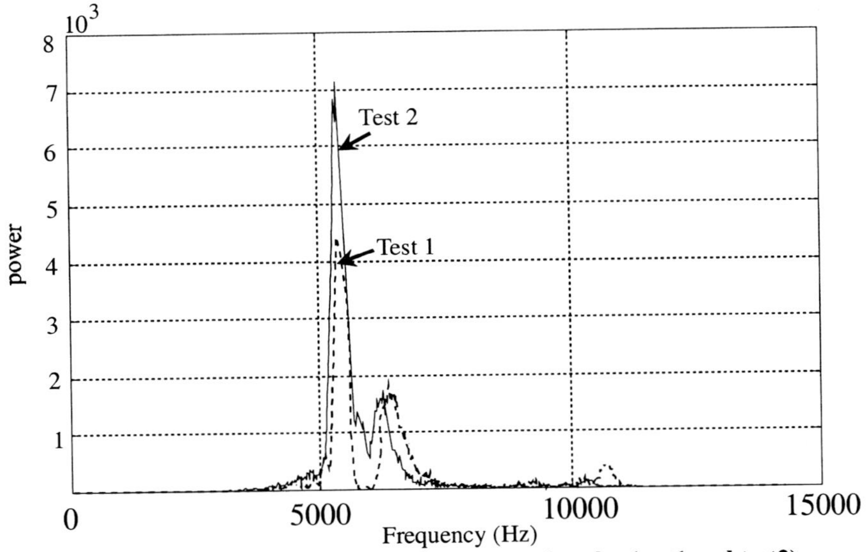

Figure 4 Frequency spectrum of the two measured cycles (test1 and test2)

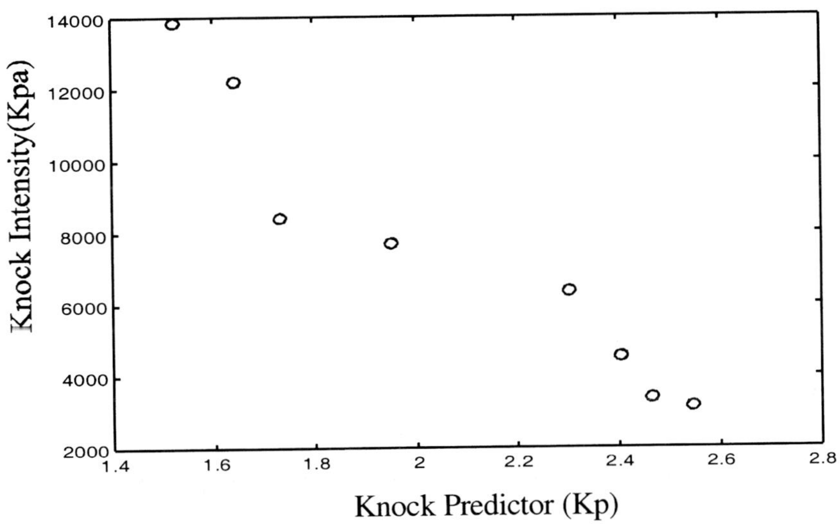

Figure 5 Variation of knock predictor versus knock intensity For CR=8, Iga=20-30° BTDC and N=1000-2000 rpm

Elastic body contact simulation for predicting piston slap induced noise in IC engine

G OFFNER
Christian-Doppler Laboratory for Engine and Vehicle Acoustics, Institute for Combustion Engines and Thermodynamics, TU Graz, Austria
H H PRIEBSCH
Christian-Doppler Laboratory for Engine and Vehicle Acoustics, Institute for Combustion Engines and Thermodynamics, TU Graz, Austria (also, AVL List, Graz, Austria)

ABSTRACT

Piston slap induced noise may contribute significantly to the noise emitted from Internal Combustion engines in a specific frequency range. There is a high interest to simulate this phenomenon in the development process of both, Diesel and gasoline engines in order to analyse measures for reducing this noise source in an early development stage of the engine. Piston slap is mainly affected by various design parameters of piston and liner, by the temperature distribution and by the combustion timing.

The paper summarises a methodology for predicting the piston to liner contact in running engines by means of MBD (Multi-Body Dynamics) and FEM. The advantages of this methodology are precise contact modelling on one hand and freedom in the model application on the other. Thus, trends caused by changes in the piston design and of combustion parameters on the contact impact can be analysed by simulation. Contact statistics e.g. peak contact pressures and their locations, etc. help to assess the importance of the contact events.

In addition to the mathematical modelling of the excitation the paper describes the transfer mechanisms of the piston slap phenomenon. Thus, the model is extended in order to analyse vibration transfer via engine structure. Results of simulation work show structure surface velocity levels and their contribution to integral levels in different frequency bands.

1. INTRODUCTION

For the development of all types of IC engines, the optimisation of piston to liner contact is of central importance. Its design affects key functions such as durability, performance and noise of the engine. Designing optimum piston / liner assemblies by means of simulation can significantly reduce development cost for prototype testing work.

Besides the primary oscillating motion, the piston performs a secondary motion due to the gap between piston and cylinder liner (briefly liner). This piston secondary or slap motion (piston slap) consists of translational and rotational components. It is caused by the reaction force and torque of the con-rod small end, which may change their direction and value during the engine cycle and allow the piston to move within the existing clearance. The amount of clearance is affected by the geometry and by the elasticity of the parts in the sliding contact.

Piston to liner contact occurs directly between piston skirt and liner and indirectly via piston rings. in the running engine, the structure in the contact area is excited radially in piston slap direction, and tangentially in sliding direction. The clearance between piston skirt and liner is formed by their shapes. Each shape is caused by the manufacturing (profile) and actual deformations due to loads (e.g. temperature, gas, assembly loads). The clearance is partly filled with oil. The amount of oil filling is depending on the oil itself, on the contacting parts (piston, rings and liner) and the engine running conditions.

The piston slap motion is affecting both the contact between the sliding parts and the excitation of the structure vibration. Thus on one hand, sliding contact between piston skirt and liner directly affects wear and mechanical friction losses of the engine. Furthermore in interaction with the piston rings, effects on blow-by and LOC (lube oil consumption) can be observed.

On the other hand, piston slap is an impact phenomenon causing engine noise and cavitation of the liner in the cooling water jacket. Piston slap induced noise is known as a very significant source of noise excitation mainly existing in the 2 kHz octave band. Its reduction is more often related to customer annoyance and subjective complaints than to meet legislation limits. Well known as a possible problem for Diesel engines, it can be observed for modern Gasoline engines, too. Due to all the important influences on engine design, piston slap phenomena have been analysed by simulation [4,7] and experimental methods [2,5,8].

Using a FEM model for the entire engine and applying the relevant forces for noise excitation (gas forces, mass forces of moving crank train and valve train part, etc.) the effect of piston slap excitation can be analysed in principle. The example in figure 1 shows the calculated surface velocity levels on an envelop mesh of a 4 cylinder Diesel engine. The results show the difference of the velocities with and w/o considering precalculated excitation forces on the cylinder liners. The liner forces were simulated by a single mass moving in the actual clearance between piston and liner due to engine operating condition at 4000 rpm, full load.

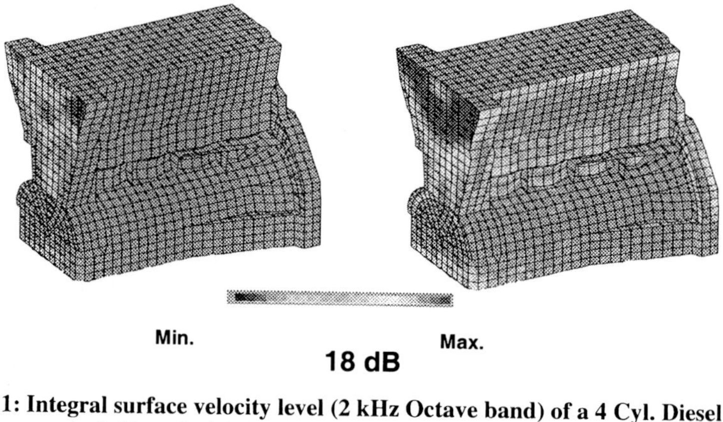

Figure 1: Integral surface velocity level (2 kHz Octave band) of a 4 Cyl. Diesel engine w/o (left) and with (right) considering liner excitation forces.

The spots on the right engine model in the frequency range 1400 – 2800 Hz indicate clearly the importance liner excitation for the structure borne noise. Despite showing the contributions of piston slap induced noise in principle, a model as described in figure 1 is not sufficient to analyse the effects of the piston impact events and their interaction with structure vibration as detailed as required for a noise optimisation.

For a very detailed analyses of all important parameters concerned, a simulation model has been developed by the authors based on the theory of elastic multi-body-dynamics and is described within this paper. For this first model, the assumption is made, that the contribution of piston rings to liner noise excitation is neglectable. For this reason, piston rings are considered in the calculation model by masses only. In the piston skirt contact the assumption is made, that radial excitation is dominating compared to sliding excitation. The effect of sliding excitation will be shown in a future publication. The resulting model of this paper enables analysis of design modifications, e.g. change of manufacturing profiles, piston design and offset, and to analyse their effects on the contact area.

2. BASIC EQUATIONS

In this chapter the formulation of representative equations describing the mathematical simulation model and the introduction of ingenious simplifications is described. Because of the complex structure of a multi-body-system, it has to be broken down into coupled systems, consisting of bodies, e.g. piston and liner, with linear elastic behaviour and connections, e.g. lubricated regions, considering the non-linear forces, acting between the connected bodies. The basic equations, used in case of simulation procedure of elastic piston liner contact are

- Equation of motion to compute global motions and vibrations of bodies
- Non-linear joint equations (e.g. Reynolds equation, spring-damper functions) to compute forces and moments, acting between connected bodies

2.1. Equation of motion

For calculation of motions each body has to be divided into a sufficiently high number of sub-bodies (partial masses). The dynamic behaviour of each of these rigid partial masses is given by the classical equation of motion for linear systems

$$\boldsymbol{M} \cdot \ddot{\boldsymbol{q}} + \boldsymbol{D} \cdot \dot{\boldsymbol{q}} + \boldsymbol{K} \cdot \boldsymbol{q} = \boldsymbol{f} \quad (1)$$

derived from the equations of momentum and angular momentum. $\boldsymbol{q} = [\boldsymbol{q}_1, \boldsymbol{q}_2, ..., \boldsymbol{q}_n]^t$ is the generalized displacement vector for n partial masses. Each element of the vector is a vector itself with three translatorial motion components and three rotational motion components ($\boldsymbol{q}_i = [u_1, u_2, u_3, \gamma_1, \gamma_2, \gamma_3]_i^t$). Damping matrices ($\boldsymbol{D}$) are calculated from given mass matrices ($\boldsymbol{M}$) and stiffness matrices ($\boldsymbol{K}$) according to the linear combination $\boldsymbol{D} = \alpha \cdot \boldsymbol{K} + \beta \cdot \boldsymbol{M}$, in the course of which α and β are functions of the structural damping and modal frequencies. The right hand side of the equation is given by adding up the external loads ($\boldsymbol{f}^{(a)}$), exciting joint forces and moments ($\boldsymbol{f}^*$) and inertia terms ($\boldsymbol{p}^*$):

$$\boldsymbol{f} = \boldsymbol{f}^{(a)} + \boldsymbol{f}^* + \boldsymbol{p}^* \quad (2).$$

External forces (e.g. gas force) and moments are determined functions given in time, calculated from given measurement data. The non-linear terms of excitational loads are given by joints, connecting one body to another (e.g. contact forces acting between piston and liner resulting from solution of Reynolds equation). In case of bodies with global motions (e.g. piston), non-linear inertia terms, detaily discussed in [9,10], also have to be considered in the equation.

2.2. Reynolds equation

Pressure distribution of the oilfilm in a lubrication region between two elastic bodies can be calculated using Reynolds equation derived from Navier-Stokes' equation and equation of continuity:

$$-\frac{\partial}{\partial x}\left(\frac{1}{12\eta}\cdot\theta\cdot h^3\cdot\frac{\partial p}{\partial x}\right)-\frac{\partial}{\partial z}\left(\frac{1}{12\eta}\cdot\theta\cdot h^3\cdot\frac{\partial p}{\partial z}\right)+\frac{w_{Liner}+w_{Piston}}{2}\frac{\partial\ (h\cdot\theta)}{\partial z}+\frac{\partial\ (h\cdot\theta)}{\partial t}=0\,. \quad (3)$$

The derivation takes as well laminar conditions, Newton fluid properties as special geometrical assumptions (pressure is constant in direction of gap height) of the lubrication region between piston and liner into account. The shear velocity part of the equation considers the axial velocity components of both connected bodies according to the stick condition:

$$\begin{aligned} w(x,0,z,t) &= w_{Piston} \\ w(x,h,z,t) &= w_{Liner} \end{aligned}\,. \quad (4)$$

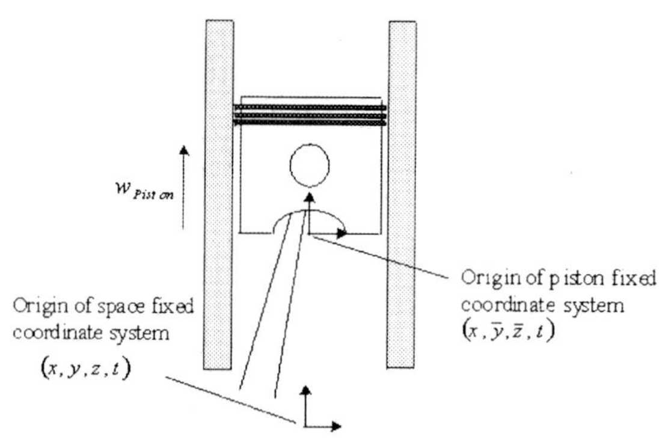

Figure 2: Coordinate transfomation to the piston fixed coordinate system

In order to get a time invariant calculation region, a coordinate transformation has to be done from the space fixed coordinate system (x,y,z,t) to the piston fixed coordinates $(x,\bar{y},\bar{z},t)$, where the coordinate in gap direction (y) has to be normalized (Fig. 2):

$$\begin{aligned} z &= \int_0^t w_{Piston}(\xi)d\xi+\bar{z} =: z_0(t)+\bar{z} \\ y &= \bar{y}\cdot\bar{h}(x,\bar{z},t) \end{aligned}\,. \quad (5)$$

So, for each function f

$$\begin{aligned} f &: R^4 \mapsto R \\ (x, y, z, t) &\mapsto f(x, y, z, t) \end{aligned} \quad (6)$$

the corresponding function in the piston fixed coordinate system $\bar{f}$

$$\begin{aligned} \bar{f} &: R^4 \mapsto R \\ (x, \bar{y}, \bar{z}, t) &\mapsto \bar{f}(x, \bar{y}, \bar{z}, t) \end{aligned} \quad (7)$$

is defined according to

$$\bar{f}(x, \bar{y}, \bar{z}, t) := f(x, \bar{y} \cdot \bar{h}, z_0 + \bar{z}, t). \quad (8)$$

Carrying out the transformation (8) for equation (3) yields to the Reynolds equation given in a piston fixed coordinate system:

$$-\frac{\partial}{\partial x}\left(\frac{1}{12\bar{\eta}} \cdot \bar{\theta} \cdot \bar{h}^3 \cdot \frac{\partial \bar{p}}{\partial x}\right) - \frac{\partial}{\partial \bar{z}}\left(\frac{1}{12\bar{\eta}} \cdot \bar{\theta} \cdot \bar{h}^3 \cdot \frac{\partial \bar{p}}{\partial \bar{z}}\right) + \frac{w_{Liner} - w_{Piston}}{2} \frac{\partial (\bar{h} \cdot \bar{\theta})}{\partial \bar{z}} + \frac{\partial (\bar{h} \cdot \bar{\theta})}{\partial t} = 0. \quad (9)$$

Due to piston functionality and manufacturing profile, no hydrodynamic contact between piston and liner will occure in piston pin area. For that reason, the hydrodynamic calculation can be reduced to the lubricated area L in direction of TS-ATS with the bounary B. Because of the limited oil supply in the contact area L and since piston skirt and liner have not exact cylindric contours (e.g. manufacturing profile of a piston skirt, deformation due to assembly and thermal load of a liner contact surface), L can be divided into an area with hydrodynamic contact (L_C) and an area with out hydrodynamic contact ($L_{\bar{C}}$) at each discrete point of time:

$$L = L_C \cup L_{\bar{C}}. \quad (10)$$

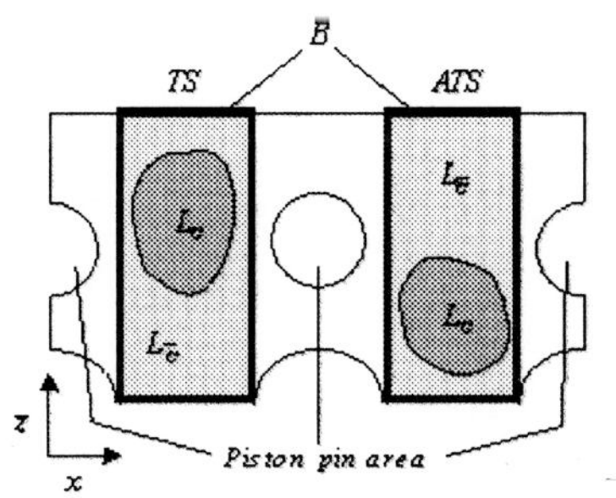

Figure 3: Calculation regions on unrolled piston skirt surface

3. SIMULATION PROCEDURE

The simulation procedure consists of three main steps:

- Preprocessing, including generation of geometries and structural matrices for each elastic body of the multi-body model using FEM, condensation, determination of external loads and generation of contact surface profile data from real messured data for connected bodies
- Multi-body dynamics (MBD) vibration analysis in time domain, considering condensed structure matrices, non-linearities occuring at connections between the elastic bodies and loads
- Postprocessing including datarecovery to the uncondensed system and contact statistics

Stiffness and mass properties as well as geometry information can be generated using a usual FE-software package (For results shown in this paper MSC-Nastran was used.). In order to enable an efficient solution of vibration equations, a reduction (condensation) of the number of degrees of freedom has to be done. The reduced set $\boldsymbol{q}_a$ includes as well static as modal degrees of freedom and can be computed via equation (11), where $\boldsymbol{G}_{ft}$ represents the transformation matrix [1, 3]:

$$\boldsymbol{q} = \boldsymbol{G}_{ft} \cdot \boldsymbol{q}_a . \quad (11)$$

By this, the number of degrees of freedom can be reduced significantly (e.g. for the piston shown in figure 5 from 25000 to 700). Substitution of (11) in the equation of motion (1) and multiplication with $\boldsymbol{G}_{ft}^t$ from left hand side results in

$$\overline{\boldsymbol{M}} \cdot \ddot{\boldsymbol{q}}_a + \overline{\boldsymbol{D}} \cdot \dot{\boldsymbol{q}}_a + \overline{\boldsymbol{K}} \cdot \boldsymbol{q}_a = \bar{\boldsymbol{f}} , \quad (12)$$

where $\overline{\boldsymbol{M}}$, $\overline{\boldsymbol{D}}$ and $\overline{\boldsymbol{K}}$ denote the condensed mass, damping and stiffness matrices and $\bar{\boldsymbol{f}}$ is the condensed force vector. The reduced matrices together with the table of degrees of freedom and the node positions are taken from FE software via an interface. Vibration analysis is performed on the reduced system only.

3.1. Time integration

The condensed equation of motion, (12), for each connected body of the multi-body system together with the equations, computing the forces and moments acting between the connected bodies (e.g. Reynolds equation) yield to a total system of high complexity. Due to the non-linear characteristic of this system, it has to be solved in the time domain. In order to minimize numerical error a direct implicit integration method (Newmarks method) considering adjusted time step size is used for time integration. In each time step both the equilibrium in the equation of motion and in Reynolds equation and the equilibrium of the total system have to be fulfilled. The total system considers the interaction of excitation loads, resulting from integration of pressure distribution $\bar{p}(x,\bar{z})$, and the function of clearance height in the lubricated region between two connected bodies $\bar{h}(x,\bar{z})$:

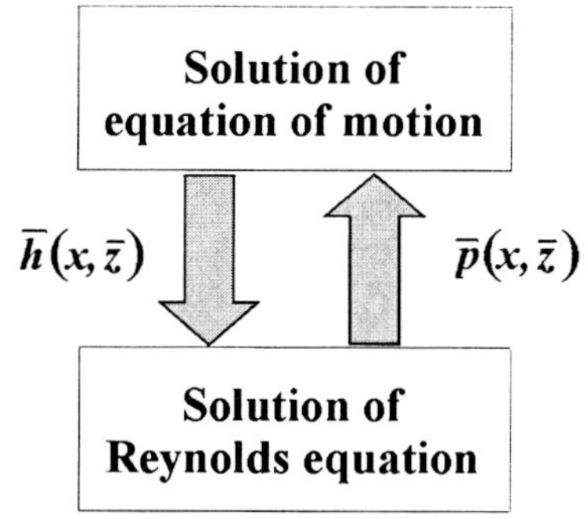

Figure 4: Solution structure of the total system

3.2. Solution of equation of motion

Equation (12) has to be solved in $\boldsymbol{q}_a$ with known external forces, the calculated excitation forces and moments and non-linear inertia terms. For calculation of the positions of the partial masses of a body, the equation of this body will be integrated. For the derivatives with respect to time of the vector $\boldsymbol{q}_a$, a direct implicit integration method is used ([1], [3]):

$$\begin{aligned}\dot{\boldsymbol{q}}_a &= a\cdot\boldsymbol{q}_a + \boldsymbol{b}_{q_a}\\ \ddot{\boldsymbol{q}}_a &= c\cdot\boldsymbol{q}_a + \boldsymbol{d}_{q_a}\end{aligned}\ . \ (13)$$

The scalar parameters a and c depend on the actual time step size and $\boldsymbol{b}_{q_a}$ and $\boldsymbol{d}_{q_a}$ are functions depending on solutions of $\boldsymbol{q}_a$, $\dot{\boldsymbol{q}}_a$ and $\ddot{\boldsymbol{q}}_a$ of former timesteps. Substitution of (13) and transformation of the resulting equation lead to a linear system for determination of the generalized displacements with the effective stiffness matrix $\hat{\boldsymbol{K}}$ and the effective load vector $\hat{\boldsymbol{f}}$:

$$\hat{\boldsymbol{K}}\cdot\boldsymbol{q}_a = \hat{\boldsymbol{f}}\ . \ (14)$$

Solution of this linear system can be done easily by factorization using the method of Cholesky [1].

3.3. Solution of Reynolds equation

The excitation forces needed for solving the equation of motion are calculated by integrating the oil film pressure in the clearance between the two connected bodies. The pressure distribution in the lubrication region L_C $(\theta = 1)$ is calculated by solving the Reynolds equation in the piston fixed coordinate system. The oil viscosity may be constant or it may depend on the oil film pressure obeying an exponential law:

$$\bar{\eta} = \eta_0\cdot e^{\alpha p}\ . \ (15)$$

The calculation is performed on an equidistant calculation grid, moved with the piston skirt surface. Because of the regular structure of the grid nodes, a finite volume method is used for

The calculation is performed on an equidistant calculation grid, moved with the piston skirt surface. Because of the regular structure of the grid nodes, a finite volume method is used for calculation. In order to get the terms resulting from the shear part of the equation, a combination of backward difference methods of second order and central difference methods of second order have to be used.

Both pressure distribution in the lubricated contact region L_C and the contact region itself, are determined iteratively. The classical SOR method (successive over relaxation) is used because of the optimal starting values for the first step of iteration that can be taken from the last time step. The relaxation parameter ω is controled considering the stability of the calculated pressure values. The usage of this relaxation method also allows the consideration of a simple cavitation algorithm: each negative pressure value is set equal to a given cavitation pressure value, that is constant [6].

4. RESULT EXAMPLES AND TREND ANALYSIS

For parametrical studies, a one cylinder 4-stroke engine with 83.0 mm stroke and speed of 3000 rev/min was simulated. Both piston skirt and liner surface data were generated by a linked profile generator. In all result examples one and a half cycles were calculated. Because of simulation beginning oscillations, the first half cycle was cut off in the shown results.

The basic Finite Element models of the piston and liner structures are shown in figure 5 and figure 6. These models are generated with a constant nominal diameter of 76.0 mm. For consideration of surface profiles, surface profile data for the piston and the liner and constant values for nominal clearance were included separately. The length of liner, of piston and of piston skirt were 180 mm, 80 mm and 54.5 mm respectively. The piston pin is placed 30.5 mm from the lower edge of the skirt with no offset. The piston pin and the connecting rod are beam-mass models. The crankshaft is modelled with one node moving along a circle with constant velocity. The piston and the piston pin as well as the pin and the connecting rod, as the connecting rod and the crankshaft node, were connected via non-linear spring-damper functions.

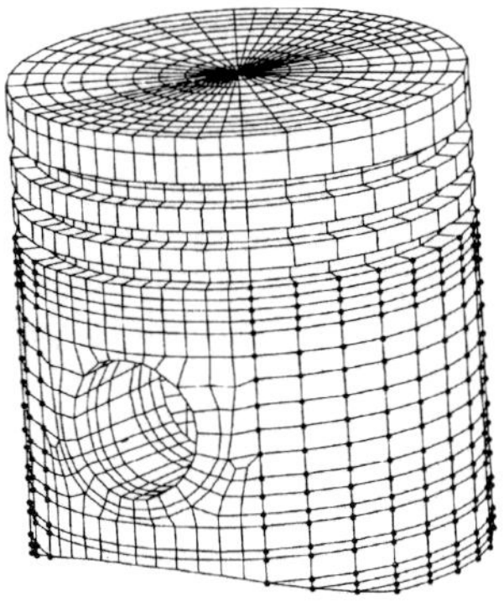

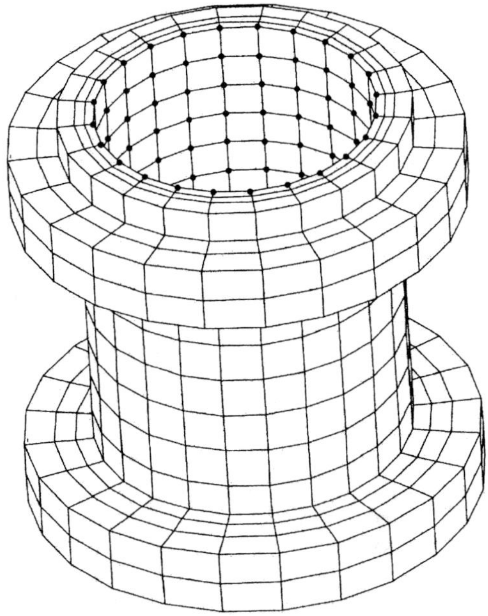

Figure 5: FE-model of piston **Figure 6: FE-model of liner**

Boundary conditions for hydrodynamic calculations were set to ambient pressure at the lower and at the top edge of the piston skirt. For all calculations a constant value of $\overline{\eta} = 1.0 \cdot 10^{-8}$ Ns/mm² was used for viscosity and a constant oil height at the liner wall was predefined. Nodes, that were considered in the solution of Reynolds equation are marked with fat dots (Fig. 5, Fig. 6). Figure 7 shows the used gas force for one engine cycle that is applied to each of eight nodes at top piston surface in piston stroke direction separately.

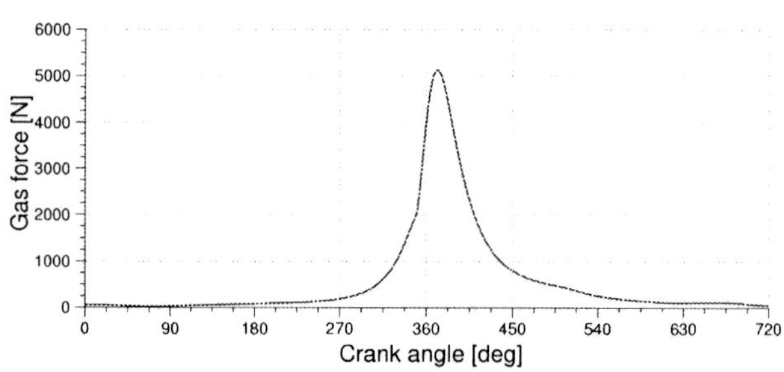

Figure 7: Gas force

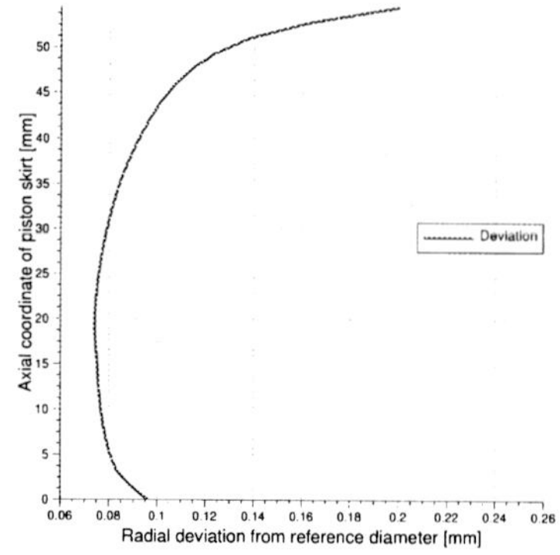

Figure 8a: Meridian of an axial symmetrical profile of a piston skirt

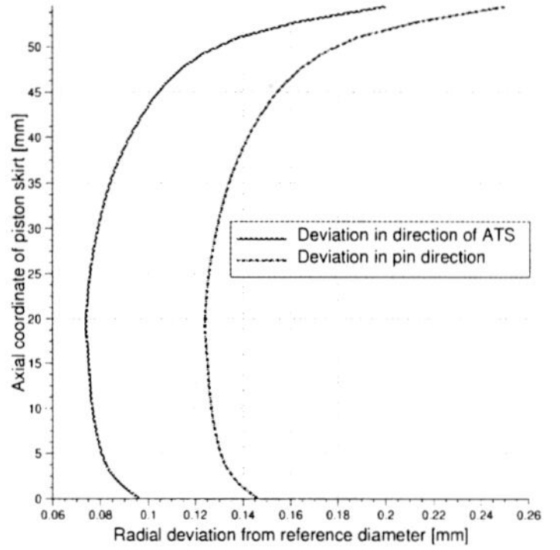

Figure 8b: Meridians of a piston skirt profile with single ovality

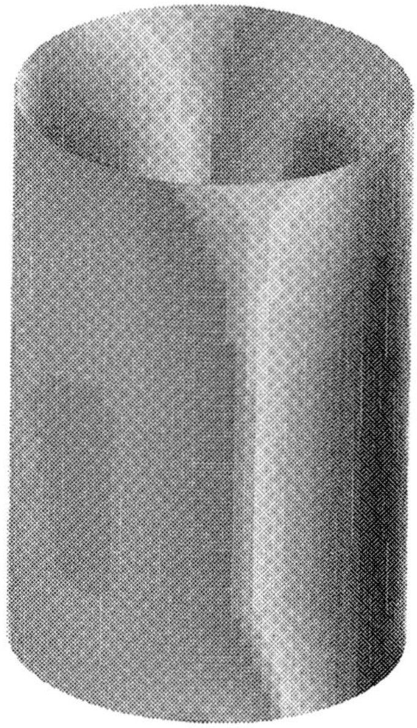

Figure 9: Liner surface profile

Two different types of piston profiles were tested in simulation. On one hand an axially symmetrical profile was used with maximum deviation of 0.2 mm at the upper edge of the skirt (Fig. 8a) , on the other hand a profile with single ovality considering a maximum deviation of 0.2 mm in the direction of the anti-thrust side (ATS) and a deviation of 0.25 mm in piston pin direction (Fig. 8b) was tested. The used liner surface profile (deviation from nominal diameter), calculated from measurement data, is shown in figure 9.

In order to characterize the friction loss and wear behaviour, the distribution of oilfilm thickness between two contacting bodies is of great importance. Besides the surface profile of the two bodies, the distribution is mainly influenced by the nominal radial clearance between the two bodies. The first study tries to figure out the influence of the nominal radial clearance between piston skirt and liner. The simulation was perfomed, using two different values of nominal clearance (10 μm, 100 μm) in addition with a piston surface profile with single ovality (Fig. 8b) and a constant oil heigth of 30 μm at inner liner wall.

Figure 10 shows the minimal gaps between the piston skirt and cylinder liner, which value is identical with the minimum oilfilm thickness below 30 μm. Differences of the two results are mainly caused by inertia effects resulting from corresponding terms in the equation of motion (1).

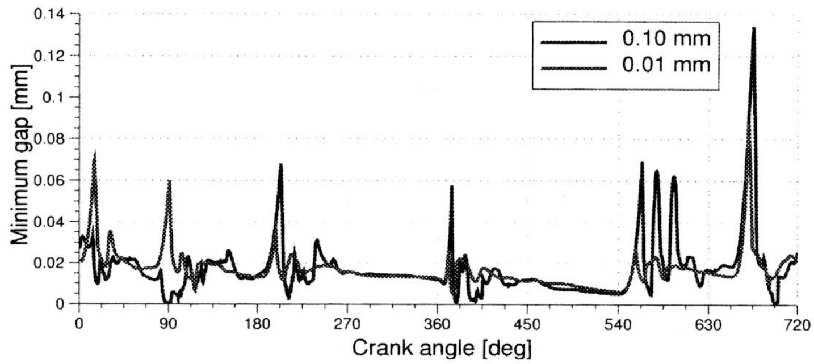

Figure 10: Minimum gap for different nominal clearance

In order to reveal the effect of the available oil quantity, figure 11 shows the effect of different values for available oil height at the liner wall on peak oil film pressure. Values of 30 μm and 130 μm were used in addition with a piston surface profile with singular ovality (Fig. 8b). The shown results underline the sensibility of the system due to oil supply. In correspondence with figure 7 the maximum oilfilm pressure values were calculated at FTDC – 7.87 MPa in the case of 130 μm and 46.72 MPa in the case of 30 μm.

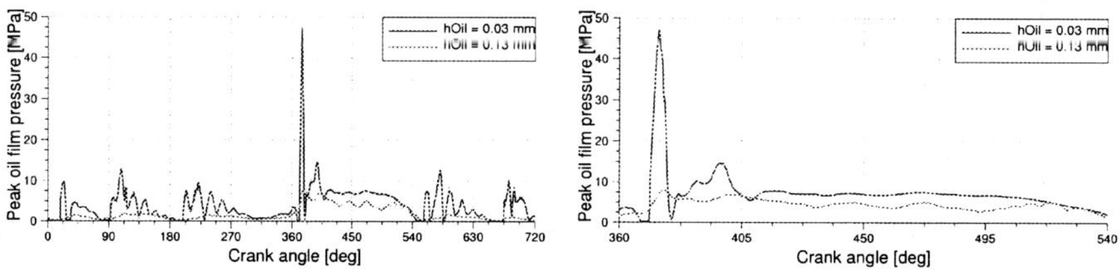

Figure 11: Peak oil film pressure for one full cycle and zoomed for the expansion stroke

In the next study, a constant oil film of 130 μm, a nominal radial clearance of 10 μm and an axialsymmetric piston skirt profile (Fig. 8a) were used in the calculation. In order to analyse the exact location of both small gap values between piston skirt and liner and the maximum oilfilm pressure values, a vertical scanning line was placed at the thrust-side of the piston skirt surface (Fig. 12).

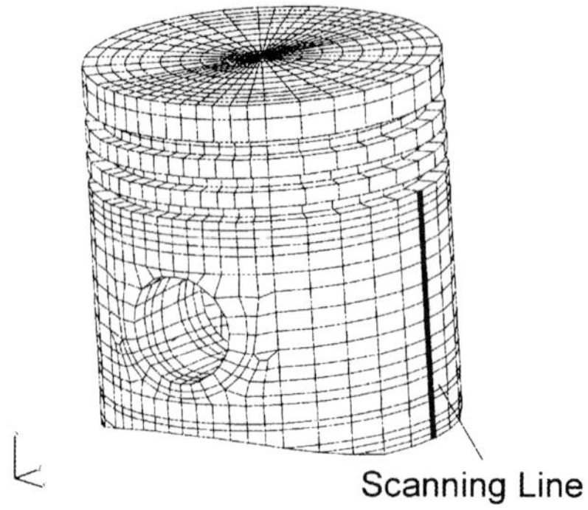

Figure 12: Scanning line at thrust-side of piston skirt surface

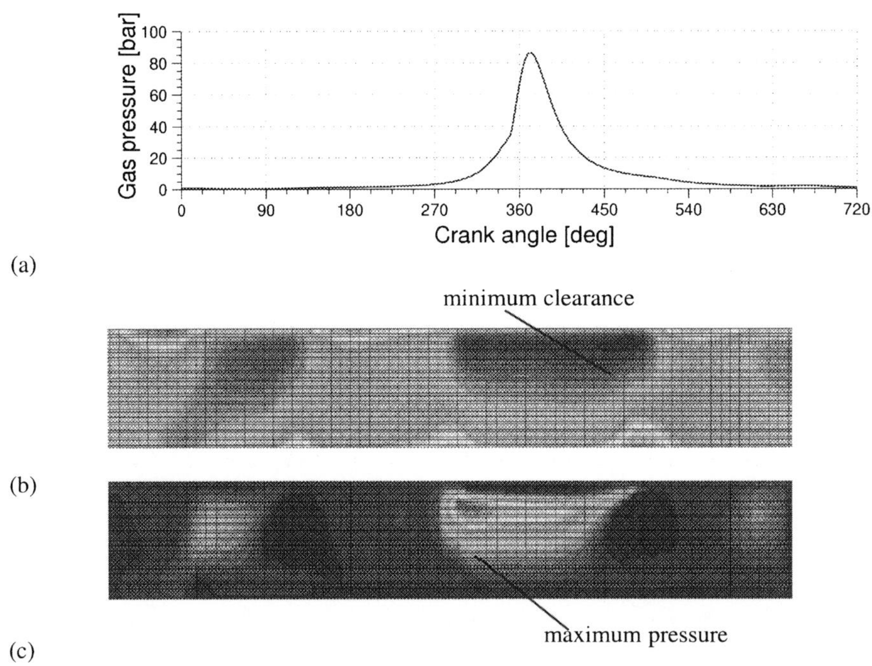

Figure 13: Gas pressure upon piston top (a), clearance between piston skirt and liner (b) and the corresponding oilfilm pressure distribution (c) at scanning line

In figure 13 the extracted results of clearance height (b) and pressure distribution (c) behaviour for one engine cycle (a) at the scanning line are shown. Via comparison of the clearance height animation over piston skirt height as well the tilting movement as the translatioal movement in direction of TS-ATS of the piston can be seen. Figure 13 (c) shows the corresponding pressure distribution. Due to the maximum gas pressure, especially after FTDC an area of small clearance height (17.5 μm) and maximum pressure (4.47 Mpa) can be located. Whereas pressure reduces significantly to a value of 2.69 Mpa the clearance value decreases slightly in the expansion stroke to a minimum value of 11.4 μm in the bottom dead center (540 deg.). This effect is caused by the additional tilting movement of the piston to the upper edge of the skirt in the last phase of the expansion stroke.

In order to be able to calculate the displacements, the velocities and the accelerations of degrees of freedom, that are not in the condensed set, (e.g. at outer cylinder wall surface), it is necessary to compute $\boldsymbol{q}$, $\dot{\boldsymbol{q}}$ and $\ddot{\boldsymbol{q}}$ of the uncondensed model too. This can be done easily, via the matrix multiplication given in equation 11.

This datarecovery was performed for the condensed results of the example discussed before. Figure 14 and 15 show the calculated distribution of integrated normal velocities of the uncondensed cylinder FE-model. The 0.5 kHz octave, the 1 kHz octave and the 2 kHz octave are shown in figure 14. The 1/3 octave band results analysed for 630 Hz, 1000 Hz and 2000 Hz are visualized in figure 15. Via comparison of the three results for both octave and 1/3 octave band, the maximum in 2 kHz octave band can be seen.

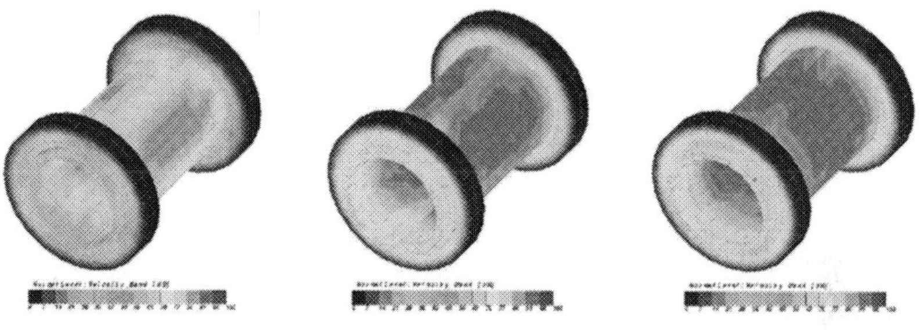

Figure 14: Integrated velocities for 0.5 kHz octave, 1 kHz octave and 2 kHz octave

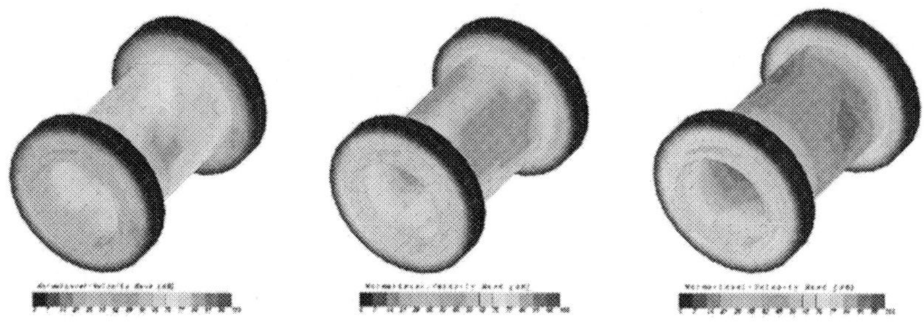

Figure 15: Integrated velocities for 630 Hz 1/3 octave, 1 kHz 1/3 octave and 2 kHz 1/3 octave

5. SUMMARY AND OUTLOOK

Beside the theoretical background of the multibody simulation tool, the total simulation procedure (building up the model, perfoming the simulation, datarecovery to the uncondensed system) is presented in this pulication. The prediction procedure is now ready to be applied for contact improvement between piston and liner structure in order to reduce noise excitation, friction and wear.

The presented results are calculated using a one cylinder 4-stroke engine model. Future aspects in this ongoing project will be the simulation of vibrations of an entire engine in order to simulate absolute levels of structure borne noise and the evaluation of the calculated results on an engine test bed. These results will be described in future publications.

ACKNOWLEDGEMENTS

The work described within this paper was funded by the Austrian Government and by the AVL List GmbH, Graz, Austria.

Nomenclature:

$\boldsymbol{f}^{(a)}$ ....External loads
$\boldsymbol{f}^{*}$ .... Excitational loads
$\bar{\boldsymbol{f}}$ .... Condensed force vector
$h, \bar{h}$ .... Clearance height
$p, \bar{p}$ .... Oil film pressure
$\boldsymbol{p}^{*}$ ....Inertia terms
$\boldsymbol{q}$ ....Generalized displacement vector
$\boldsymbol{q}_a$ .... Condensed displacement vector
t ....Time
w ....Axial velocity component
w_{Liner} ....Axial velocity of liner
w_{Piston} .... Axial velocity of piston
x .... Circumferential direction
$y, \bar{y}$ ....Gap direction
$z, \bar{z}$ ....Axial direction
$\boldsymbol{G}_{ft}$ ....Transfomation matrix
$\boldsymbol{K}$ .... Stiffness matrix
$\bar{\boldsymbol{K}}$ ... Stiffness matrix of condensed system
$\boldsymbol{D}$ .... Damping matrix
$\bar{\boldsymbol{D}}$ .. Damping matrix of condensed system
$\boldsymbol{M}$ ....Mass matrix
$\bar{\boldsymbol{M}}$ .... Mass matrix of condensed system
$\eta, \bar{\eta}$ ....Lubricant viscosity
$\theta, \bar{\theta}$ ....Fill ratio

REFERENCE

[1] Bathe, K.-J.: *Finite-Elemente-Methoden*, Springer Verlag, 1990

[2] DeLuca, J.C.: *The Influence of Cylinder Lubrication on Piston Slap*, Doctor-Thesis, Universidade Federal de Santa Catarina, Florianopolis, 1998

[3] *Excite Reference Manual (Version 5.0), AVL LIST GmbH,* Graz, 1999

[4] Kageyama, H.; Hara, S.; Kawabata, Y.: *Study of the Simulation of Piston Skirt Contact*, JSAE 9430095, Elsevier Science, 1994

[5] Kamp, H.; Spermann, J.: *New Methods of Evaluating and Improving Piston Related Noise in Internal Combustion Engines*, SAE Paper 951238, Detroit, 1995

[6] Krasser, J.: *Thermoelastohydrodynamische Analyse dynamisch belasteter Radialgleitlager*, Dissertation, Institut fuer Mathematik/C, TU Graz, 1996

[7] Knoll, G.; Peeken, H.; Lechtape-Grüter, R.; Lang, J.: *Computer Aided Simulation of Piston and Piston Ring Dynamics*, ASME ICE-Vol. 22, p 301, 1994

[8] Künzel, R.: *Die Kolbenbewegung in Motorquer- und Motorlängsrichtung, Teil 2: Einfluß der Kolbenbolzendesachsierung und der Kolbenform*, MTZ 56, S 534, 1995

[9] Priebsch, H. H.; Affenzeller, J.; Kuipers, G.: *Prediction Technique of Vibration and Noise in Engines*, Proceeding, IMechE Conference Quiet Resolutions, 1990

[10] Priebsch, H. H.; Krasser, J.: *Simulation of Vibration and Structure Borne Noise of Engines – A Combined Technique of FEM and Multi Body Dynamics* to be published at CAD-FEM USERS' MEETING, Bad Neuenahr – Ahrweiler, 1998

Multi-body dynamics for the assessment of engine induced inertial imbalance and torsional-deflection vibration

D ARRUNDALE and **S GUPTA**
Mechanical Dynamics Limited, UK
H RAHNEJAT
Department of Mechanical Engineering, University of Bradford, UK

Abstract:

Use of multi-body dynamics is increasing in the area of powertrain NVH analysis. This growth is much owed to a need for vehicle refinement in the area of engine induced vibration and the often associated structure-borne noise. In recent years the trend has been in the development of engines with high output power to weight ratio, through use of materials of lighter construction and with a greater power torque. Thus, component flexibility has been playing a greater role in system dynamics. The paper presents a multi-body approach for the determination of sources of powertrain Noise, Vibration and Harshness (NVH) by paying particular attention to crankshaft system elastodynamic behaviour that contributes significantly to half-engine order vibrations.

Keywords: Engine Dynamics, Elasto-multibody dynamics, Half engine order vibrations, Flywheel nodding motion

1. Introduction:

In recent years an increasing use is made of materials of lighter construction, both for enhanced engine efficiency by a corresponding reduction in mechanical losses, and for an improved NVH performance. The realisation of the latter was perceived to occur as a by-product of reduced inertial imbalance. The selection of materials for powertrain components has been largely based upon their fatigue endurance limits alone, not paying much attention to issues of structural dynamics as an NVH concern. Parallel to these developments much attention has been paid to improved thermodynamic performance of engines as thermal losses account for 60-65% of all losses in an engine. As a result of the improved combustion processes a greater power torque is now applied to lighter engine components, yielding larger torsional-deflection and lateral bending oscillations. With improved ride comfort from an NVH perspective and a perceived reduction in aerodynamic noise in intermittent driving conditions in congested areas, the driver concern has shifted towards a plethora of powertrain induced vibration and the associated structure-borne noise. The implication of this has been an assortment of palliative measures introduced by the industry at a significant cost, an approach which is unlikely to be sustainable. Root-cause identification and solution of most powertrain NVH problems calls for detailed modelling of the internal combustion engine using a multi-body dynamic approach.

The analysis comprises simultaneous solution for large displacement dynamics of engine components such as piston and connecting rod motions, together with infinitesimal elastic response of , for example, crank-pins and support bearings.

Zeischka et al (1) have highlighted a multi-body elastodynamic model of the crankshaft and the engine block, making use of finite element models to represent the elastic behaviour of the various components. The hydrodynamic action of the supporting journal bearings were taken into account, using impedance charts. These give the journal reactions as a function of the Sommerfeld number. The authors have demonstrated that vibration characteristics of four stroke, four cylinder in-line internal combustion engines can be accurately predicted. These characteristics comprise even order harmonics of engine speed (i.e. even engine orders) due to inertial imbalance introduced by reciprocating motion of pistons and the articulated motions of connecting rods, even if one is to assume a rigid crankshaft system. Analytic solutions for verification of this is provided in references (2,3). Component flexibility causes torsional-deflection oscillations of the crankshaft system at half engine order multiples. Analytic verification for this is provided in reference (3) with experimental evidence portrayed in references (4-6). These half engine order multiples are particularly troublesome and induce a variety of drivetrain noise and vibration problems (see for example clutch in-cycle axial vibration (7-11), and driveline impact induced elastodynamic response, referred to as clonk (12-15)).

Katano et al (16) have also obtained the dynamic forces generated in an engine which induce resonant conditions in the crankshaft system, using a multi-body dynamic approach. Lacy (17) has studied torsional vibration of a four cylinder gasoline engine, using a multi-body model. In his model the crankshaft nodes were connected to the main journal bearing housing by an oil film module, having linear and rotary stiffness and damping. This oil film module was initially reported by Kikuchi (18) and has been employed by Lacy with a constant journal eccentricity, resulting in an axi-symmetric oil film constraint. However, transient conditions that are prevalent in engine dynamics can lead to small perturbations that render this assumption void, as in reality journal eccentricity alters and the hydrodynamic oil film reaction occurs as a result of combined lubricant entraining and squeeze film motions (19,20).

Hitherto, most of the multi-body models reported in literature assume linearised journal bearing reactions, represent power torque fluctuations by simplified induced torque upon the crankshaft system and are devoid of realistic assembly constraints between the various engine components. A detailed multi-body model of a single cylinder four stroke engine, comprising inertial components, assembly constraints and finite width hydrodynamic journal bearings, subjected to a numerically calculated combustion force has been reported by Boysal and Rahnejat (21,22). They have included the necessary detail to investigate the secondary tilting motion of the piston and the combined torsional vibration and conical whirling motion of the crankshaft. The level of detail has included the evaluation of main journal bearing hydrodynamic restoring reactions by solving the Reynolds equation for combined entraining and squeeze film motions, calculation of friction torque, determination of piston slapping action against the cylinder bore, piston friction and piston compression ring to cylinder wall elastohydrodynamic reaction using an extrapolated finite line oil film expression, first obtained by Rahnejat (23). The authors have shown that inclusion of such detail can lead to a considerable computation effort, requiring several hours of CPU time. However, for the simulation study to be of any practical use in industry this level of detail can be viewed as almost essential. The authors did not include the effect of component elasticity in their model, apart from the elastic contact deformation of mating parts.

When component elasticity is included in the model, the degrees of freedom are increased significantly. Therefore, time for solution of the problem increases and a realistic approach calls for omission of some detail which is deemed as not essential, given the primary purpose of a given study. To determine the NVH characteristics of the crankshaft one may omit secondary tilting motion of the pistons and ignore the likely unstable whirl of the journals, but include the elastic distortion of the latter. All other modelling details should be retained. This approach is reported here.

2. Model Description:

The multi-body model in this investigation represents a six cylinder four stroke in-line engine (see figure 1). The inertial components include the pistons, the connecting rods, the flywheel, the cam gear, the engine block and the crankshaft. The list of parts in the model with their masses is given in table 1. In this table, column 1 gives the part number for each component. Note that certain parts are repeated a number of times in a multi-cylinder engine as in this investigation.

The crankshaft is modelled as a continuous system, comprising point mass-inertial elements interconnected by three dimensional elastic field elements. Therefore, each crank-pin or web is represented as two concentrated inertial element, referred to in the table as crank-pin a and crank-pin b. The crankshaft is made of SG cast iron with a modulus of elasticity of 169 GPa, a modulus of rigidity of 66 GPa and a Poisson's ratio of 0.27. The elastic field elements in the crankshaft system are each represented by a 6X6 stiffness matrix. The crankshaft, as a continuous system, undergoes elastic deformation, giving the induced shear forces and bending moments in terms of relative displacements between two ends of a crank-pin or web. The bearing shells' stiffness is added in the same manner.

The stiffness of the oil film in the journal bearings is in series with the corresponding thin shell stiffness element and at a considerably higher value. This has been shown to be true under elastohydrodynamic conditions, where the oil film roof ripple oscillations are a fraction of the contact deflection of the mating elastic members under load (see for example Mehdigoli et al (24)). Therefore, the stiffness of the thin shell bearing is governed by the shell thickness. Therefore, the need for the inclusion of an oil film module is alleviated. The same is of course not true if hydrodynamic conditions prevail. However, most modern cars employ thin shell bearings. The oil film contributes to friction torque, acting upon the crankshaft system. This is included in the model by a spline function fitted to experimentally measured data. The friction torque is initially overcome by the starter motor torque which is also included in the model.

The counter-balance masses are added to the crankshaft system in order to eliminate or reduce the primary imbalance at engine order. These masses are also discretely positioned in the same manner as described above.

The various inertial components in the model are assembled by specified holonomic constraint functions, a combination of a number of which represent a practical joint. A list of the joints in the multi-body model is given in table 2. In this table, column 1 gives an identification number for a given joint. Column 2 gives the location of the assembly site. Column 3 indicates the type of joint employed.

The combustion force in each cylinder is obtained by numerical solution for the rate of change of cylinder pressure as a function of heat release during the combustion process and the heat transfer rate to the cylinder head and bore walls in each step of simulation. The formulation and method of solution for this is outlined in references (3,21,22). Friction force is applied between each piston and cylinder wall, as also outlined in the same references. The cylinders' firing order for this engine is 1-4-2-6-3-5, with piston 1 denoting the closest piston to the flywheel position in this model. In this four stroke engine a cylinder fires every 120 degrees of crankshaft rotation. Therefore, the phase angle vector for this arrangement is given as:

$$\{\phi_j\}_{j=1-6} = \{0, {}^{4\pi}\!/_3, {}^{8\pi}\!/_3, {}^{2\pi}\!/_3, {}^{10\pi}\!/_3, 2\pi\}^T \qquad [1]$$

Before embarking upon the method of multi-body formulation and simulation, attention has to be paid to the validation of the numerical results. The approach advocated by Rahnejat (3) involves simplified analytic solution for power torque and the induced forces and moments acting upon the crankshaft system. This enables the verification of the numerically obtained vibration spectra. Experimental evidence for vibration spectra of six cylinder engines also exist (25).

Sources contributing to the spectra of vibration include the combustion power torque, the induced inertial imbalance torque and combined torsional-deflection of the crankshaft system due to component flexibility. Simplified analytic solutions for all of these sources of vibration are provided in reference (3). A brief description of these is provided here for the specific six cylinder four stroke engine under investigation.

3. Simplified Analytic Solutions :

The induced inertial torque applied to the crankshaft for a multi-cylinder engine can be obtained analytically as (3,26):

$$T_i = \frac{1}{2} m_t r^2 \omega^2 \sum_{j=1}^{n} \left\{ \frac{r}{l} \sin(\omega t + \phi_j) - \left[1 + \frac{r}{l}\right] \sin 2(\omega t + \phi_j) - \frac{r}{l} \sin 3(\omega t + \phi_j) \right\} \qquad [2]$$

Note that higher order terms, being a product of ascending powers of crank radius to the connecting rod length have been omitted in the above equation as the ratio $\frac{r}{l} \approx 0.3$.

Now replacing for the phase angles from equation [1] for $n = 6$, the above equation simplifies to:

$$T_i = -3m_t\omega^2 \frac{r^3}{l} \sin 3\omega t \qquad [3]$$

This indicates that the dominant contribution in induced imbalance torque is the third engine order, a fact that has been corroborated by much experimental evidence, for example see March and Croker (25).

The power torque due to the cylinders' combustion process for a multi-cylinder engine is given as (2,3,26):

$$T_p = \sum_{j=1}^{n} P_{jk} e^{i\frac{k}{2}(\omega t - \phi_j)} \qquad [4]$$

Where the power torque is given in terms of all k th harmonics of the fundamental combustion frequency, these being multiples of half-engine order for a four stroke engine. Assuming no cylinder-to-cylinder combustion variation, it follows that: $P_{jk} = P_k$ for all values of $j = 1 - 6$. Replacing for the phase shift vector from equation [1] into equation (4) it can be shown that (26):

$$T_p = 6P_k e^{i(3k\omega t)} \text{ for whole multiples of } k \qquad [5]$$

This means that the fundamental power torque contribution is at the third engine order, with all other whole multiples of 3ω having a non-zero contribution.

The "roughness", referring to half-engine order multiples, observed in experimentally obtained spectra relate to torsional-deflection response of the elastic crankshaft system. The analytic proof for multi-cylinder engines is provided in reference (3). For a six cylinder engine the resultant torsional deflection of the cylinder head is given by:

$$\theta_k = \frac{6BP_k}{GI} e^{i\frac{k}{2}\omega t} \quad \text{for} \quad k = 3m \qquad [6]$$

and : $$\theta_k = -\frac{4iBP_k}{GI} e^{i\frac{k}{2}\omega t} \quad \text{for} \quad k = 3m - 2 \qquad [7]$$

Note that: $$\theta_k = 0 \quad \text{for} \quad k = 3m - 1 \qquad [8]$$

In the above equations $m = 1,2,3,\ldots.$

It can be observed that half engine order multiples remain, with the highest contributions occurring when $m = 1$ in both equations [6] and [7], at $1\frac{1}{2}$ and $\frac{1}{2}$ engine orders. Other half order multiples at $3\frac{1}{2}$ and $4\frac{1}{2}$ engine orders exist.

Contributions due to second, third and sixth engine orders also remain. Therefore, the induced inertial imbalance, the power torque and the torsional deflection oscillations exhibit a significant number of engine order responses.

These analytic solutions form a guideline for verification and understanding the numerical non-linear multi-body dynamic analysis.

4. Multi-body Formulation:

The numerical multi-body model is created in the ADAMS software (a trade mark of Mechanical Dynamics Inc.). The method of formulation and solution is described by Orlandea (27) and Rahnejat (3) . A brief description of this is provided here.

The equations of motion for each part in the assembly of parts in the engine model, described in section 2 are given by the Lagrange's equation for constrained systems as:

$$\frac{d}{dt}\left(\frac{\partial K}{\partial \dot{q}_j}\right) - \frac{\partial K}{\partial q_j} - F_{q_j} + \sum_{k=1}^{n} \lambda_k \frac{\partial C_k}{\partial q_j} = 0 \qquad [9]$$

where $\{q_j\}_{j=1\to 6} = \{q_t, q_r\} = \{x, y, z, \psi, \theta, \phi\}^T$ is the vector of generalised co-ordinates, with the rotational components given in the Euler's body 3-1-3 frame of reference.

The n constraint functions for the different joints in the engine model are represented by a combination of holonomic and non-holonomic functions as:

$$\begin{bmatrix} C_k \\ \dot{q}_j \dfrac{\partial C_k}{\partial q_j} \end{bmatrix} = 0, j = 1 \to 6, k = 1 \to n \qquad [10]$$

The generalised forces in equation [9] act in the direction of the generalised co-ordinates and preserve the virtual work done by the original force. The original forces include weight of parts, the applied directed forces and moments, and reactions introduced by restraining elements in the system such as bushings and elastic fields. The virtual work for the gravity force acting in the centre of mass of a part is given as:

$$\{\delta W\} = \{\delta q_t\}^T m\{g\} \qquad [11]$$

And for an applied forces F acting at a point P in $\{e\}, \{\beta\}$ local frame of reference :

$$\{\delta W\} = \{\delta q_t\}^T \{F\} + \{\delta \beta\}^T \{\hat{e}\}[T_{ig}]\{F\} \qquad [12]$$

where:

$$[T_{ij}] = \begin{bmatrix} \cos\psi\cos\phi - \sin\psi\cos\theta\sin\phi & -\cos\psi\sin\phi - \sin\psi\cos\theta\cos\phi & \sin\psi\sin\theta \\ \sin\psi\cos\phi + \cos\psi\cos\theta\sin\phi & -\sin\psi\sin\phi + \cos\psi\cos\theta\cos\phi & -\cos\psi\sin\theta \\ \sin\theta\sin\phi & \sin\theta\cos\phi & \cos\theta \end{bmatrix}$$

and $j=g$, denotes the ground frame of reference.

The coefficient of $\{q_t)^T$ gives the generalised force, whilst the second term provides the torque about the centre of mass of the part with the infinitesimal rotation $\{\delta\beta\}^T$ expressed in terms of the local part frame of reference. The torque contribution (i.e. the second term in equation [12]) can be given in terms of the global Euler frame of reference as: $[E]^T\{\hat{e}\}[T_{ig}]\{F\}$, where $[E]$ is the transformation from the Euler-axis frame to the local part frame:

$$[E] = \begin{bmatrix} \sin\theta\sin\phi & 0 & \cos\phi \\ \sin\theta\cos\phi & 0 & -\sin\phi \\ \cos\theta & 1 & 0 \end{bmatrix}$$

Also note that: $\vec{e}\Lambda\vec{F} = \{\hat{e}\}\{F\}$ and $\{\delta\beta\} = [E]\{\delta q_r\}$.

Similarly, the virtual work done by an applied torque is given by:

$$\{\delta W\} = \{\delta q_r\}^T[E]^T\{e\}\{T\} \qquad [13]$$

The generalised torque components are given as the coefficients of the term $\{\delta q_r\}^T$.
In a similar manner it can be shown that bushing and field forces (such as Eulerian beams) produce generalised forces and moments of the following form:

$$\{F_{q_i}\}_{j=1\to3} = \{F\} \quad \text{and} \quad \{F_{q_i}\}_{j=4\to6} = [E]^T([T_{ig}]\{T\} + \{\hat{e}\}[T_{ig}]\{F\}) \qquad [14]$$

The reaction forces and moments, introduced by the imposed constraints, are obtained in the same manner in terms of the Lagrange multipliers, λ_k where the following relation for infinitesimal changes should hold:

$$\lambda_k\delta C_k = 0 \qquad [15]$$

where C_k is a holonomic constraint function. These are the primitive functions that ensure positional or orientation conditions. Certain combinations of these form a physical joint. The most common types are the at-point or point coincident constraint, the inplane and in-line joint primitives, perpendicular axes and prescribed angular orientation such as the imposition of parallel axes condition. Coupling constraints may also be imposed, relating the position (by a holonomic constraint) or velocity (by a non-holonomic constraint) of parts with respect to each other.

The joints in the engine model are listed in table 2. As an example, generalised constraint forces for a revolute joint are described here. A revolute joint consists of an at-point and two axes perpendicularity holonomic relations. Using the Bernoulli's virtual work principle for an at-point constraint, it can be shown that:

$$\delta W = (\{\delta q_t\}_i^T + \{\delta q_r\}_i^T [E]_i^T \{\hat{e}_i\}[T_{ig}] - \{\delta q_t\}_j^T + \{\delta q_r\}_j^T [E]_j^T \{\hat{e}_j\}[T_{jg}])\{\lambda\} \qquad [16]$$

Therefore, the generalised force for both parts *i* and *j* are given as the coefficients of the terms $\{\delta q_t\}_{i/j}^T$, in other words: $\{\lambda\}$. The generalised constraining torque is provided by the coefficient of the terms $\{\delta q_r\}_{i/j}^T$, or : $[E]_{i/j}^T \{\hat{e}_{i/j}\}[T_{i/j,g}]\{\lambda\}$.

For the perpendicular axes there are clearly no generalised constraining forces, and the generalised constraining moments on parts *i* and *j* are given as: $[E]_{i/j}^T \{\hat{a}_{i/j}\}[T_{ij}]\{a_{j/i}\}\{\lambda\}$. Note that for a co-directed *z* axes at a revolute joint the vectors $\{a_i\}$ and $\{a_j\}$ yield the conditions: $\vec{z}_i.\vec{x}_j = \vec{z}_i.\vec{y}_j = 0$.

The formulated generalised forces in case of body forces, applied forces and the constraining reactions can be implemented in equation [9]. The differential-algebraic equation set can now be represented as follows:

$$\begin{bmatrix} \left[\frac{1}{\delta t}\frac{\partial K}{\partial \dot{q}_j} + \frac{\partial K}{\partial q_j}\right] & \begin{bmatrix} \left[\frac{\partial C_k}{\partial q_j}\right]^T & 0 \\ 0 & \left[\frac{\partial C_k}{\partial q_j}\right]^T \end{bmatrix} \\ \left[\frac{\partial C_k}{\partial q_j}\right] & [0] \\ \left[\frac{\partial C_k}{\partial q_j}, \dot{q}\frac{\partial}{\partial q_j}\left(\frac{\partial C_k}{\partial q_j}\right)\right] & [0] \end{bmatrix} \{\delta q_j, \delta\lambda_k\}^T = \{F_{q_j}\} \qquad [17]$$

The matrix on the left hand side is referred to as the Jacobian matrix.

The applied field element matrices are given as three dimensional Eulerian beams:

$$\{F,T\}^T = \begin{bmatrix} \frac{EA}{L} & 0 & 0 & 0 & 0 & 0 \\ 0 & \frac{12EI_{zz}}{L^3} & 0 & 0 & 0 & \frac{-6EI_{zz}}{L^2} \\ 0 & 0 & \frac{12EI_{yy}}{L^3} & 0 & \frac{6EI_{yy}}{L^2} & 0 \\ 0 & 0 & 0 & \frac{GI_{xx}}{L} & 0 & 0 \\ 0 & 0 & \frac{6EI_{yy}}{L^2} & 0 & \frac{4EI_{yy}}{L} & 0 \\ 0 & \frac{-6EI_{zz}}{L^2} & 0 & 0 & 0 & \frac{4EI_{zz}}{L} \end{bmatrix} \{e,\beta\}^T - [D]\left\{\frac{de}{dt},\frac{d\beta}{dt}\right\}^T$$

[18]

where the first matrix is the stiffness matrix and $[D]$ is the structural damping matrix, given usually as a percentage of the former (in the range 1%-5% for the crankshaft system).

The method of solution is described in references (3,27,28).

5. Simulation studies:

A multi-body engine model thus established can be employed for some practical simulation studies. These may include verification studies in respect of conceptual designs, prior to physical prototype testing. These can encompass parametric studies, for example in relation to the determination of the required starter motor torque for a given resistive friction torque produced by supporting crankshaft engine bearings. The analysis can be extended to bearing selection and crankshaft structural flexibility for optimisation studies in striking a balance between reducing powertrain weight, whilst guarding against noise and vibration concerns emanating from the lowering of torsional deflection and bending modes of the crankshaft system.

The most important step in such studies is the initial model validation against experimental spectra and known simplified analytic solutions, such as those highlighted in section 3. This paper presents simulation results, related to this important step in multi-body modelling of internal combustion engines. To carry out this task the model, highlighted in section 4 is subjected to a simulation study for a period of 2 seconds with 2048 steps of simulation. The validation process is carried out in the frequency domain. To avoid aliasing the spectral content up to a frequency of approximately 250 Hz can be relied upon with such a sample size, according to the Nyquist criterion (i.e. half the actual sampling rate). Therefore, the applied combustion force time history in each cylinder must be either monitored experimentally or evaluated numerically in a correct manner to match the simulation study. This leads to a given engine speed for a specified starter motor torque and a corresponding resistive friction torque. In the current study an experimental measured combustion time history is employed in all cylinders. Figure 2(a) shows the input cylinder combustion force time history. The starter motor torque applied at the onset of simulation for a period of 0.1 seconds, from a magnitude of 1000 Nm in the clock-wise direction is shown in figure 2(b). Figure 2(c) provides the spectral content of the applied cylinder combustion force, showing the prominence of the half engine order contribution at 12.5 Hz for the four stroke process, and all its higher harmonics up to the $5\frac{1}{2}$ engine order at 137.5 Hz.

6. Results and Discussion:

The interest in this simulation is to ascertain the torsional-deflection modes of the crankshaft system. For the conditions depicted in figure 2: (a)-(c), the crankshaft clock-wise rotation commences from rest to reach a steady state mean angular velocity of 157 rad/s, as shown in figure 3(a). The small perturbations at the steady state conditions have in fact a peak-to-valley oscillation amplitude of 1.5 rad/s (i.e. a fluctuation of approximately 14 rpm, representing a deviation of $\pm 0.5\%$) from the nominal steady state conditions. Such conditions are in fact fairly typical in four stroke gasoline engines, leading to small torsional deflection contributions which are nevertheless quite significant in a number of ways. Firstly, they lead to NVH concerns, progressively viewed in the industry as a key refinement issue. Secondly, their action, coupled with the bending modes in-plane and out-of-plane of the crank-throw lead to conical whirling motion of the flywheel, particularly with lighter engine materials and thin bearing shell deformation. This problem can lead to an assortment of clutch and drivetrain NVH problems (for example see Kelly et al (7-11)). The spectral content of the crankshaft rotational speed is shown in figure 3(b). Note that the main contributions occur at the 1.5, 3rd, 4.5, 6th and 7.5 engine orders. This is in-line with the analytic solutions in section 3, indicating a verification of the numerical method. These results also agree with a similar 6 cylinder in-line spectral content, reported by March and Croker (25).

The conical motion of the flywheel has been investigated by Kelly et al (7-11) as a major clutch system axial vibration issue, with its root cause found to be the firing of cylinders nearer to the flywheel position. This problem leads to tactile vibration at the clutch pedal during the engagement and disengagement processes, as well as noise in the footwell area. In their case, they have shown both experimentally and by numerical prediction that the problem in 4 cylinder diesel engines occur with the firing of the 3rd and particularly the 4th cylinders. In the case of the 6 cylinder gasoline engines this problem manifests itself to a much lesser extent, owing to the lower combustion forces in gasoline engines. However, the problem still remains as a concern to powertrain engineers. The so-called "flywheel nodding motion" is as a result of elasto-multibody dynamic of the flexible crankshaft system. This is shown in the zoomed plot in figure 4(a) for the case under investigation here. The peak- to- valley amplitude of vibrations is approximately 0.02 mm. This value is a fraction of the case in diesel engines; found to be in the region of 0.1 mm for 4 cylinder 1.8 diesels by Kelly (29). Nevertheless, the flywheel motion with this amplitude and at the fundamental combustion frequency of half engine order represents a repetitive Dirac-type impact function into the drivetrain system that induces an assortment of noise and vibration problems. Figure 4(b) shows the spectrum of flywheel nodding motion, indicating the half engine order contribution and all its harmonics, as anticipated. Furthermore, figure 5 illustrates that the problem is particularly poignant with the firing of the 1st cylinders being the closest cylinder to the flywheel position in the model denotations. The firing order is indicated by cylinder numbers in the figure. The flywheel nodding acceleration is superimposed upon the cylinder firing order, showing a maximum peak-to-valley oscillation of approximately 32g for cylinder 1 closest to the flywheel position.

Increasingly crankshaft bearings with thin shell configurations made of materials of low elastic moduli are employed in modern engines as a part of a drive in the construction of light weight engines. One adverse effect of this approach is the manifestation of torsional deflection and bending motions of the crankshaft system at these locations, resulting in the aforementioned nodding motion of the flywheel, a consequence of which is the clutch whoop problem. The flexure of the bearing shell is quite complex and leads to elastohydrodynamic regime of lubrication in the support bearings under transient conditions. This problem has been highlighted by Rahnejat (20), who reports on a solution of the thin shell engine bearing problem under combined entraining and squeeze film motion. The method of solution for the Reynolds' hydrodynamic equation and the elastic film shape is the column method, which assumes that the deflection at any location in the shell is dominated by the lubricant pressure element acting directly upon it. In reference (20) the solution is undertaken for the lubricant reaction acting in the vertical direction, whilst the shell deformation is due to an aligned pressure distribution, ignoring the small slope along the width of the bearing. However, due to the conical whirl of the crankshaft and the flywheel shell deformation takes place in the radial and along its axial direction. The resulting three dimensional pressure distribution is skewed, yielding high pressures at the extremities of the shell width when the shaft undergoes dominant vertical oscillations due to fluctuations in the combustion gas force. The slope of deformation is worst for the shell bearing nearest to the flywheel. Figure 6(a) shows the misaligned three dimensional pressure distribution at the maximum cylinder combustion force, resulting in a combined vertical and moment loading of the shell bearing with a magnitude of 10 KN. The shell is 3 mm in thickness, the radius of the journal is 30 mm and the lubricant film thickness under static equilibrium is 30 μm. The maximum pressure has occurred in the direction of the sloping crankshaft towards the flywheel location, indicated by the edge of the shell (i.e. the zero axial position). This is more clearly seen in figure 6(b), being the pressure isobar plot, with the maximum pressure of 16 MPa. The circumferential direction is the direction of entraining motion (i.e. shown as unwrapped in all these figures). Figure 6(c) shows the corresponding lubricant film thickness with a minimum value of 4 μm in the vicinity of the maximum pressure value.

Conclusion:

The multi-body formulation method presented here is generic and has been shown to be capable of modelling complex multi-body systems. The results obtained for a powertrain problem show the use of such numerical techniques during the design evaluation process in practical applications. The numerical predictions have shown to conform well with both simplified analytic solutions and with experimental rig and vehicle data.

Acknowledgements:
The authors wish to express their gratitude to Mechanical Dynamics International and Ford Motor Company for their financial support and collaboration in this research project.

Nomenclature:

B	: Cylinder-to-cylinder bore distance
C_k	: k^{th} holonomic constraint function
$[D]$	: Structural damping matrix
E	: Modulus of elasticity in equation (18)
F	: Applied force
F_q	: Generalised force in Euler frame of reference
G	: Modulus of rigidity
g	: Acceleration of free fall
J	: Second area moment of inertia
j	: Cylinder identity number
K	: Kinetic energy
k	: k^{th} constraint or harmonics of the combustion frequency
l	: Length of beam element in equation (18), otherwise connecting rod length
m	: Mass
m_t	: Mass of translational imbalance
n	: Number of constraints in equation (9), otherwise number of cylinders
P_j	: Combustion force
q	: Generalised co-ordinates
r	: Crank radius
T_i	: Induced inertial torque
T_p	: Power torque
t	: Time
W	: Work done
x,y,z	: Cartesian frame of reference
ϕ_j	: Cylinder firing phase angle
λ	: Lagrange's multiplier
θ_k	: Angle of twist
ψ, θ, ϕ	: Euler angles
ω	: Crankshaft angular velocity

References:

(1)- Zeischka, J., Mayor, L. S., Schersen, M. and Maessen, F., "Multi-body dynamics with deformable bodies applied to the flexible rotating crankshaft and the engine block", ASME 94 Fall Tech. Conf., Lafayette, USA.

(2)- Nakada, T. and Tonosaki, H., "Excitation mechanism of half order engine vibrations", IMechE Conf. Trans., C487/017/94, MEP, 1994, pp. 1-7.

(3)- Rahnejat, H., Multi-body Dynamics: Vehicles, Machines and Mechanisms, Co-publishers: PEP (UK) and SAE (USA), July 1998.

(4)- Dixon, J., Rhodes, D. M. and Phillips, A. V., "The generation of engine half orders by structural deformation", IMechE Conf. Trans., C487/032/94, MEP, 1994, pp. 9-17.

(5)- Kinoshita, M., Sakamoto, T. and Okamura, H., "An experimental study of a torsional/bending damper pulley for an engine crankshaft", SAE Trans., Pap. 891127, 1989, pp. 23-33.

(6)- Soma, N., Nakada, T. and Iwahara, M., "Engine shaking measuring techniques using angular velocity fluctuation of flywheel and its application", JSAE Trans., 43, 1990, pp. 114-118.

(7)- Kelly, P. and Rahnejat, H., "Clutch pedal dynamic noise and vibration investigation" in Rahnejat, H. and Whalley, R. Multi-body Dynamics: Monitoring and Simulation Techniques, MEP Publications, 1997, pp. 23-31.

(8)- Rahnejat, H., Centea, D., Kelly, P., "Non-linear multi-body dynamic analysis for the study of in-cycle vibrations (whoop) of cable operated clutch systems", Proc. 30th ISATA, Florence, Italy, June 1997, pp. 245-252.

(9)- Kelly, P., Rahnejat, H., Biermann, J. W. and Hagerodt, B., "Combining design of experiments and modelling techniques to resolve complex clutch pedal noise and vibration" IMechE Conf. Trans. on European Conf.on Vehicle Noise and Vibration, MEP Publications, May 1998, pp. 297-309.

(10)- Kelly, P. Rahnejat, H. and Biermann, J. W., "Multi-body dynamics investigation of clutch pedal in-cycle vibration (Whoop)", IMechE Conf.Trans. on Multi-body Dynamics: New Techniques & Appl., C/553, PEP, Dec. 1998.

(11)- Kelly, P., Rahnejat, H. and Biermann, J. W., "A parametric study of clutch pedal whoop: Numerical prediction and experimental Verification", Proc. 32nd ISATA, Vienna, Austria, June 1999.

(12)- Menday, M. T., Rahnejat, H. and Ebrahimi, M., "Clonk: an onomatopoeic response in torsional impact of automotive drivelines", Proc. Instn. Mech. Engrs., Part D: J. Automobile Engng., 213, D4, 1999.

(13)- Krenz, R., "Vehicle response to throttle tip in/tip out", SAE Pap. No. 850967, 1985.

(14)- Biermann, J. W. and Hagerodt, B., "Investigation of clonk phenomenon in vehicle transmissions- measurement, modelling and simulation", in Rahnejat, H. and Whalley, R. (Eds.), Multi-body Dynamics: Monitoring and Simulation Techniques, MEP Publications, 1997.

(15)- Arrundale, D., Hussain, K., Rahnejat, H. and Menday, M. T., "Acoustic response of driveline pieces under impacting loads (Clonk)", Proc. 31st ISATA, Dusseldorf, June 1998, pp. 319-331.

(16)- Katano, H., Iwamoto, A. and Saitoh, T., "Dynamic behaviour of internal combustion engine crankshafts under operating conditions", IMechE Conf. Trans., C430/049, MEP Publications, 1991, pp. 205-209.

(17)- Lacy, D. J., "Computers in analysis techniques for reciprocating engine design", IMechE Conf. Trans., C14/87, MEP Publications, 1987, pp. 55-68.

(18)- Kikuchi, K., "Analysis of unbalance vibration of a rotating shaft system with many bearings and discs", Bull. JSME, 13, 1970.

(19)- Hamrock, B. J., Fundamental of Fluid Film Lubrication, McGraw-Hill Publishers, 1994.

(20)- Rahnejat, H., "Multi-body dynamics: Historical evolution and application", Proc. Instn. Mech. Engrs., Part C: J. Mech. Engng. Sci., Special Millennium Issue, 214, C1, 2000.

(21)- Boysal, A. and Rahnejat, H., "Torsional vibration analysis of a multi-body single cylinder internal combustion engine model", J. Appl. Math. Modelling, 21, 1997, pp. 481-493.

(22)- Boysal, A. and Rahnejat, H., "ADAMS generic engine: A tribo-elastodynamic analysis tool", Proc. 96 ADAMS Int. Users' Conf., Mechanical Dynamics Inc., Yipsilanti, USA, 1996, pp. 1-20.

(23)- Mostofi, A. and Gohar, R., "Oil film thickness and pressure distribution in elastohydrodynamic point contacts", Proc. Instn. Mech. Engrs., Part C: J. Mech. Engng. Sci., 24, C4, 1982, pp. 173-182.

(24)- Mehdigoli, H., Rahnejat, H. and Gohar, R., "Vibration response of wavy surfaced disc in elastohydrodynamic rolling contact", Wear, Vol.139, 1990, pp.1-15.

(25)- March, J. P. and Croker, M. D., "Present and future perspectives of powertrain refinement", IMechE Conf. Trans. European Conf. on Vehicle Noise and Vibration, C521/023/98, 1998, pp. 23-40.

(26)- Arrundale, D., Elasto-multibody dynamic analysis applied to noise, vibration and harshness (NVH) in powertrain systems, MPhil Thesis, University of Bradford, 1998.

(27)- Orlandea, N. V., "A study of the effects of the lower index methods on ADAMS sparse tableau formulation for the computational dynamics of multibody mechanical systems", Proc. Instn. Mech. Engrs., Part K: J. Multi-body dynamics, 213, K1, 1999.

(28)- Orlandea, N. V. and Coddington, R., "Reduced index sparse tableau formulation for improved error control of the original ADAMS program", Mechanics in Design, 1, Toronto, 1996, pp. 219-228.

(29)- Kelly, P., Multi-body dynamic analysis of engine induced automotive clutch pedal vibration, PhD Thesis, University of Bradford, 1999.

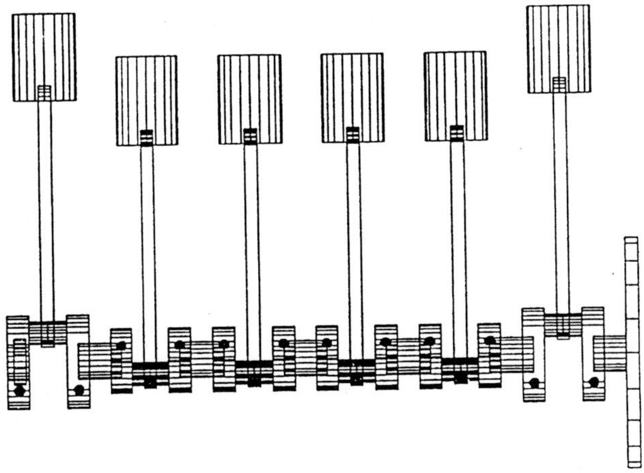

(a)- Side on view

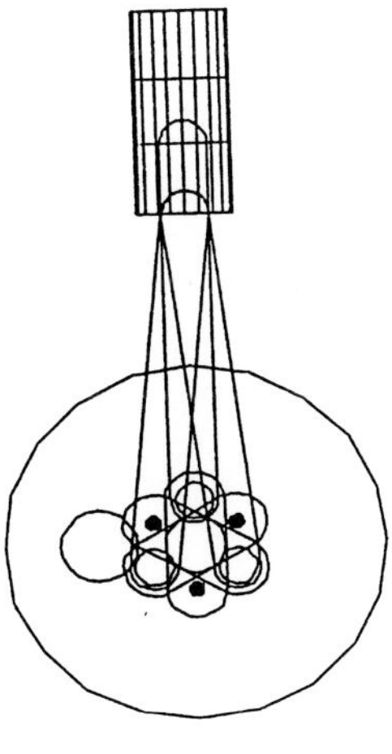

(b)- End on view

Figure 1: Multi-body model of a six cylinder internal combustion engine

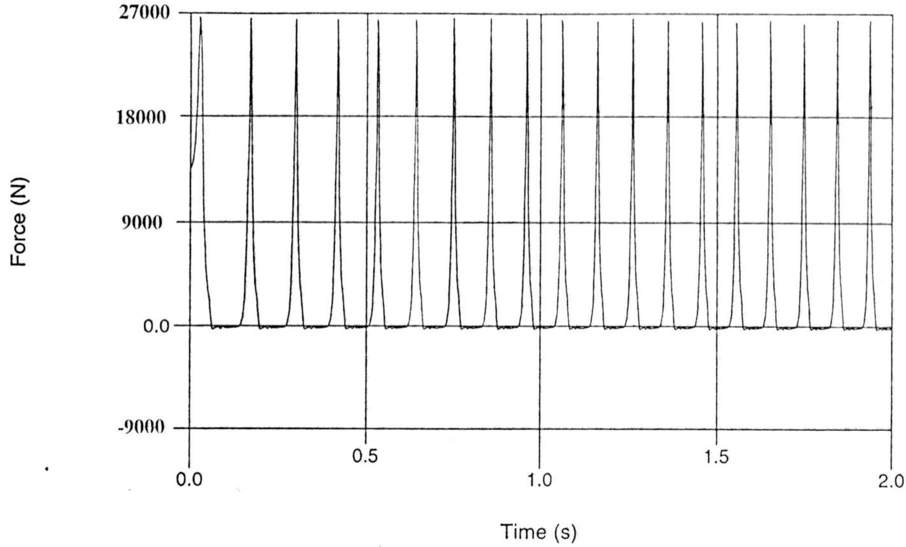

Figure 2(a): Cylinder combustion force time history

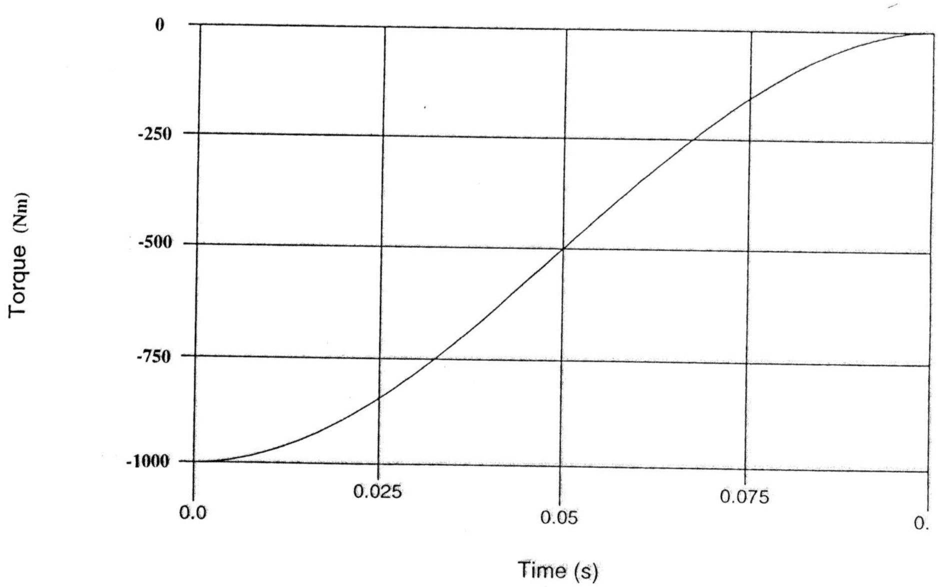

Figure 2(b): Starter motor torque

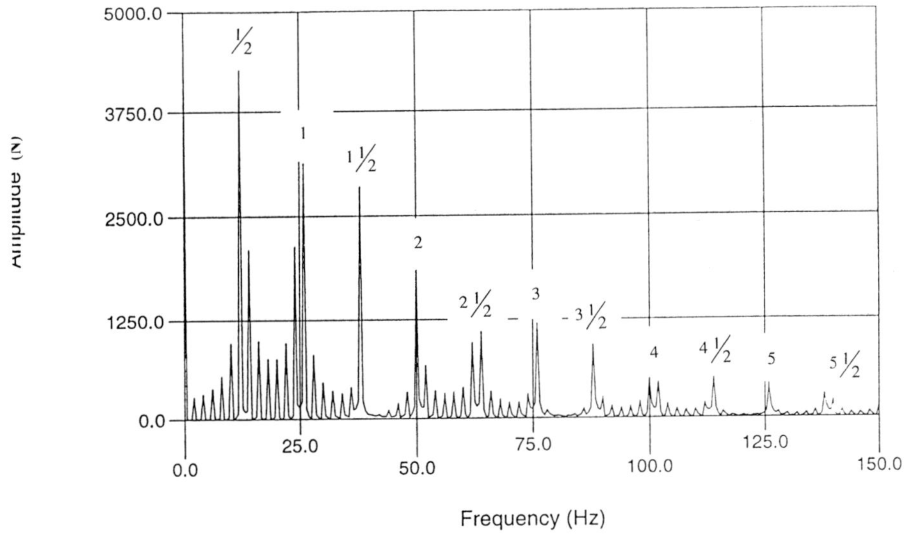

Figure 2(c): Spectral composition of the cylinder gas force

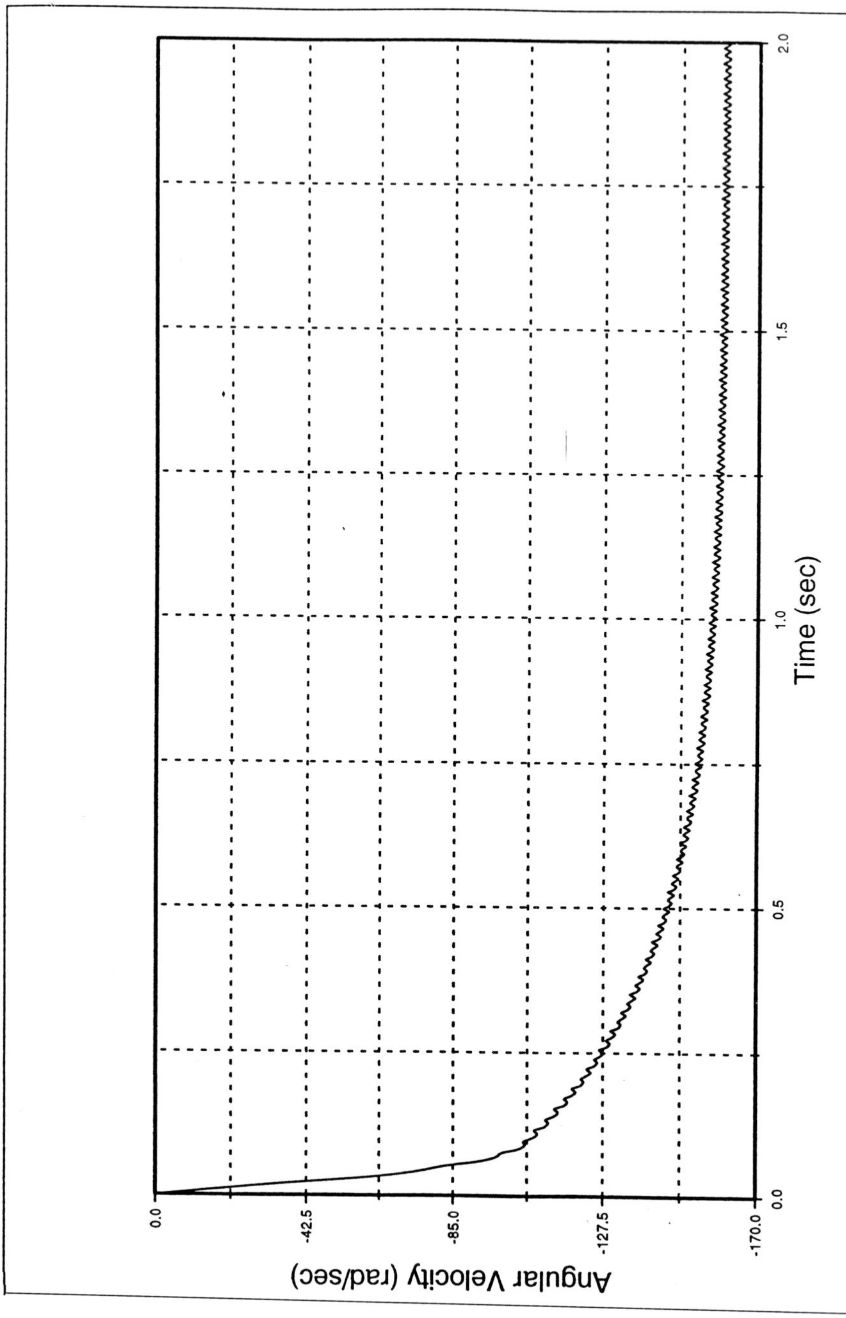

Figure 3(a): Ramp in crankshaft angular velocity

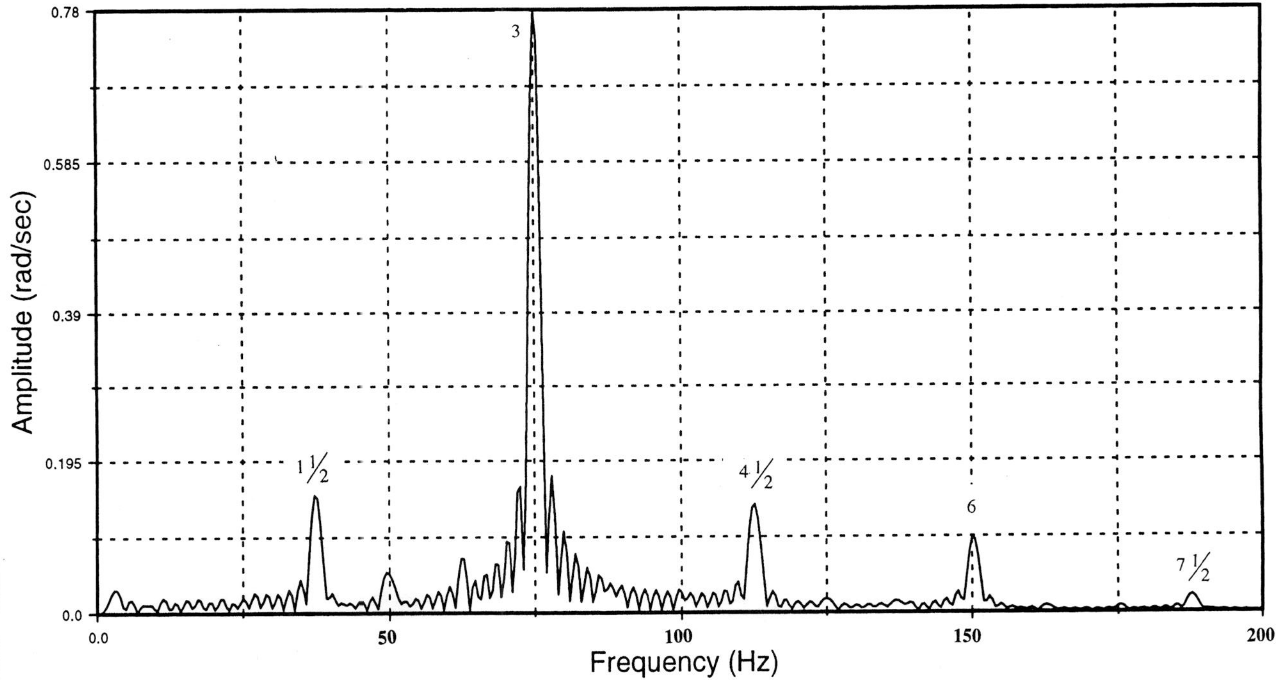

Figure 3(b): Spectral composition of crankshaft torsional vibration

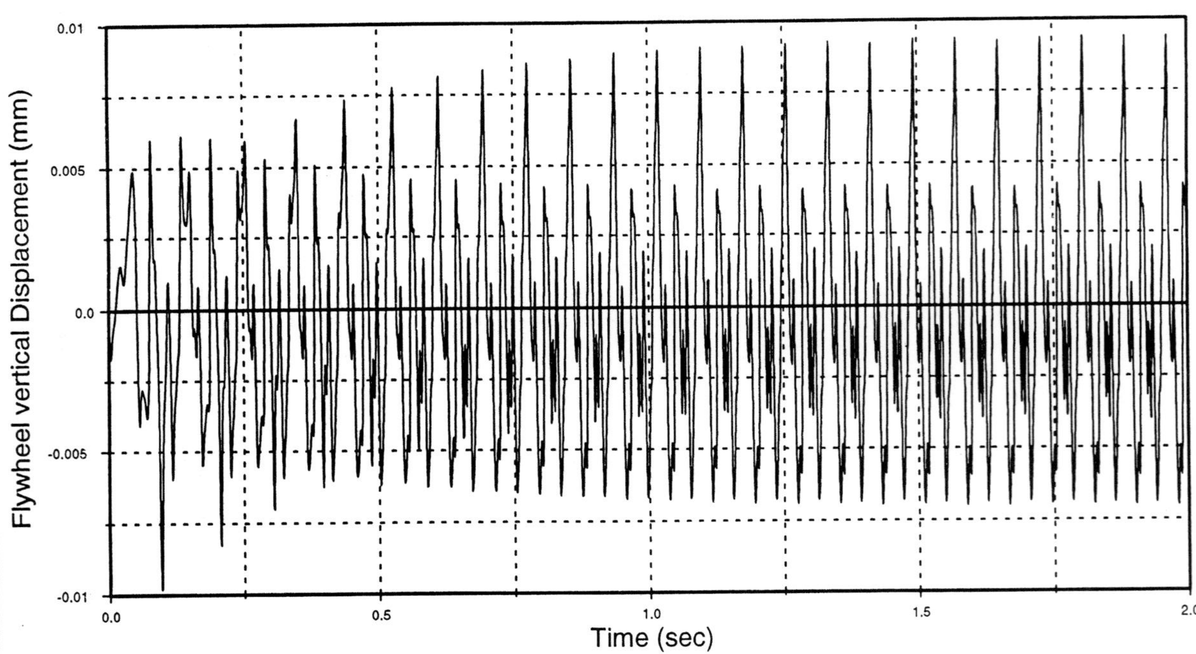

Figure 4(a): Time history of flywheel nodding motion

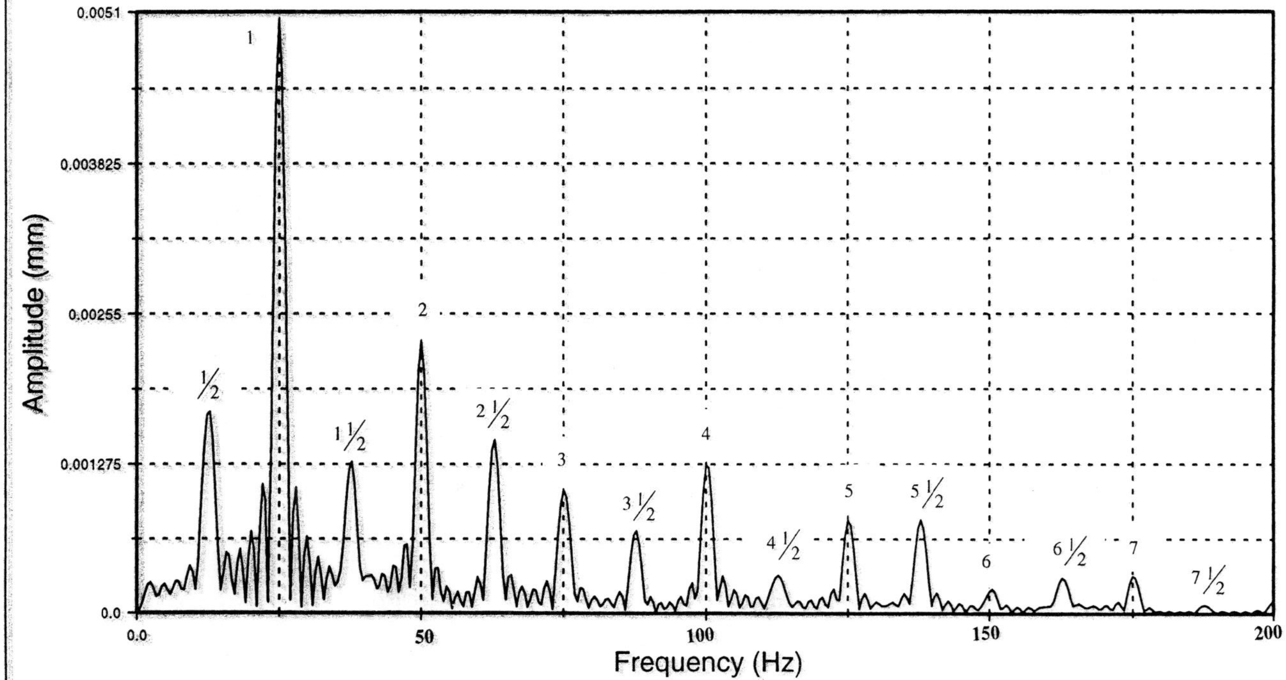

Figure 4(b): Spectrum of flywheel nodding motion

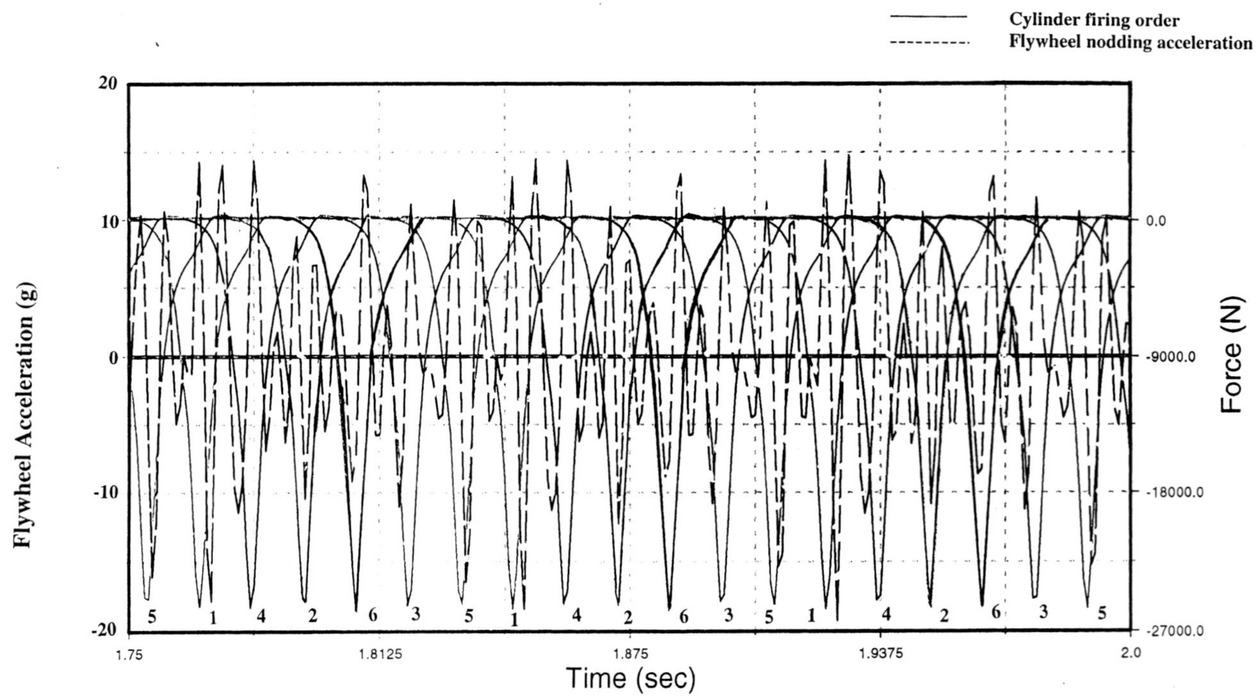

Figure 5: Flywheel nodding motion with cylinder firing

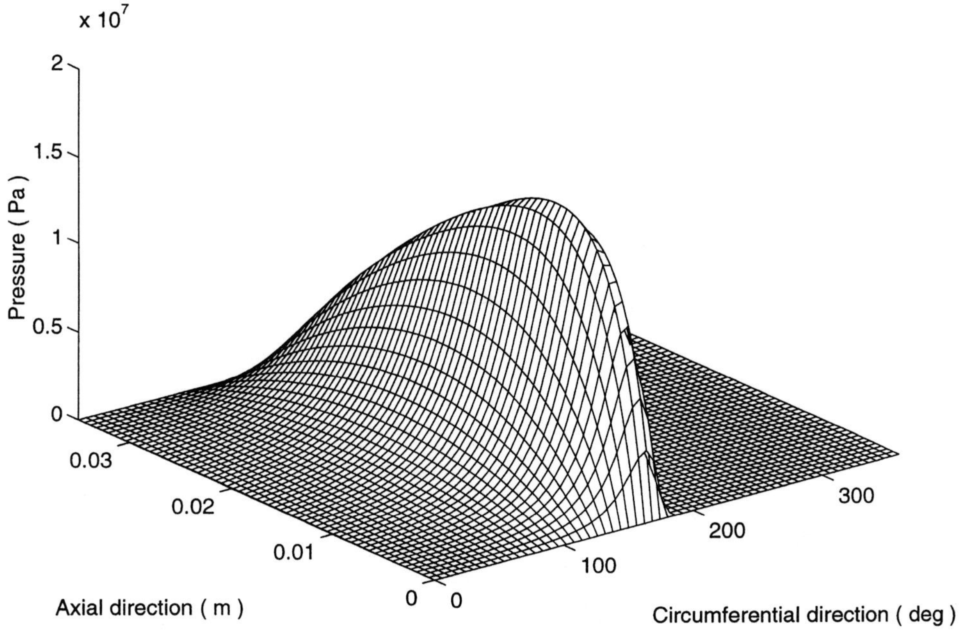

Figure 6(a): Three dimensional elastohydrodynamic pressure distribution

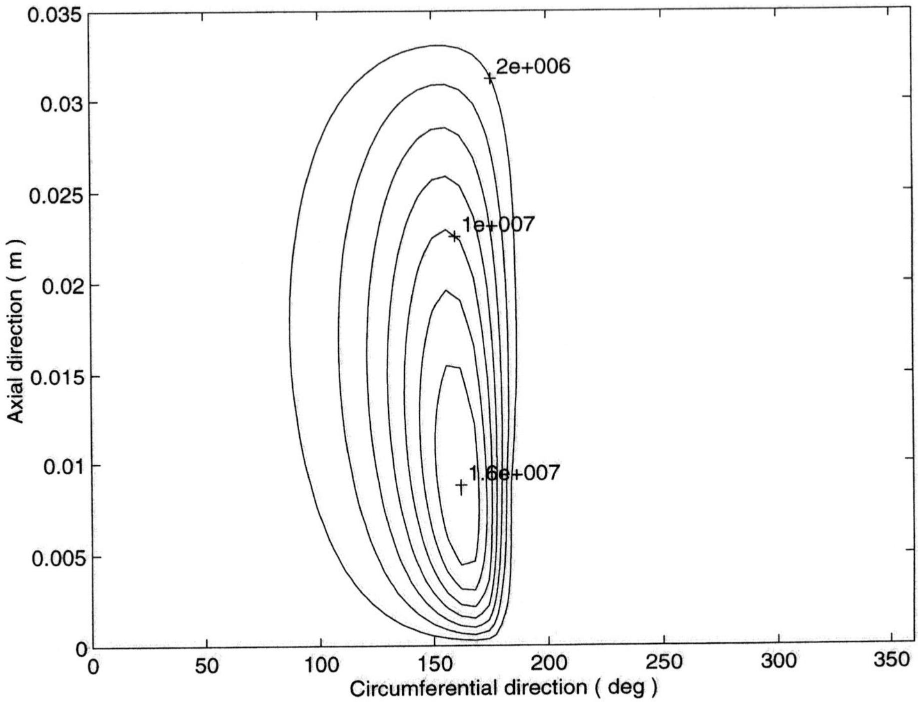

Figure 6(b): Pressure isobar plot for the thin shell engine bearing

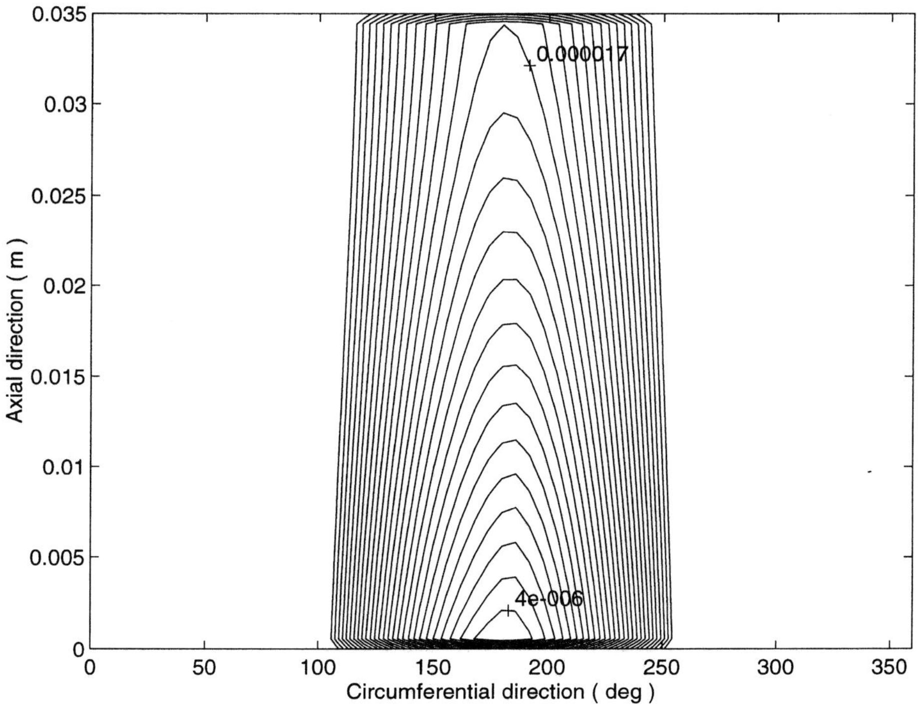

Figure 6(c): Oil film contour for the thin shell bearing

Part Number	Description	Mass (kg)
1	Ground	-
2	Flywheel	19.61
3, 8, 9, 10, 11, 12, 13	Crankshaft (7 pieces)	10.81
58	Block	-
59	Cam gear	0.784
	One per cylinder	
40, 41, 42, 43, 44, 45	Piston	9.19
34, 35, 36, 37, 38, 39	Connecting rod	3.422
5, 24, 25, 26, 27, 28	Crank pin a	0.294
6, 29, 30, 31, 32, 33	Crank pin b	0.294
4, 14, 15, 16, 17, 18	Web a	3.435
7, 19, 20, 21, 22, 23	Web b	3.435
46, 47, 48, 49, 50, 51	Balance mass a	1.8125
52, 53, 54, 55, 56, 57	Balance mass b	1.8125

Table 1: List of Parts in the Six Cylinder Model

Joint No.	Description	Joint Type
1	Engine Block to Ground	Fixed
2	Crankshaft to Block	Revolute
3	Flywheel to Crankshaft	Fixed
4 - 9	Pistons to Block	Translational
10 - 15	Connecting rods to Pistons	Revolute
16 - 57	Webs, Pins, Masses, etc. to Crankshaft	Fixed
58	Cam to Block	Revolute
Prim 1 - 5	Connecting rods to Crankshaft	In-line
Coupler 1	Dummy cam to Crankshaft	Coupler

Table 2: List of Constraints in the Six Cylinder Model

Powertrain Systems

The establishment of realistic multi-body clutch systems NVH targets using rig-based experimental techniques

P KELLY
Ford Werke AG, Koeln, Germany
A REITZ and **J-W BIERMANN**
Institut fur Kraftfahrwesen Aachen, Germany

1. ABSTRACT

The importance of Noise, Vibration and Harshness (NVH) in vehicle development process is important in improving the customers subjective perception of a quality image As for many complex NVH problems, there are many factors that either directly or indirectly influence the multi-body dynamic clutch system. One such NVH phenomenon is clutch whoop. This noise and vibration issue occurs on vehicles with diesel engine during the process of disengaging and engaging the clutch.

The first Multi-body Dynamics Conference introduced whoop and a number of research establishments have further improved the knowledge of the topic. This paper expands the boundary by showing how a specially built test rig, developed at Aachen University, enables the clutch systems NVH behaviour during disengagement and engagement to be determined in much more detail. One unique feature of this paper is how the influence of the torsional and bending aspects of the flywheel can be separated.

The reactions of the clutch system in a real vehicle can be predicted on this rig by testing the vibration characteristic for different combinations of torsional and bending excitations. This development provides a time efficient method to reduce clutch NVH and to establish realistic targets.

2. INTRODUCTION

Good subjective NVH behaviour was used by expensive cars in the past to give them a competitive advantage. Today a mid price car has impressive NVH. This development has meant that issues that where not considered important before have been addresses and resolved. The clutch is a good example of this trend. This paper will show some of the areas where the clutch has an influence and how these are now well understood, and goes on to explain the NVH phenomena that are still being investigated.
Development tools to investigate clutch NVH involve vehicle measurements, rig and CAE analysis. For simple investigations vehicle tests are normally used. However, to gain a true understanding of the system influences for such things as target setting, vehicle tests are not suitable. To truly understand the multi body excitation sources that occur in a powertrain a rig or CAE tools is required.

The investigation on a special NVH rig driven by an electric engine with a low noise and vibration levels, offers the possibility to separate axial and torsional influences. To achieve early program clutch targets a special powertrain subsystem rig was developed.

3. POWERTRAIN SYSTEM NVH PHENOMENA

Figure 1 shows the main NVH phenomena that are influenced more or less by the clutch system. Mostly all described phenomena are effected by both, vibration and noise aspects. On a test bench it is possible to investigate these aspects in more detail.

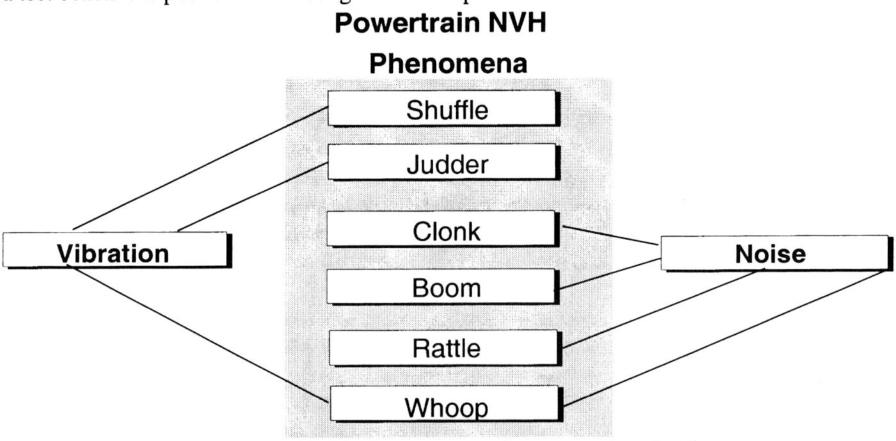

Figure 1 Main NVH phenomena relevant to the clutch system

Shuffle is a low frequent longitudinal vibration issue effecting the whole vehicle. It is caused by input load changes in the frequency range between 2 and 8 Hz. Shuffle is influenced by the clutch system in because the use of a dual mass flywheel has a significant beneficial effect. A rig simulation of the shuffle is not possible to date.

Clutch judder is also a low frequency back and forward motion of a vehicle in the frequency range 5 and 20 Hz. This is caused by the torsional vibrations of the driveline which occur during the clutch engagement process, usually in the take off process and is especially pronounced on rear wheel drive vehicles with diesel engines. Take-off judder is essentially a total powertrain issue but can be influenced by the frictional characteristics of the clutch linings [1].

Clonk is a hard metallic noise impulse of 20 to about 100 ms duration that can occur as a load change reaction. Other factors related to the speed of clutch engagement and the backlash in the system. The use of a dual mass flywheel also has beneficial effects. For the analysis of clonk on vehicles with front wheel and rear wheel drive special test benches exist [2].

Rattle and boom are also effected by the characteristic of a dual mass flywheel. Rattle is a broad-band noise of the transmission caused by the meshing of the loose gears induced by the 2nd or 3rd engine order. Boom is as well excited by engine orders and perceived in the vehicle interior when a body or cavity mode is excited. Investigations on idle rattle and in gear rattle on rigs have been successfully performed on specific transmission and driveline test benches [3].

The so called clutch whoop occurs during disengaging and engaging the clutch. This is one of the few NVH features that has a distinct noise and vibration aspect. The effects are low frequency pedal vibrations that increase during the pedal travel with a accompanying unpleasant noise. The first known paper on this topic was at the first International Symposium on Multi-body Dynamics [4]. The root cause of whoop is clarified by results from basic vehicle measurements described in the next section.

4. VEHICLE INVESTIGATIONS

Using five vehicles from different manufacturers the objective was to investigate the whole clutch system to increase the NVH behavioural understanding of whoop. Figure 2 shows the range of vibration measurements at various engine speeds between idle and 4000rpm during clutch disengagement and engagement. Each time the vehicle was stationary and the transmission was in neutral.

The vibration of the clutch pedal and the release bearing were measured as well as the torsional vibration at the flywheel and the transmission input shaft. The picture shows that the trend in pedal vibration increases with engine speed. Surprisingly there is a large variation in pedal vibration performance across the range of similar cars. The vehicle actually investigated in this paper was the average performer.

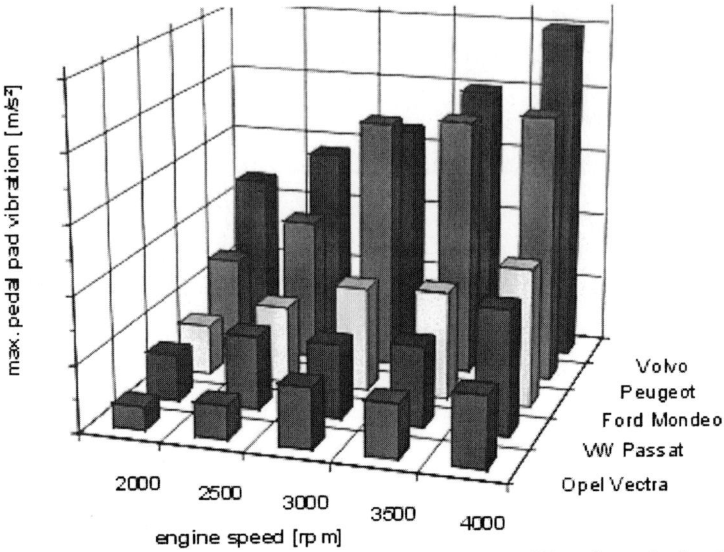

Figure 2 Shows the range of vibration measurements of Benchmarked vehicles

Figure 3 shows the clutch pedal vibration during disengagement and engagement at an engine speed of 2000 rpm. The left side presents the spectrum of the structure borne noise. At the beginning of the measurement the clutch is free. After 0.8s the "off point" (also known as the disengagement point). is passed. Between 1.1s and 1.2s the clutch is fully disengaged before the clutch engages. The amplitudes of the vibration vary considerably with the position of the clutch pedal. The spectrum clearly shows the engine orders 1, 1.5, 2 and 2.5 increase during the disengagement process. The 2nd engine order is the dominant carrier frequency and reaches its maximum after the disengagement point.

At the right side of figure 3 the auto correlation of the clutch pedal vibration is presented. The auto correlation is a function that shows the relationship of a signal to itself. The y-axis represents the time τ after that the signal is repeated. Thus the auto-correlation helps to detect the modulation frequency of a signal. The signal is repeated after τ=60 ms. This is equivalent to a frequency of 16.7 Hz which is the half engine order at 2000 rpm. This shows that the pedal vibration during the disengagement and engagement process is modulated at half engine order.

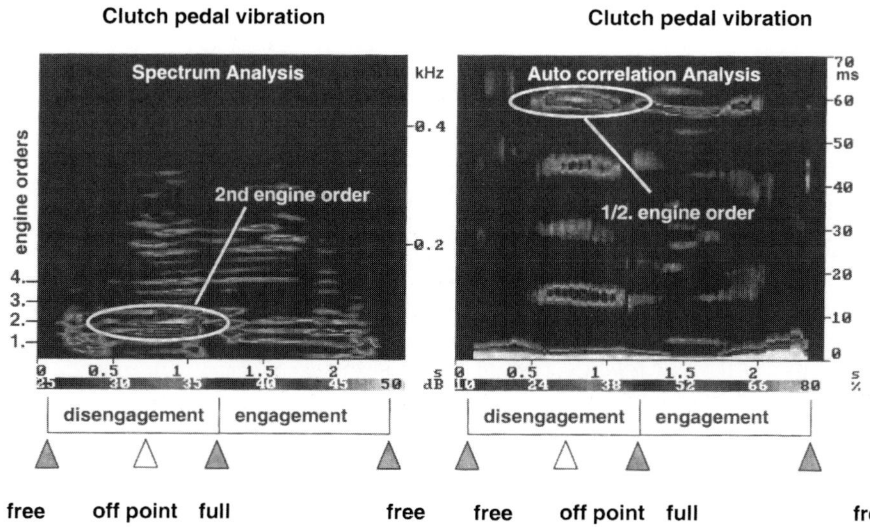

Figure 3 Clutch pedal vibration during disengagement and engagement, n=2000 rpm

The 2nd engine order is excited by the torsional vibrations of the 4 cylinder diesel engine. To determine the root cause of the half engine order in the vehicle the flywheel modes are measured at three positions.

Figure 4 shows a measurement example taken at an engine speed of 2000 rpm.

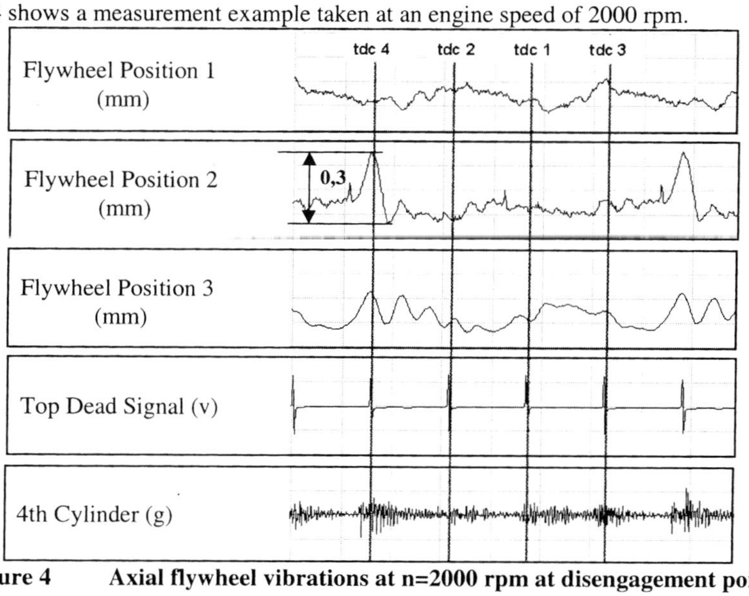

Figure 4 Axial flywheel vibrations at n=2000 rpm at disengagement point

Figure 4 shows a firing cycle of the 4 cylinder diesel engine at the disengagement point of the clutch. It can be seen that the flywheel is excited mainly at the top dead centre (tdc) of the fourth cylinder (tdc4). When the 4th cylinder fires the crankshaft bends. The effect of this bending is an axial flywheel vibration that leads to an impulsive excitation to the clutch [5]. The maximal amplitudes from the measured vehicle is approximately 0.3 mm at flywheel position 2.

As a result of the vehicle measurements Figure 5 shows the amplitude range of the critical flywheel modes.

axial vibration, 1/2. engine order

- frequency: upto 35 Hz (4000 rpm)
- maximal amplitude: 0.3 mm

torsional vibration, 2nd engine order

- frequency: upto 135 Hz (4000 rpm)
- amplitude 1500 rad/s² peak

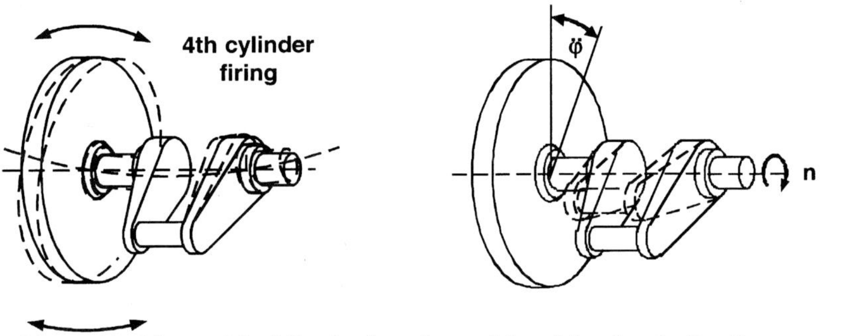

Figure 5 Most critical flywheel modes, axial and torsional vibrations

The amplitude range of the axial flywheel vibration with half engine order is up to 0.3 mm. This depends on the bearing concept of the crankshaft. Depending on the engine speed, the 2nd engine order excitation with the transmission gear set in neutral position leads to amplitudes up to 1500 rad/s² peak at the measured vehicles with diesel engine.

The vehicle measurements demonstrate the complexity of flywheel excitation and its transfer to the clutch pedal. Investigations on a test bench will help to gain a deeper understanding also for CAE modelling.

5. TEST RIG DEVELOPMENT

Using the information from the vehicle measurements it was possible to develop a rig to provide the same outputs. However, the new aspect here was to design the rig so that the axial and torsional vibration inputs could be separated analysed.

Also to be included in the specification of the rig :

- Simulation of clutch pedal vibration in idle position with constant engine speeds during engaging and disengaging the clutch.
- Assembly of the whole clutch system from pedal to flywheel to investigate the vibration transfer behaviour.
- A low noise and vibration drive by using an electric motor.

Figure 6 shows the concept of the test bench. The clutch system is assembled completely with transmission, hydraulic system and pedal box. An electric motor is used as the power input. The axial flywheel vibrations are excited by a connecting rod and an eccentric shaft. This shaft is powered by a belt drive with a speed ratio of 2:1.

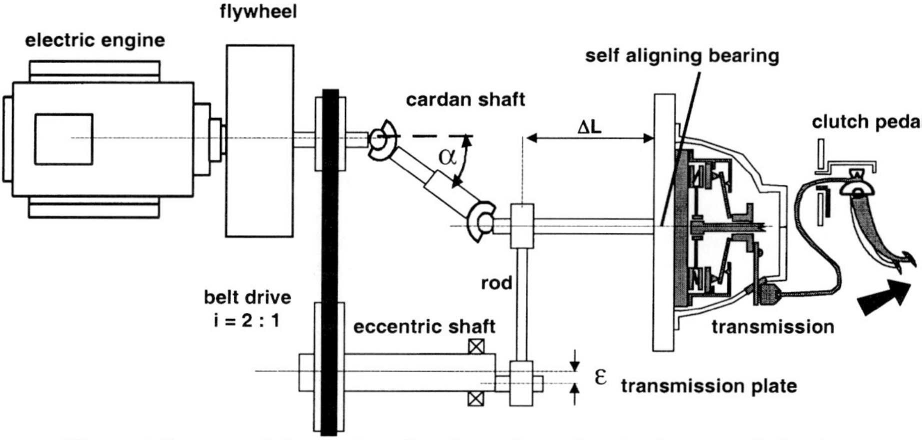

Figure 6 Concept of the test bench to investigate the clutch system behaviour

A self-aligning bearing mounted in the transmission plate allows the rocking motion of the flywheel. The distance ΔL between self aligning bearing and connection rod relates to the amplitude of the axial flywheel excitation. If the crankshaft deflection of one engine type is known, the axial excitation of the flywheel can be adjusted to suit. Clearly the larger the distance ΔL the smaller the amplitude becomes.

To simulate the 2nd engine order a cardan shaft is used. Rotating with a bending angle α cardan joints produce torsional vibrations with a 2nd order. If two cardan joints are assembled in such a way that both have the same angle and both joints rotate in one plane, the irregularities are therefore neutralised and the output has no torsional vibration. On the other hand if two cardan joints are assembled at 90° to each other then the 2nd order is amplified. This effect is used in the test rig.

The torsional vibration depend on the bending angle α as shown in figure 6. Thus by a variation of this angle the amplitude of the 2nd order can be increased on the rig and the excitation frequency can be adjusted by changing the speed of the electric motor.

The complete system consists of a pedal box mounted on a solid block and all the actuation parts including the actual transmission and clutch. The sideshafts were also added to reproduce completely the in-vehicle situation. Measurements of the vibration are recorded at suitable points along the whole system. To ensure that the repeatability of the results the room and transmission oil temperature were controlled, and the pedal actuation stroke together with application time were controlled by a hydraulic push rod. The transmission was kept in idle which also meant that the driveline loads were not required to be controlled. With this rig concept it was possible to simulate separately axial and torsional excitation. A combined axial and torsional excitation was also feasible.

The following test conditions were investigated:
•rotation (800 to 4000 rpm)
•axial flywheel vibration (±0.0 to ±0.5mm)
•axial flywheel vibration + rotation , 1/2. engine order
•torsional vibration, 2nd engine order (0 to 1500 rad/s²)
•axial excitation (1/2. engine order) + torsional vibration (2nd engine order)

6. RESULTS OF RIG TESTS

The examples shown describe the possibility of analysing the process in the system by means of the test rig. Figure 7 represents the effect of the torsional vibration on the clutch pedal vibration as a function of the pedal travel during the disengagement of the clutch. In this case at a flywheel speed of 2000rpm. Interestingly the clutch pedal vibration increases actually during the pedal stroke. If the individual curves are compared, it becomes clear that the amplitude of the 2nd engine order has a influence on the pedal vibrations. The higher the torsional excitation of the flywheel, the larger the acceleration of the pedal. The picture shows the trend of larger torsional values lead to increasing pedal vibrations. At values above 800 rad/s² the vibration levels rise significantly.

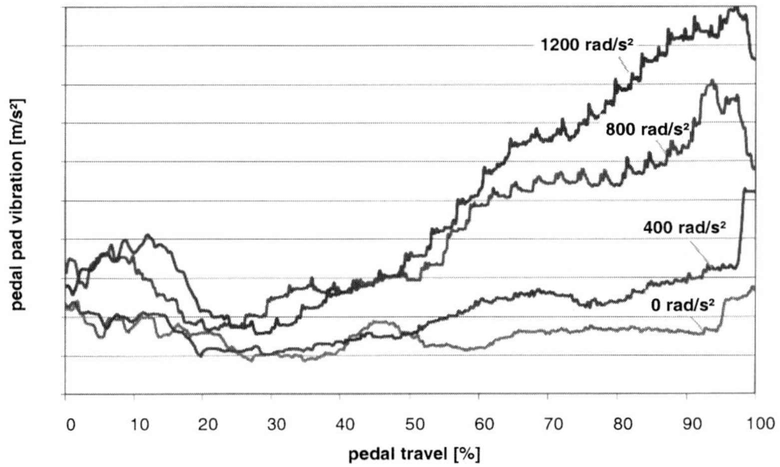

Figure 7 Effect of the second engine order on the pedal pad vibration at 2000 rpm

This test is not possible in the vehicle because the excitation of the clutch pedal is influenced by other sources and transfer paths. Therefore the analysis on the rig provides valuable information and can be used for further CAE-simulation.

A comparison between different amplitudes of axial excitation is now possible. The lowest clutch pedal vibration is possible without axial or torsional excitation when the flywheel rotates continuously. In this condition the measured values predominantly are those produced through the dynamic slip stick effect from the friction lining. The pedal vibrations are about twice as high when the flywheel rotation is superposed by an axial excitation. The level is equivalent to a torsional excitation with 400 rad/s². In vehicles with diesel engine the 2nd engine order excitation at 2000 rpm is up to 500 rad/s². This illustrates that at conditions with a very low torsional excitation the axial flywheel vibration is dominant for the clutch NVH. The higher the amplitudes of the 2nd engine order for example at higher engine speeds the more the clutch system behaviour is effected by torsional aspects. With this knowledge gained from the excitation experiment it is possible to propose targets for a reduction of the pedal vibration. The values of these target can be measured directly from the test rig measurements. The rig has uses in a parameter design study for the tuning of the clutch system components. As an example figure 8 shows the result of a rig test and a vehicle measurement to evaluate a clutch design manufactured to reduce pedal vibrations. To protect the confidentially requirements, the different designs are referred to as letters A to C. In figure 8 it is easy to see which design leads to the best improvement. The highest pedal vibration is measured with design A which is the original clutch system. With designs B and C vibration level improvements are clearly shown.

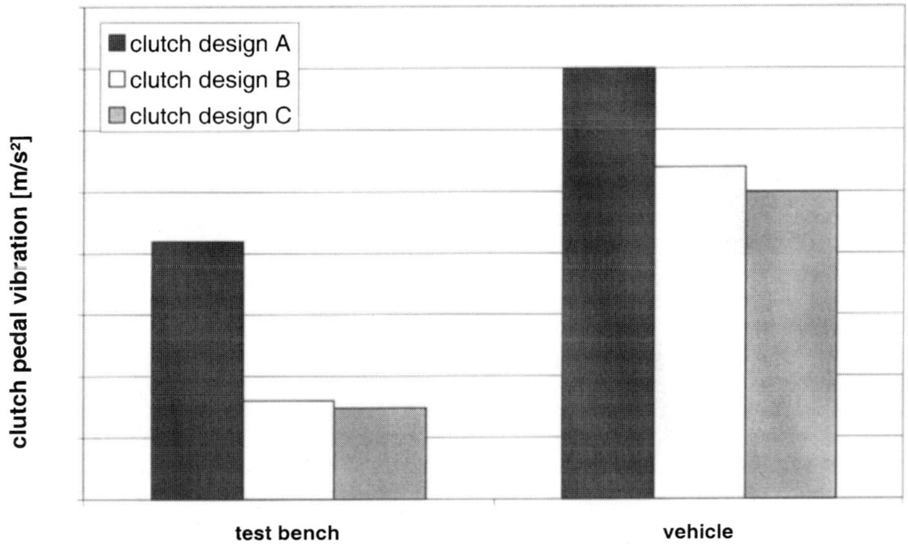

Figure 8 Improvement effect of different clutch designs, test bench and vehicle measurement

In the vehicle the test rig tendencies are verified.. The trend is again confirmed even though the amplitudes of the clutch pedal vibration in the vehicle are higher than on the rig. This is because the vehicle has other additional sources and transfer paths that were not present on the rig.

7. SUMMARY

It can be seen that the test bench presented in this paper enables to determine the clutch systems NVH behaviour during disengagement and engagement in more detail. The behaviour of the clutch system in a real vehicle can be predicted in general on a test bench by measuring the characteristic for each kind of excitation. So it is possible to find a way to reduce the vibration and noise and to propose good NVH targets.

REFERENCES

[1] Drexl, H. J., "Clutch judder – causes and counter measures", Technische Konferenz SITV 1990

[2] Hagerodt, B., "Untersuchung zu Lastwechselreaktionen frontgetriebener Personenkraftwagen",Dissertation am Institut für Kraftfahrwesen Aachen , Schriftenreihe Automobiltechnik, Forschungsgesellschaft Kraftfahrwesen, Aachen, 1998

[3] Reitz, A., Biermann, J. W., Kelly, P., "Special testbenches to investigate driveline related NVH phenomena", 8th Aachen Colloquium, "Automobile and Engine Technology", 1999

[4] Kelly, P., and Rahnejat, H., "Clutch Pedal dynamic noise and vibration investigation", International Symposium on Multi-body Dynamics - Modelling and Simulation Techniques, MEP, Bradford, UK, pp. 23-31, Mar 1997

[5] Kelly, P., Biermann, J. W., Hagerodt, "Clutch pedal dynamic noise and vibration investigation using Taguchi methods", 6th Aachen Colloquium, "Automobile and Engine Technology", 1997

Measured torsional damping levels for two spur gearbox rigs

S J DREW and **B J STONE**
Department of Mechanical and Materials Engineering, The University of Western Australia, Australia
B A LEISHMANN
Formerly at UWA, now a PhD Student at Cambridge University

ABSTRACT
AC servo-drive torsional exciters were used to measure system torsional damping levels of two spur gearbox rigs while rotating. A laser torsional vibrometer and torque telemetry measured angular velocity and dynamic torque. Rig #1 was a servo-drive and single reduction gearbox, lightly-loaded using an electric brake. Rig #2 consisted of back-to-back gearboxes with a recirculating torque. The servo-drives were part of the systems under investigation. Previous papers have included measured and modelled frequency response data, operating deflected shapes and receptance modelling details. This paper presents modal torsional damping data obtained from frequency response functions.

1 INTRODUCTION

Gearboxes are a vital and commonly used component in rotating machinery drive trains. They exhibit an inherent coupling between shaft torques, gear tooth forces and shaft bearing forces, resulting in a coupling of the torsional and transverse system dynamics and vibration. In addition, gear profile errors and the time varying torsional stiffness associated with the meshing action of the gear teeth generate dynamic torques and forces within the gearbox, that are transmitted to the remainder of the machine. Gear vibration has been the subject of a very significant research effort for much of the twentieth century, with particular interest in the prediction and measurement of gear tooth forces and the design of low noise gears.

There is an increasing focus on the coupled vibration of gears and to a lesser extent on the effect of system dynamics. It is also recognised that the level of torsional damping in rotating machinery is generally low and that torsional resonance can thus result in significant dynamic torsional stresses (1, 2). However, only a very limited amount of measured torsional damping data have been published for rotating machinery. Most of the data that are available are modal damping levels for the few lowest natural frequencies of machines (normally below 100 Hz). Torsional damping sources include the driver (eg electric drive or turbine), couplings, gearboxes, shafts and driven machines.

Previously published work includes some torsional damping data for machine trains containing gears, while operating. For many of these studies, the required torsional excitation of the system was achieved by altering the shaft speed until the gearmesh frequency coincided with the torsional natural frequency of interest. By adjusting the shaft speed and hence gearmesh frequency (in small increments), the torsional response of the system was measured in the vicinity of the natural frequency. The torsional damping level for that natural frequency was then calculated (eg using the half-power method). This method of torsional excitation has a number of significant limitations. Firstly, there is an implicit assumption that the gearmesh excitation torque remains constant in the vicinity of the resonance, which may not be true. Secondly, once the numbers of gear teeth are selected, the torsional excitation frequency is not independent of the machine speed. Thus, the measurement of torsional damping levels at various machine speeds is impractical. Thirdly, the torsional excitation amplitude is dependent on the gear parameters (eg tooth stiffness, contact ratio and profile errors) and cannot readily be controlled. Therefore, torsional damping measurements at different torsional excitation levels are also impractical. In addition, the phase of the torsional excitation cannot be easily controlled.

In some recent work, Drew (3) described the torsional testing of two spur gearbox rigs using an AC servo-drive torsional exciter to conduct torsional *modal damping* measurements for the two rigs. Use of the servo-drive exciter enabled the torsional excitation amplitude and frequency to be controlled using a signal generator, independent of the shaft speed and gear characteristics. Angular velocity measurements were obtained using a Dantec laser torsional vibrometer (LTV). Dynamic torque measurements were obtained using a quartz torque sensor and torque strain gauges, in conjunction with two FM radio telemetry systems. Frequency and time domain receptance modelling were also undertaken to predict the response of the rigs to torsional excitation and for comparison with the experimental results, as described by Drew and Stone (4) and Leishman *et al.* (5, 6).

This paper presents experimental torsional modal damping results for the two rigs tested by Drew (3). The first was a lightly loaded, single reduction spur gearbox rig consisting of an AC servo-drive, a gearbox and a small DC motor used to provide a mean brake torque of 4 Nm. In a previous paper by Drew and Stone (4), the measured servo-drive characteristics were used with a frequency domain torsional model of the proposed gearbox rig to predict the system response. At that stage, it was anticipated that a friction brake would be used, however, this later proved to be unsatisfactory and a DC motor (brake) was used for the actual rig. Section 4 of this paper presents measured torsional vibration results for this low-load rig. Experimental angular velocity and torque transfer functions using random (white noise) excitation were measured. The first three torsional natural frequencies and corresponding modal damping levels were determined from the frequency response data.

The second rig was a back-to-back gearbox rig, consisting of the AC servo-drive and two gearboxes torsionally loaded against each other to provide a mean locked-in torque of 40 Nm. Fixed-sine torsional excitation was used to measure torsional damping levels for the ninth torsional natural frequency of the system (1540 Hz) and these results are presented in Section 5.

2 DESCRIPTION OF THE SINGLE REDUCTION GEARBOX RIG

A schematic diagram of the test rig is shown in Figure 1 and gearbox details are presented in Table 1. Also shown in Figure 1 is the boundary defining the system under investigation. In structural modal analysis, the stimulus is almost always defined as the force input to the system, measured at the point where the vibration shaker is connected to the structure under investigation. Dynamic response is normally measured using an accelerometer, or more recently with a laser Doppler vibrometer (LDV). In contrast, for a rotating machine, the drive forms an integral part of the torsional system. Therefore, the torsional system under investigation included the complete servo-drive (servo-amplifier and motor), in addition to the gearbox and the DC load motor. System stimulus (input) was defined as the speed control input voltage to the servo-amplifier. System response (output) was defined as the angular velocity of the shaft (at various positions), plus the dynamic torque at the input and output of the gearbox. The familiar half-power technique was used to estimate measured damping levels.

Table 1. Gearbox specifications for the single reduction rig (3)

Item	Value
Pinion, number of teeth	49
Wheel gear, number of teeth	83
Gearbox ratio	1.6939:1
Facewidth (mm)	20 mm
Pressure angle (degrees)	20°
Module (mm)	2 mm
Contact ratio	1.8
Pinion pitch diameter (mm)	104 mm
Pinion outside diameter (mm)	176 mm
Wheel gear pitch diameter (mm)	108 mm
Wheel gear outside diameter (mm)	180 mm
Shaft centre distance	140 mm
Gearbox bearing type	Taper roller

A Dantec laser torsional vibrometer (LTV) was used to measure the shaft angular velocity at several positions, as shown in Fig. 1. In addition, two torque telemetry systems were used to measure system torques. A Kistler Quartz Torque Transducer was fitted to the input of the gearbox, to measure dynamic torques. An associated charge amplifier, batteries and telemetry transmitter were attached nearby. Strain gauges, a telemetry transmitter and battery were also fitted to a machined section of shaft between the gearbox output shaft and the DC load motor, to measure both static and dynamic torques. An HP 35670A Dynamic Signal Analyser and a Macintosh PC fitted with a National Instruments data acquisition card and LabVIEW software were used for data acquisition and signal processing.

A lumped mass, frequency domain receptance model was used to predict the frequency response and deflected shapes of the lightly loaded gearbox system. The model used is shown in **Figure 2** and was the simplest model required to estimate the first few torsional natural frequencies of the system. Inertia, stiffness and damping values used in the model were adjusted to match the measured and predicted natural frequencies. Figure 2 also shows the predicted frequency response of the servo-motor to a dynamic torque applied to the motor shaft and **Figure 3** shows the predicted deflected shapes at the three torsional natural frequencies.

3 DESCRIPTION OF THE BACK-TO-BACK GEARBOX RIG

A schematic diagram of the back-to-back gearbox rig is shown in Figure 4 and gearbox details are included in Table 2. An AC servo-drive was used to drive the experimental rig at a mean speed of 480 r.p.m. with a locked in torque of 40 Nm. The recirculating section consisted of two commercial gear casings that had been modified to suit the chosen gear parameters (gear ratio, centre distance and contact ratio). Suitable shafts and gears were manufactured, with mild steel shaft diameters of at least 20 mm used throughout. Further details for the test rig are included in reference (5).

Experimentally, the torsional system boundary included the servo-amplifier, servo-motor, connecting shafts, couplings and both gearboxes. System input (stimulus) was defined as the speed control input signal to the servo-amplifier and the system outputs were the angular velocities measured at ten measurement positions. An HP35670A Digital Signal Analyser and a Dantec LTV were used to obtain frequency response functions at eleven measurement positions on the rig (5). Measured torsional frequency response function data have been published (5, 6) and typical data for one measurement position are also shown in Figure 4.

Table 2. Gearbox specifications for the back-to-back gearbox rig (5)

Item	Slave gearbox	Test gearbox
Pinion, number of teeth	52	39
Wheel gear, number of teeth	60	45
Gearbox ratio	1.1538:1	1.1538:1
Facewidth (mm)	25	6
Pressure angle (degrees)	20°	20°
Diametral pitch (teeth/inch)	16	12
Design contact ratio	1.773	1.726
Pinion pitch diameter (mm)	82.55 mm	82.55 mm
Pinion outside diameter (mm)	85.73 mm	86.78 mm
Wheel gear pitch diameter (mm)	95.25 mm	95.25 mm
Wheel gear outside diameter (mm)	98.43 mm	99.48 mm
Shaft centre distance	88.9 mm	88.9 mm

4 RESULTS FOR THE SINGLE REDUCTION GEARBOX RIG

A random noise voltage input to the servo-amplifier was used to generate broad band torsional excitation of the gearbox rig while running at 1200 r.p.m. The DC load motor provided a small constant torque of 4 Nm, which prevented the gears from disengaging during the tests. Measured frequency response functions of angular velocity versus voltage input and of dynamic torque versus voltage input are presented in Figure 5.

The torsional natural frequencies were well separated and the system natural frequencies were estimated as the frequencies of maximum FRF magnitude. Similar results for the three torsional natural frequencies were obtained from all five measurements and are included in Table 4, however, the variation was larger than would normally be expected. This was attributed mainly to the use of random excitation, the effect of laser speckle harmonics and also the presence of genuine torsional vibration; all of which contributed to increased noise levels in the data. The use of random excitation with the servo-drive enables measurements to be undertaken quite rapidly. Normally, fixed-sine measurements are normally required with the torsional exciter in order to obtain highly accurate data. This is particularly the case at higher frequencies (above approximately 200 Hz). Even so, reasonably consistent estimates of torsional damping were obtained from the FRF data (using the half power points), as shown in Table 5 and Figure 6(a).

Table 4. Measured torsional natural frequencies for the lightly loaded rig (3)

Mode	Posn 1 (LTV)	Posn 3 (LTV)	Posn 4 (LTV)	Posn 3 (Torque)	Posn 4 (Torque)	Average
1	7	7.3	7	7.5	7.25	7.2 Hz
2	66	62.8	63.2	65.2	66.1	64.7 Hz
3	155	155.2	152.4	155.2	154.2	154.4 Hz

Table 5. Torsional modal damping levels for the lightly loaded rig (3)

Mode	Posn 1 (LTV)	Posn 3 (LTV)	Posn 4 (LTV)	Posn 3 (Torque)	Posn 4 (Torque)	Average
1	14.3 %	15.1 %	16.1 %	13.1 %	13.8 %	14.4 %
2	4.4 %	4.8 %	5.3 %	3.9 %	5 %	4.7 %
3	1.8 %	1.8 %	1.9 %	1.7 %	1.2 %	1.7 %

5 RESULTS FOR THE BACK-TO-BACK GEARBOX RIG

The first nine torsional natural frequencies of the back-to-back rig were measured (5) and these are shown in Table 6. Detailed damping measurements for the ninth natural frequency of the back-to-back gearbox rig were also conducted using fixed sine excitation (4) and a summary of the results obtained is presented in this section. The locked in torque was 40 Nm (46 Nm for the low speed shaft) and the input shaft speed was 500 r.p.m. In reference (5), the ninth torsional natural frequency was at 1548 Hz. After those experiments, the rig was partially dismantled and the reassembled, leading to a small change in the natural frequency from 1548 Hz to 1540.5 Hz in reference (4). The fact that this change in natural frequency was so small (0.5%) indicates that the system torsional characteristics were quite consistent, even after a partial disassembly.

Frequency response data were measured at each end of the gearbox loop couplings (Positions 5, 6, 9 & 10 in Figure 5), using fixed sine excitation. A frequency increment of 0.5 Hz was chosen to obtain eight or more frequency intervals between the two half power points and thus a maximum error of 1% in taking the maximum FRF value as the true peak for the resonance. Suitable windowing and averaging resulted in coherence values (ie >0.98 for most data), with significant confidence in the data obtained.

Table 6. Measured torsional natural frequencies for the back-to-back rig (5)

Mode #	Natural frequency (Hz)
1	11 Hz
2	111 Hz
3	272 Hz
4	595 Hz
5	720 Hz
6	982 Hz
7	1080 Hz
8	1303 Hz
9	1548 Hz

For the purpose of the analysis it was assumed that the damping was viscous and linear for the particular test conditions adopted. Also, since the natural frequencies were well separated, SDOF behaviour in the vicinity of the natural frequency was also assumed. Three methods of damping estimation were used:

(1) Half power method; using the nearest data points for the half-power points.

(2) Interpolated half power method; using linear interpolation (between the two closest points), to determine the half-power points.

(3) SDOF circle fit method. Normalised Nyquist diagrams for the four measurement positions each yielded 36 damping estimates, one for each combination of data pairs containing a data point above and below the natural frequency.

Figure 6(b) and Table 7 present the measured torsional damping results. The results in Table 7 are shown to three significant figures. However, it is considered that a maximum of two significant figures is more realistic, giving a best estimate of 0.17±0.02% for the actual torsional damping ratio. A comparison of the results for the three calculation methods indicates that the simplest (the half-power technique) proved to be as good an estimator of the torsional damping level as the more sophisticated SDOF circle fit method. The interpolated half-power technique produced similar results, although possibly underestimating the damping levels slightly.

Table 7. Measured torsional damping levels at the ninth torsional natural frequency (1540 Hz) for the back-to-back gearbox rig (3)

Method	Average damping (% critical)	Standard deviation (% critical)
Half power method	0.162	0.000
Interpolated half-power	0.156	0.005
SDOF circle fit	0.167	0.020

6 DISCUSSION AND CONCLUSIONS

6.1 Single reduction spur gearbox rig

The first three torsional natural frequencies of the rig were estimated from the experimental data, by picking the frequency of maximum FRF amplitude for each resonance. Measured average natural frequencies were 7.2±0.3 Hz, 64.7±1.9 Hz and 154.4±2 Hz.

Torsional damping levels for the three natural frequencies were also calculated from the measured FRF data, using the half-power method and an assumption of viscous damping. The average damping values obtained for the three natural frequencies were 14%, 4.7% and 1.7%, respectively.

Variations in measured values for torsional natural frequency and damping were attributed mainly to the use of random excitation (rather than fixed sine excitation), and to the relatively coarse frequency resolution used for the FRF data. Satisfactory results were obtained below 200 Hz.

6.2 Back-to-back gearbox rig

Detailed measurements of the torsional *modal damping* levels for the ninth torsional natural frequency of the back-to-back gearbox rig (approx. 1540 Hz) were undertaken. Four measurement positions were used, with fixed-sine excitation and a fine frequency resolution. With suitable averaging, high coherence levels were achieved. All four measurements resulted in a natural frequency measurement of 1540.5 Hz (± 0.5 Hz).

There were three main assumptions for the damping calculations. The first assumption was that the frequency response in the region of the natural frequency was equivalent to that of a single degree-of-freedom system (SDOF behaviour). Secondly, that the modal damping was viscous. Thirdly, that the system behaviour was linear for the operating conditions that were tested.

Torsional damping levels were measured for the 1540.5 Hz natural frequency using; (i) the half-power method, (ii) the interpolated half-power method, and (iii) the SDOF Circle Fit method. Damping estimates were averaged over four measurement positions inside the gearbox loop. For the operating conditions tested, the torsional damping ratio of the ninth torsional natural frequency of the back-to-back gearbox rig (1540.5 Hz), was 0.17±0.02% of critical. It was considered that the torsional damping measurements undertaken were very demanding due to the very low damping levels. However, the damping measurements obtained were very consistent.

6.3 Conclusions

Results obtained clearly demonstrate that accurate torsional *modal damping* measurements were possible using a torsional exciter and laser torsional vibrometer on rotating gearbox rigs. This is considered to be a significant outcome. The torsional modal analysis techniques used allow the torsional characteristics of gearbox systems under realistic speed and load conditions to be obtained readily and accurately. Very low system torsional damping levels were measured and illustrate the capacity for significant stresses associated with torsional resonance problems.

7 REFERENCES

1. H. R. Simmons and A. J. Smalley, 1984. *Transactions of the ASME, Journal of Engineering for Gas Turbines and Power*, **106**, 946-951. Lateral gear shaft dynamics control torsional stresses in turbine-driven compressor train.
2. B. J. Maher and R. J. Davey, 1996. *First Australasian Congress on Applied Mechanics, Melbourne*, **2**, 795-799. Torsional analysis of a twin drive lime kiln.
3. S. J. Drew 1999 *The measurement and excitation of torsional vibration.* PhD dissertation, The University of Western Australia. (Submitted for Examination).
4. S. J. Drew and B. J. Stone 1995 *Proceedings of the Second International Conference on Gearbox Noise, Vibration and Diagnostics, London*, 171-181. Excitation of torsional vibration for rotating machinery using a 1.7 kW AC servo-drive.
5. B. A. Leishman, S. J. Drew and B. J. Stone 2000 *Proc. Instn. Mech. Engrs. Part K: Journal of Multi-Body Dynamics.* Torsional vibration of a back-to-back gearbox rig: Part 1 - Frequency domain modal analysis. (Accepted for publication).
6. B. A. Leishman, S. J. Drew and B. J. Stone 2000 *Proc. Instn. Mech. Engrs. Part K: Journal of Multi-Body Dynamics.* Torsional vibration of a back-to-back gearbox rig: Part 2 - Time domain modelling and verification. (Accepted for publication).

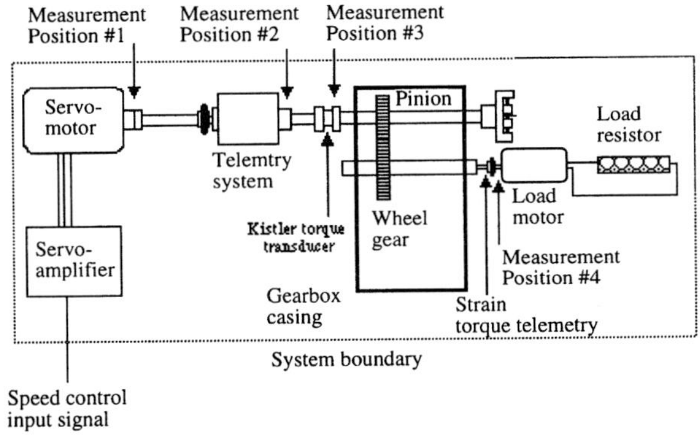

Figure 1. Schematic diagram of the lightly loaded gearbox rig, showing torque and LTV measurement positions, system boundary and system input (3).

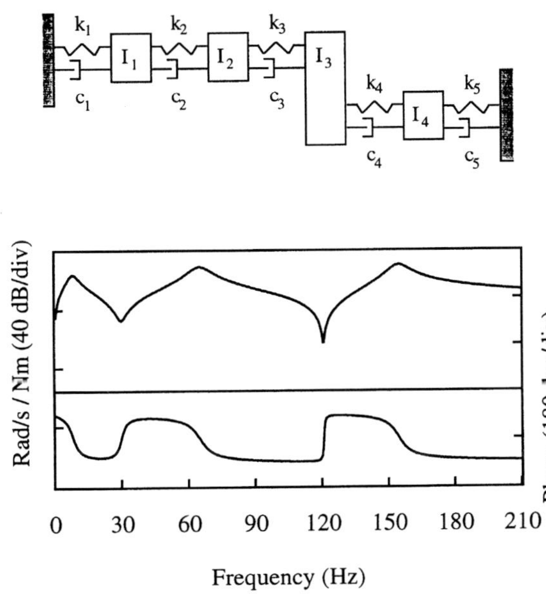

Figure 2. Lumped mass receptance model of the lightly loaded gearbox rig and predicted torsional mobility at the motor (3).

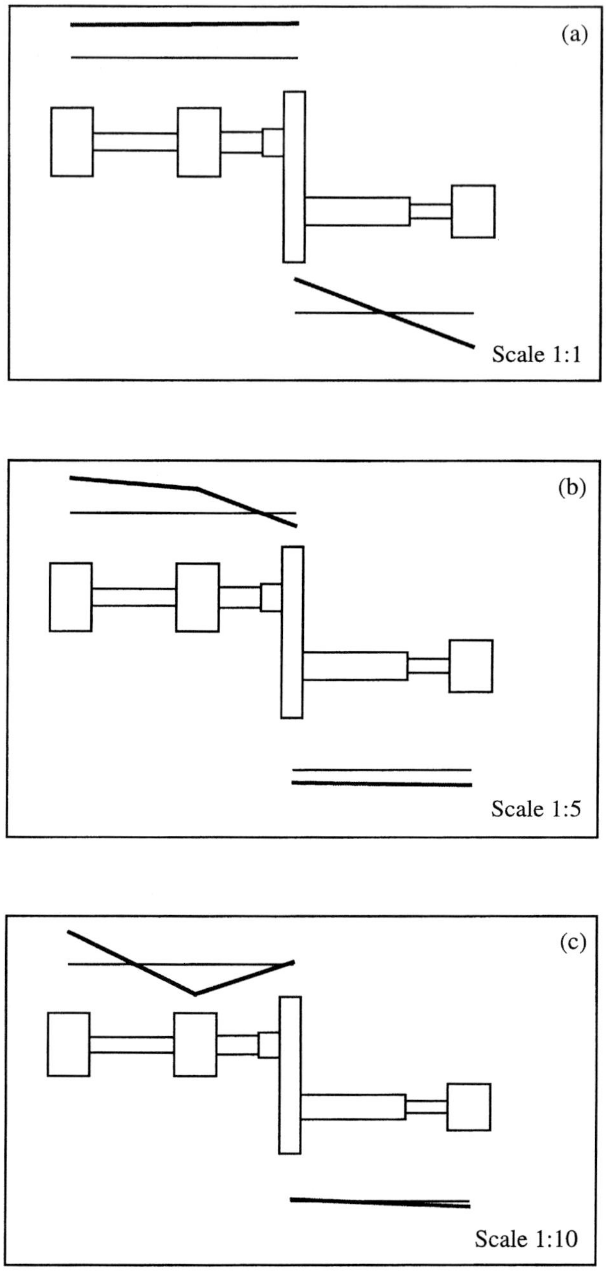

Figure 3. Predicted torsional deflected shapes for the lightly loaded rig (3): (a) First mode (7.2 Hz); (b) second mode; (c) third mode (154 Hz).

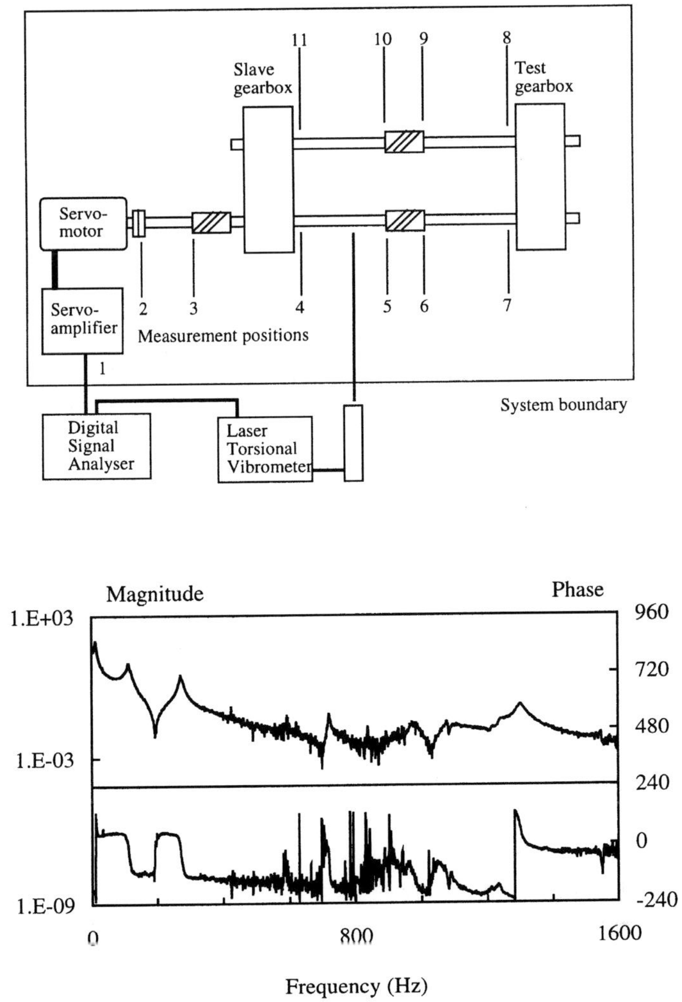

Figure 4. Schematic diagram of the back-to-back gearbox rig and a typical measured torsional frequency response (after (5)).

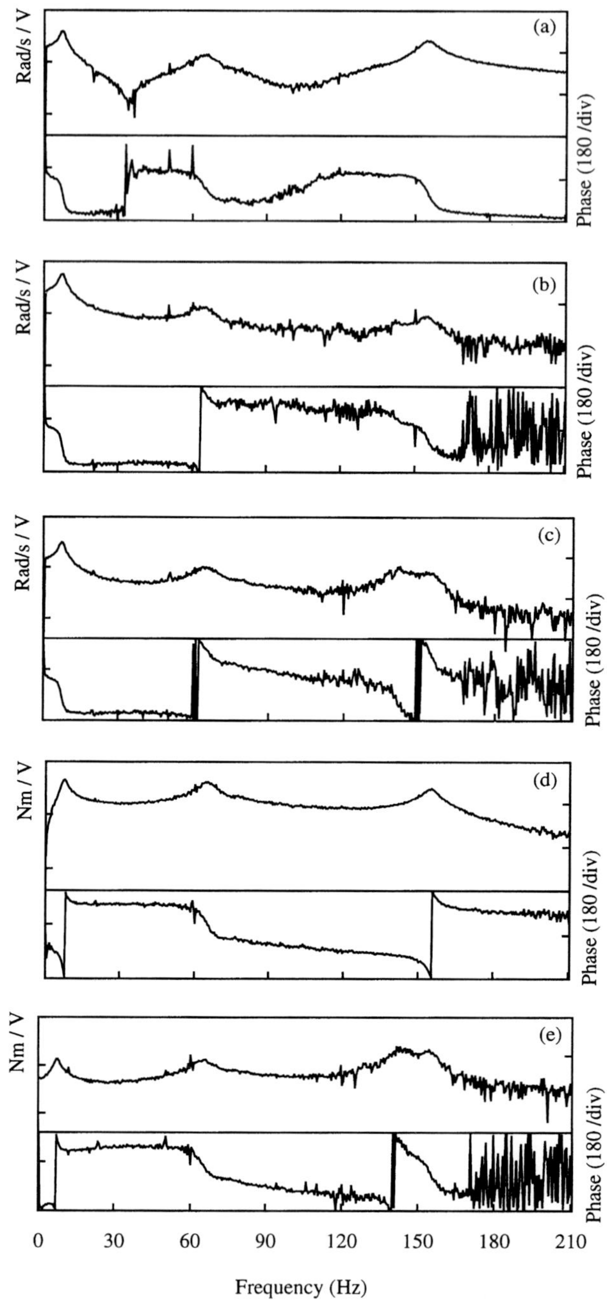

Figure 5. Measured torsional frequency response for the lightly loaded rig (3).

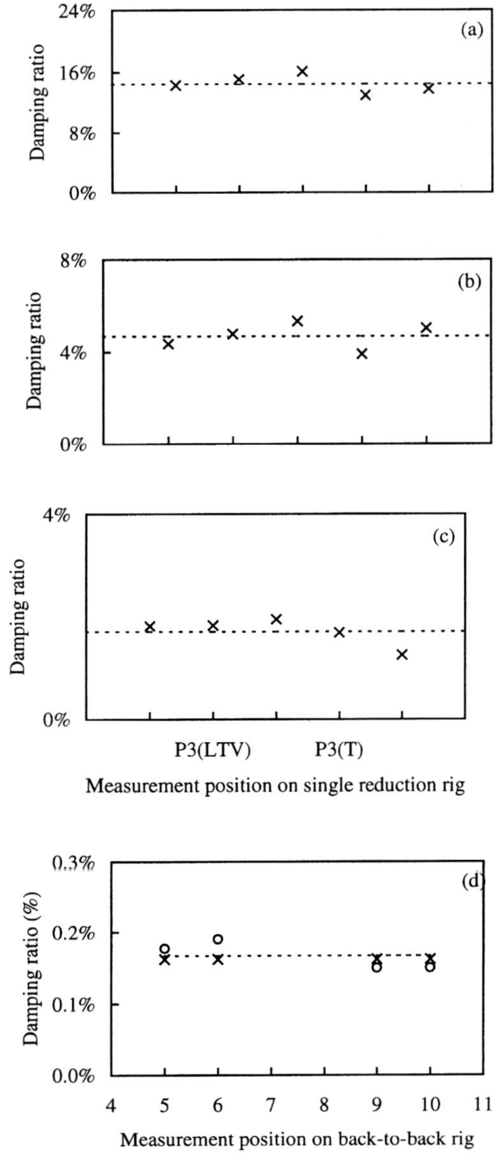

Figure 6. Measured torsional damping levels for the two gearbox rigs (3). (a) Single reduction rig (7.2 Hz); (b) single reduction rig (64.7 Hz); (c) single reduction rig (154.4 Hz); and (d) back-to-back rig (1540 Hz). - - - , Average value; o , SDOF circle fit; x , half-power method.

Test-bench investigations of CV-joints regarding NVH behaviour

S RICHTER and **J-W BIERMANN**
Institut fur Kraftfahrwesen Aachen, Germany

SYNOPSIS

The drivetrain of a passenger car represents a multi-body system, which is of some importance for the NVH-behaviour of the complete vehicle. Many noise and vibration phenomena perceived inside the vehicle are caused by parts of the drivetrain. Dimensionally small but NVH-relevant parts of the drivetrain are the inboard drive-shaft joints. These are plunging joints, which influence the NVH-performance in two different ways. They play an active and a passive role in the vibration transfer of the drive-shaft. The Institut für Kraftfahrwesen Aachen has developed a test-bench and a measurement device to analyse these phenomena. With this test configuration it is possible to measure both the plunging forces of the joints and the axial forces that occur when the joints are rotating under a bending angle. The NVH-performance of the joints depends on many different parameters such as the design, the grease and the operating point of the joints, i.e. the bending angle, the torque and the speed. In order to gain a deeper understanding of the influence of these parameters, several investigations have been carried out at the Institut für Kraftfahrwesen Aachen, and this paper provides detailed information of the studies.

1. INTRODUCTION

The customer demands on passenger cars are growing rapidly. These demands are not only concerning the driving performance of the vehicles, they are also concerning the comfort. The driving comfort not only consists of the propulsion power and the driving stability of the car, it also consists of the control comfort and the Noise Vibration and Harshness (NVH) behaviour. The latter becomes more and more important for the car manufacturer. Most vibrations perceived in the passenger compartment are normally caused by the chassis or the drivetrain of the vehicle. This paper deals with the vibrations which are generated by the drivetrain and especially by the CV-joints.

The CV-joints are playing two different roles in the NVH-behaviour of the vehicle. On one hand, they have a passive role, this is the vibration isolation. On the other hand, they have an active role, which means they generate forces leading to vibrations inside the car.

Against this background a test bench has been developed at the Institut für Kraftfahrwesen Aachen, which permits to analyse the different NVH-phenomena of the joints. This test configuration allows the measurement of both the plunging forces and the axial forces of the joints. In addition, the efficiency of the joints can be determined, but this is not subject of this paper.

The plunging forces and the generated axial forces of the joints are depending on many different parameters. The investigations presented in this paper deal with the variation of these parameters and their influence on the forces.

2. NVH PHENOMENA

2.1 Shudder

The first NVH-phenomenon which is provoked by the CV-joints is the "Shudder" effect. The mechanism causing this noise is shown in Fig. 2-1.

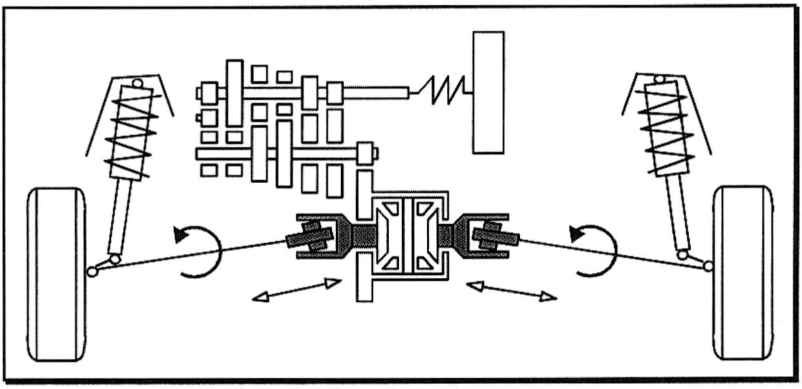

Fig. 2-1: Excitation of the "Shudder"-noise

The Shudder is a result of the generated axial forces of the CV-joints, which occur when the drive shaft is rotating with a bending angle. Due to these periodic forces the drive shafts begin to oscillate in their axial direction. These vibrations are transmitted via the outboard joints to the suspension and applied over the shock-absorber strut to the body. The Shudder-noise is then radiated into the passenger compartment. In this case the joints play an active role in the NVH-behaviour of the complete vehicle. The generation of the axial forces is shown in Fig. 2-2.

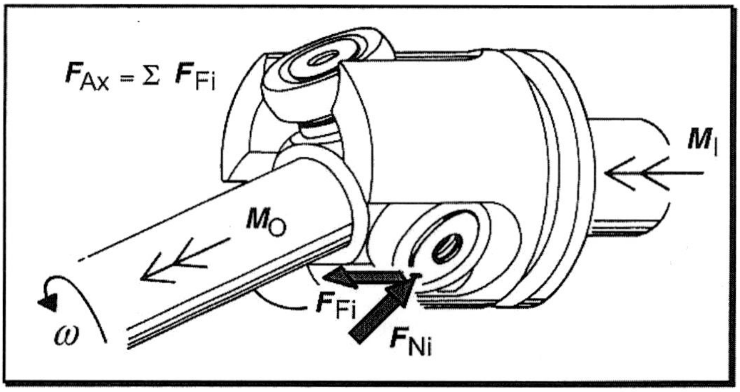

Fig. 2-2: Generation of axial forces

The axial forces occur only in driving condition of the vehicle, this means that the drive shafts are rotating. The second precondition is that the joint is running with a bending angle, which leads to an oscillating movement of the balls in the housing ball bores. The input torque (M_I) equals the output torque (M_o) and is transmitted over the three balls to the output side of the joint. The torque transmission causes pressure forces (F_{Ni}) between the balls and the ball bores. These forces show in the normal direction of the ball bore. Due to the pressure forces, internal friction effects appear (F_{Fi}). The two occurring sources of friction are sliding of the balls on their spider trunnions and combined sliding and rolling of the balls in their housing ball bores. These friction forces are directed along the trunnion axes and result in a cyclic axial force (F_{Ax}) along the spider rotational axis.

2.2. Idle boom

The second NVH-phenomenon which is associated with the CV-joints is the "Idle-boom" effect. The circumstances leading to this noise are shown in Fig. 2-3.

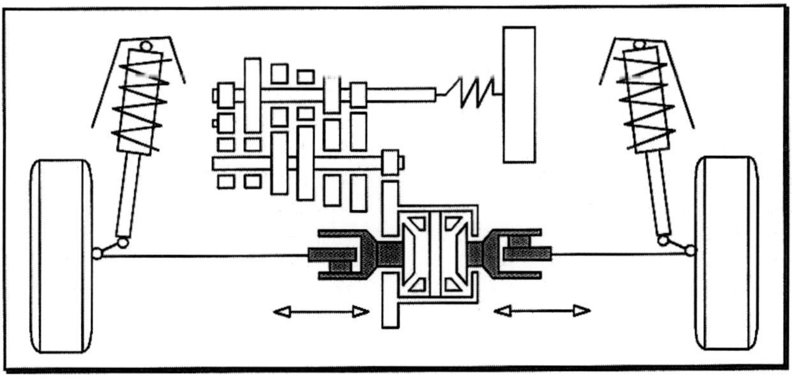

Fig. 2-3: Excitation of the "Idle-Boom"-noise

The idle-boom is caused by the plunging forces of the CV-joints. Similar to the shudder effect a vibration excitation of the drive-shafts takes place, which leads to the noise inside the car, These vibrations are also transmitted by the outboard joints to the suspension and applied over the shock-absorber strut to the body. In this case the joints play a passive role in the NVH-behaviour of the complete vehicle, because the excitation is not produced inside the joints. They only have a vibration isolating function, as it is described in Fig. 2-4.

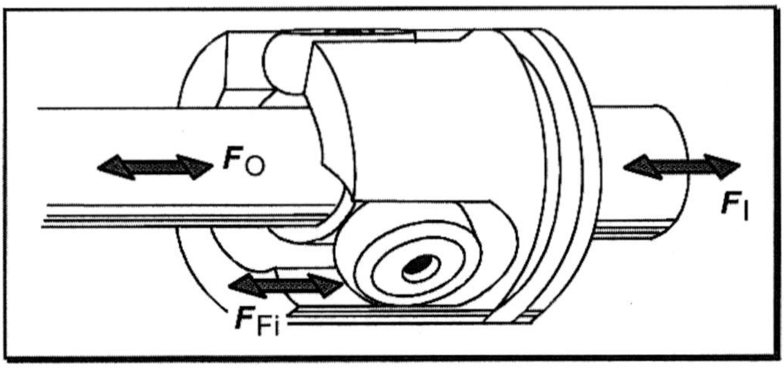

Fig. 2-4: Build-up of plunging forces

The plunging forces are a result of the length adjusting function of the joints. The length adjustment between the inboard and the outboard joint is required in many cases. One case is the compression and the rebound of the chassis, which happens during driving over undulated roads. Another case, which leads to plunging forces and to the idle boom is front-wheel driven car with an automatic gearbox running in idle condition. The unbalanced inertia forces of the engine lead to an oscillating motion of the whole engine. Because the wheels are not oscillating, the length of the drive-shafts, which are linked to the engine in a fixed manner, have to be adjusted. Due to the automatic gearbox the joints are loaded with a torque even when the car is standing still in idle condition. Due to the torque normal forces appear between the balls and the housing ball bores, which lead to internal friction. This means that the input forces (F_I) due to the oscillation of the engine are transmitted by the joint to the Body. The amplitude of the transmitted output forces (F_o) depends on the friction forces (F_{Fi}).

3. TEST BENCH

The Test bench and the measuring system, which have been developed at the Institut für Kraftfahrwesen Aachen for the analysis of these phenomena will be described in the following chapter. A front view of the test bench is shown in Fig. 3-1.

Fig. 3.1: Front view of the test bench for investigations of CV joints

The basic modul of the test bench is the drive train, with a special measuring shaft. The test bench is driven by an electric motor. The speed and the torque are adapted to the desired level by shifting the different gears of a reducing gearbox. The plunging joint is fixed with a flange to the output side of the gear. The original side shaft, which is equipped with additional flanges, is used for the connection to the measuring shaft. Data of the forces, that occur in the axial direction of the shaft, the torque and the speed are acquired with the measurement shaft. A telemetry system sends the data to the recording system outside the test room. The desired bending angle is realised with an adjustable device, which allows a movement of the angle drive at which the fixed joint is attached. Thus bending angles between 0° and 16° are possible. The speed is increased with a second gear box to reach the level required by the eddy-current brake. This set-up gives the possibility to simulate nearly every driving conditions of the joints.

The configuration of the test bench differs slightly for the different measurement tasks. These set-ups are shown in the following Fig. 3-2 and Fig. 3-3.

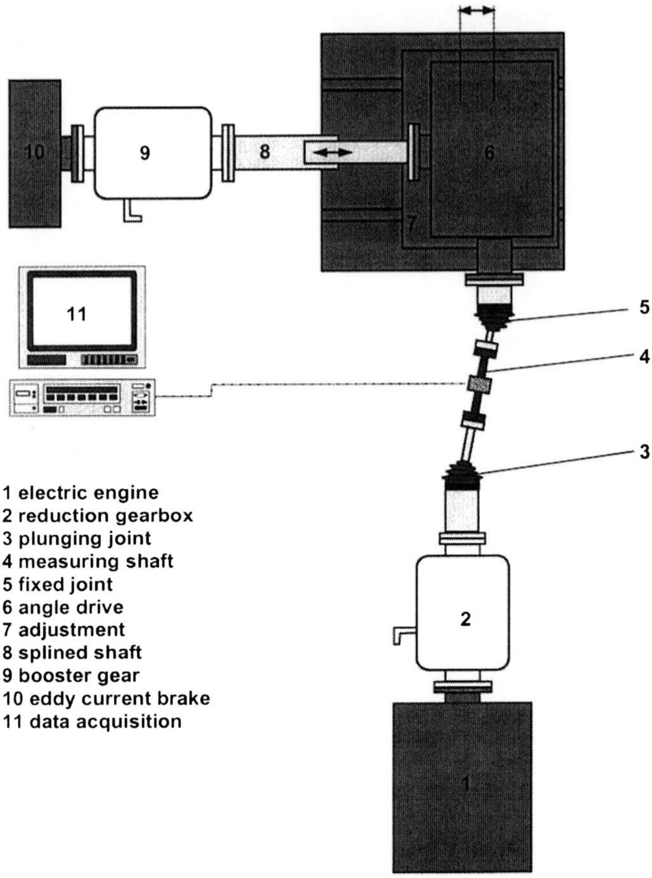

Fig. 3-2: Test bench set-up for the measurement of the generated axial forces

The configuration of the test bench corresponds to the set-up described above. This top view shows in a much clearer way how the required bending angle is build up. Moving the angle drive to the right results in the desired bending angle, but it also demands an extension of the drive train. The length adjustment of the drive train is possible due to the facility of the plunging joint to vary its axial extension. In this case the measurements are carried out in running conditions, which means that the drive train is rotating at a preselected speed and load torque. With this test bench set-up the generated axial forces leading to the shudder noise can be simulated.

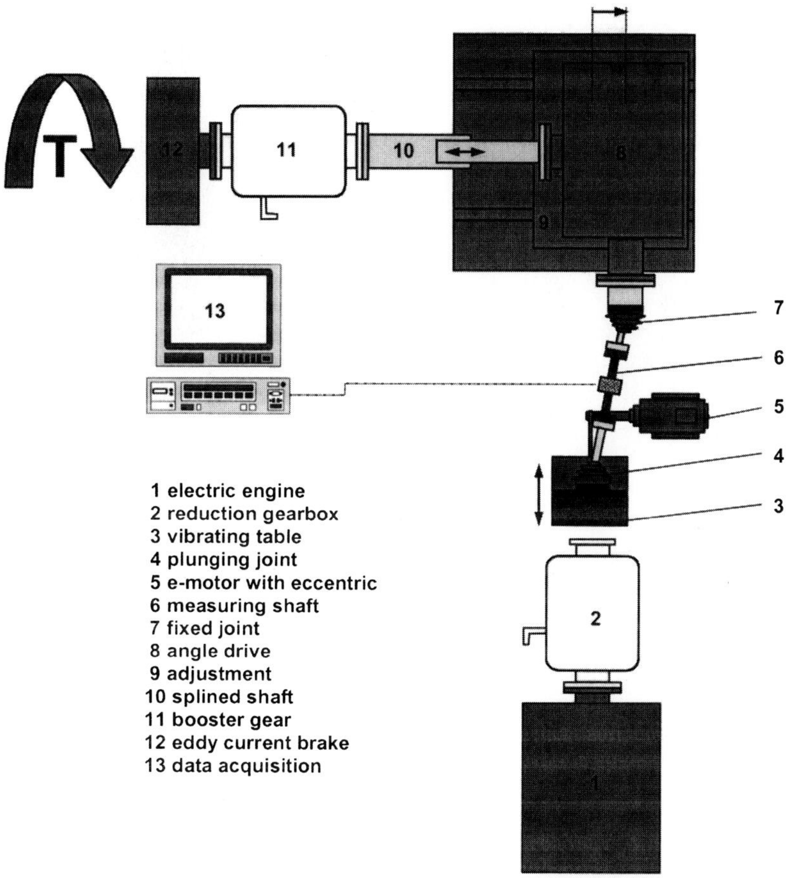

Fig. 3-3: Test bench set-up for the measurement of the plunging forces

For the measurement of the plunging forces a different configuration of the test bench is necessary. In this case the drive train is not rotating, the connection flange between the reduction gearbox (2) and the plunging joint (4) is removed. Instead, the joint is linked to a vibrating table shown in Fig. 3-4. The vibrating table is driven by an additional electric motor (5) with an eccentric at the end of the drive shaft. This results in an axial displacement of the joint, with both a variable amplitude and frequency. The non-rotating joint is loaded with a torque, which is applied with a lever system mounted on the eddy current brake. This configuration simulates the standing vehicle and the conditions leading to the idle boom noise.

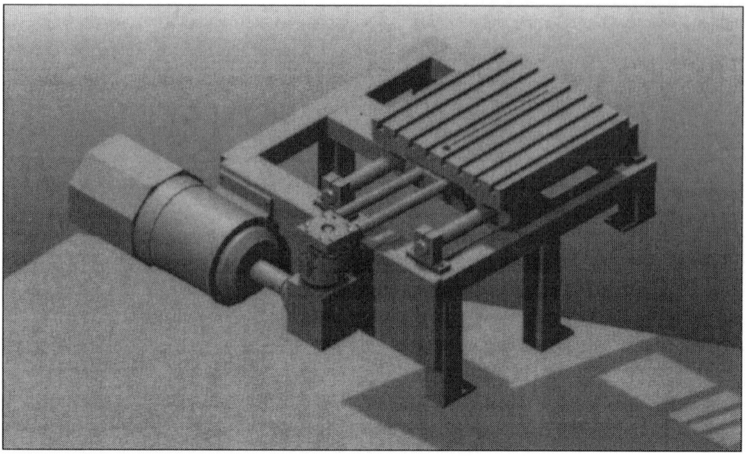

Fig. 3-4: Vibrating table

Another very important part of the test bench is the used measurement shaft. This special device is patented by the Institut für Kraftfahrwesen Aachen and is shown in Fig. 3-5.

Fig. 3-5: Measurement shaft

With this measurement shaft all data of interest can be measured nearly directly at the place of their formation. These data are on one hand two of the parameters of the operating point, the speed and the torque. On the other hand they consist of the forces occurring in the direction of the shaft rotation axis. The torque and force measurements are based on the strain gauge principle. Both values can be recorded independently of each other, which is very important since torque values are high and the axial force values are small. The speed of the shaft is measured by an inductive sensor. Another part of the shaft is the telemetry system required to send the data to the data acquisition system.

4. RESEARCH RESULTS

In the following chapter some results of various investigations carried out at the Institut für Kraftfahrwesen Aachen are presented. In Fig. 4-1 the generated axial forces of joints of different types are shown.

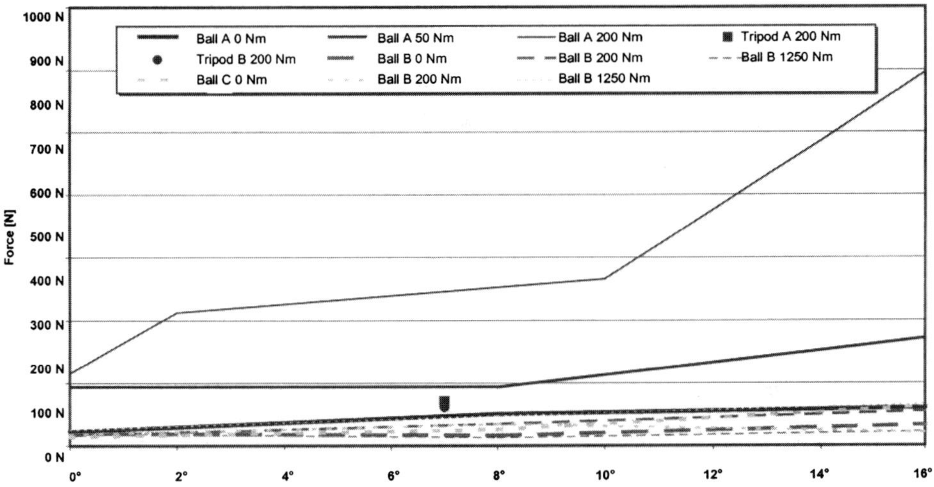

Fig. 4-1: Generated axial forces of different joint types

The chart illustrates the bending angle dependent axial forces of the joint, These results of diverse investigations give the ability to compare joints of a different design under different operating points The design can be divided into tripod-plunging joints and ball-plunging joints. Since the rotating speed of the joints has not a considerable influence on the axial forces, it is not mentioned explicitly. The lines and dotes on the chart represent the different joint types and also their different load torque.

It is apparent that the level of the axial forces depends firstly on the load torque of the joints and secondly on the bending angle. The increase of the torque leads to considerable higher axial forces. A higher bending angle also results in higher axial forces, but the rate is smaller compared with a higher torque. A further important parameter regarding the axial force is the design of the joint. The axial forces generated by a tripod-plunging joint are much higher than the forces generated by a ball-plunging joint.

The following Fig. 4-2 shows results of measured plunging forces for different joints.

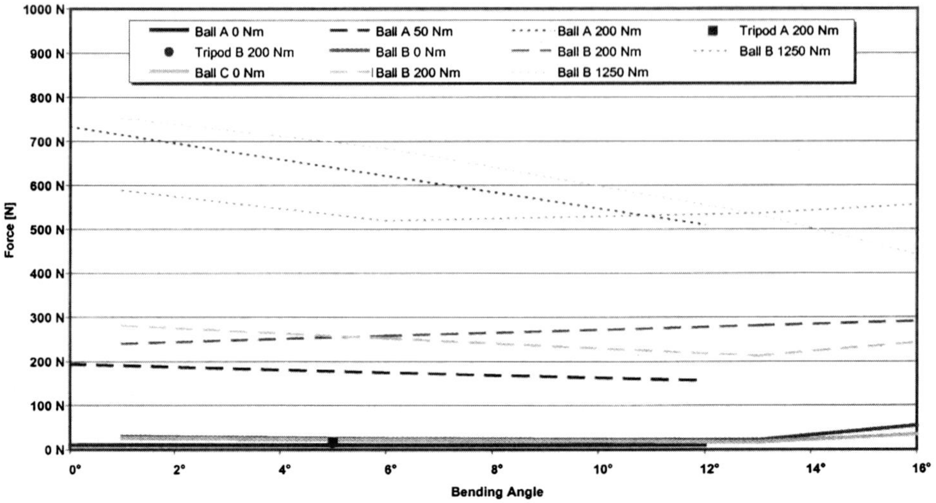

Fig. 4-2: Plunging forces of different joint types

The chart also illustrates the bending angle dependent plunging forces of investigations. The parameters of the different lines and dots are the joint design and the load torque. It is apparent that the plunging forces show a nearly complete dependence on the load torque. The bending angle of the joints only plays a subsidiary role for the plunging forces. It is significant for the different joint designs, that the ball-plunging joints have much higher plunging forces than the tripod-plunging forces

These results show that the choice of a joint for a passenger car is not a simple process. On one hand there are ball joints with extremely low generated axial forces but also very high plunging forces. On the other hand it is possible to use tripod joints, which cause very low plunging forces, but much higher generated axial forces. Against this background it becomes obvious, that the used joint can only be a compromise between the plunging and the generated axial forces. It is only possible to recommend tripod-plunging joints for front-wheel driven cars with an automatic gearbox to avoid the idle-boom effect.

Low-frequency torsional vibration of vehicular driveline systems in shuffle

A FARSHIDIANFAR, M EBRAHIMI and **H RAHNEJAT**
Department of Mechanical and Medical Engineering, University of Bradford, UK
M T MENDAY
Powertrain Systems Engineering, Ford Engineering Research Centre, Dunton, Uk

ABSTRACT

Vehicle drivelines with manual transmissions are exposed to varying dynamic engine torques during drive conditions. Beside the excitations through the cyclic irregularity of the combustion engine, there are additional load change excitations (during throttle tip-in and tip-out). With these load changes, transient vibrations and noise are induced through the contact changes in free-travelling single components of the driveline. Stiffness, backlash and friction in the driveline contact mechanism play an important role in the generation of untoward responses. To study these effects, lumped parameter models of key elements of the driveline can be used to approximate the physical behaviour of the overall system.

This paper presents a global model of a vehicle driveline for torsional vibration analysis. Two different methods have been applied for solution of the equations of motion; the Transfer Matrices Method (TMM), and the step-wise time integration of the Newton-Euler equations of motion. The results from these analyses have been compared with each other, as well as with experimental findings.

NOTATIONS

B_1, B_2, B_3	coefficients for viscous damping
c_1, c_2, c_3	coefficients for viscous damping for shafts and clutch
J_1, J_2, J_3	polar moments of inertia
J_d	polar moment of inertia for the differential
K_1, K_2, K_3	torsional stiffness components
T_1	input torque

T_2, T_3, T_4	transmitted torques
T_{c1}, T_{c2}, T_{c3}	coefficients of coulomb damping
T^L	torque to the left of the disk
T^R	torque to the right of the disk
$\theta_1, \theta_2, \theta_3$	angular displacements
$\dot{\theta}_1, \dot{\theta}_2, \dot{\theta}_3$	angular velocities
$\ddot{\theta}_1, \ddot{\theta}_2, \ddot{\theta}_3$	angular accelerations
ω	natural frequency of the system

1 INTRODUCTION

There is a progressive trend towards reduction of vehicle weight with an aim to improve running performance and reduce fuel consumption per unit travel distance and enhance overall efficiency. In recent years the emphasis has been put on criteria such as engine power, top speed and acceleration, as well as braking performance. Other priority factors include safety, fuel consumption and comfort, along with manufacturing cost-reductions. However, these requirements produce a negative effect in terms of noise and vibration. One of the most important problems concerning noise and vibration inside the vehicle is that associated with the driveline system. This accounts for a sizeable portion of noise and vibration for vehicles ranging from automobiles to heavy trucks. Whenever these vibrations are transmitted to the chassis, they cause deterioration in passenger comfort. Overcoming these problems can lead to better customer satisfaction.

Drivelines are complex multi-body non-linear dynamic systems, which have many modes of vibration and are excited by a large number of sources of excitation. These include road excitation inputs and engine NVH (Noise, Vibration and Harshness) characteristics. A typical driveline NVH spectrum contains a significant number of frequencies, ranging from a few Hertz to several kHz [1].

After the drive torsional impulse is applied to the driveline, the powertrain system oscillates with the fundamental torsional mode of the driveline and several response cycles may occur. Vehicle shunt is the lowest frequency of the longitudinal oscillation of the driveline; a fore and aft motion. This is coupled to the first rigid body torsional mode of the entire powertrain, referred to as shuffle, and usually in the range 2-8 Hz. Shuffle can be excited by any form of impact loading of the powertrain system as a load change reaction, for example in bump riding of front and rear suspensions, having different longitudinal compliance. The driver can, therefore, discern the longitudinal tip-in and back-out motion. It also occurs with sudden application of throttle in hill climbing or a sharp reduction in power along descending slopes (throttle tip-in and tip-out), particularly with large trucks and lorry drivelines, especially with manual transmissions. (Tip-in is the fast application of throttle demand and tip-out or back-out is an abrupt throttle release.)

A number of investigators have studied shuffle and driveline vibration. Rooke *et al.* [2] have shown that shuffle is minimised if an iteration method is used to find the optimum input pulse shape to minimise the response and to optimise the active system parameters. This process works only if backlash is absent. However, it should be noted that a certain amount of transmission backlash is required for gear meshing contacts, and that presence of backlash is in fact unavoidable. Hawthorn [3] used a lumped parameter model to investigate back-out shunt/shuffle. The model was then used in parameter optimisation studies and he concluded that:

- A ramp time of 0.3 seconds was critical to minimise shuffle excitation.
- There was a lash threshold. Lashes above this threshold cause further increases in the shuffle response. Below this threshold the shuffle was not affected.
- Stiff clutch springs with compliant driveshafts gave a low initial response but a slow decay rate, and conversely, soft clutch springs with stiff driveshafts gave a high initial response and a rapid decay rate.

Fothergill and Swierstra [4] have concluded that modification to clutch and driveshaft stiffness characteristics was effective in raising the modal frequency and, thus, reducing the subjective discomfort. The human body is particularly sensitive to longitudinal oscillations in the shuffle frequency range 2-8 Hz [5]. They suggested that the only way to avoid a compromise solution was to use a closed loop feedback control system which modified the excitation forces as soon as the shuffle response was detected.

Tobler [6] has stated that:

- Clonk and shuffle could not be improved by engine mounting.
- The first few degrees of lash produced the worst degradation.
- Lash reduction alone would not fix clonk, it was one ingredient in an overall clonk fix.
- Clutch slipping strategy eliminated shuffle and could improve clonk dependent on the tuning.
- Tip-in from a slight drive compared to tip-in from coast / over-run showed a significantly different response, indicating that lash was a predominant cause of clonk, and also worsened shuffle.

It is clear that these sources have used simplified lumped parameter models, some with non-linearities, to achieve a shuffle solution. But in this paper a modular approach has been developed for driveline simulation, in which any driveline could be simulated from a number of individual modules.

2 DESCRIPTION OF MODEL

The purpose in modelling any physical system is to predict and then to optimise the system performance, once sufficient confidence with the model has been established. The model should reasonably represent the characteristics of the system under investigation. A multi-body dynamic system may be represented by a series of inter-connected masses, springs and dampers. The masses are lumped at discrete stations and inter-connected to approximate the dynamic behaviour of the system. The system responds over time to a given input.

Figure 1a shows a schematic layout of a complete driveline for a conventional rear wheel drive vehicle. The inertia of the differential is reduced to the input shaft of the gearbox according to the second gear ratio of $n_1 = \frac{1}{2.08}$, using the following relationship:

$$J_3 = n_1^2 J_d \tag{1}$$

The inertia of the road wheels and tyres are also reduced to the input transmission shaft location, using the same second gear ratio and the differential ratio $n_2 = \frac{1}{4.11}$:

$$J_3 = n_1^2 n_2^2 J_d \tag{2}$$

One can use the same approach for the stiffness coefficients [7]. A branched driveline system reduces to common speeds by one to one gears, as shown in Fig. 1b. Figures 1c and 1d show the equivalent driveline system and the driveline model respectively.
Since there are a large number of different driveline configurations, resulting from in-line and transverse engines, front and rear wheel drives, direct and indirect gearboxes, it is decided to adopt a modular approach in the simulation study. The concept behind this approach is that any driveline can be simulated from a number of individual modules (e.g. clutch, gearbox, ..). In the modular approach to simulation of vehicle driveline, the system is broken down into a series of simple items or modules. The equations of motion for each module can be developed, according to the formulation method employed. This enables rapid generation of a simulation program for an entire driveline system.

This modular approach in system development allows a vehicle driveline to be represented by a mathematical model and the effects of changing system parameters to be ascertained. The first step in any simulation study is to validate the model. The validation can result from either experimental recording of noise and vibration or theoretically using a simplified analytical approach (e.g. natural frequency investigation). In the case, where a problem exists in a production vehicle the former is possible, but where the program is being used for design / prediction of the driveline characteristics, the latter approach is necessary.

In this paper, both validation procedures have been adopted. Since in the experimental rig the rear wheels were fixed to the ground, the driveline system model was considered under the same condition (Figs. 1c and 1d).

The simulation model consists of individual modules which represent the engine flywheel, clutch pressure plate, clutch disk, gearbox, differential and the rear axle. In this model the gearbox is represented as a single module, incorporating the gearbox case inertia which also allows for torque change reactions. Similarly, the differential unit incorporates a differential/rear axle case inertia in two perpendicular planes, together with a torque reaction. The following assumptions have been made, when obtaining the equations of motion:

- The driveline system is considered after clutch engagement.
- Inertia of other parts than those of J_1, J_2, J_3 are negligible.
- The gearbox is considered as a lumped system with mass moment of inertia J_g.

The inertia of the gearbox, J_2, and the inertia of the differential, J_3 were referred to the clutch. The governing equations of motion for the driveline system model shown in Fig. 1d are:

$$T_i = J_i\ddot{\theta}_i + B_i\dot{\theta}_i + T_{i+1} + T_{ci}[\pm \mathrm{sgn}(\dot{\theta}_i)] \tag{3}$$

where: $i=1,2,3$

The transmitted torque between the various sub-systems is found from the stiffness conditions that exist between the pre-loaded parts of the driveline in contact with each other. The clutch stiffness characteristic is modelled as a linear spring (without a pre-damper) and then as a non-linear spring as shown in Fig. 2, which has a pre-damper with hardening characteristics. Therefore, the transmission torque for each part can be expressed as:

$$T_i = k_i(\theta_{i-1} - \theta_i) + c_i(\dot{\theta}_{i-1} - \dot{\theta}_i) \tag{4}$$

where: $i=2,3,4$ and $\theta_4 = \dot{\theta}_4 = 0$

Taking Laplace Transforms of equations (3) and assuming that all initial conditions are zero gives:

$$\theta_i = \frac{1}{s(J_i s + B_i)}\{T_i - T_{i+1} - T_{ci}[\pm \mathrm{sgn}(\theta_i s)]\} \quad (5)$$

where: $i=1,2,3$

From the driveline system model, shown in Fig. 1, together with equations (3) to (5) a complete simulation model of the driveline system can be developed, as shown in Fig. 2.This simulation model has been constructed in MATLAB using the SIMULINK toolbox. The SIMULINK toolbox enables a model to be constructed, using existing block functions held in its library. Each block contains elements of inertia, damping, stiffness, friction and backlash. The parameter values may be interactively selected and revised. Values input in the driveline model were mainly derived from experimentally measured data or have been calculated, where in this case Table 1 gives the model's constant values.

3 TRANSFER MATRIX METHOD (TMM)

The transfer matrix method is applied to the Holzer-type problems, where a large system can be broken down to a number of inter-connected sub-systems with stiffness and damping properties. The formulation is in terms of a *state vector*, which is a column matrix of the displacements and internal forces; the *point inertial matrix*, which contains the physical properties of the sub-system, and a *field matrix*, which describes the elastic properties of the subsystem. In terms of these quantities, the calculations are made to proceed from one end of the system to the other, the natural frequencies being established by satisfying on appropriate set of boundary conditions.

Figure 3 shows a part of torsional system with one of the sub-systems isolated. The nth section consist of the disk J_n with displacement θ_n and the spring of stiffness k_n, whose ends have displacements θ_n and θ_{n+1}. When necessary the quantities to the left and right of an element are designated the subscripts L and R.

For the disk J_n, the equation of motion is:

$$T_n^R - T_n^L = J_n \ddot{\theta}_n \quad (6)$$

which for harmonic motion becomes:

$$T_n^R = -J\omega^2 \theta_n + T_n^L \quad (7)$$

Since the displacements on either side of J_n are the same, one arrives at the identity

$$\theta_n = \theta_n^R = \theta_n^L \quad (8)$$

Equations (7) and (8) can now be assembled into a single matrix equation

$$\begin{Bmatrix} \theta \\ T \end{Bmatrix}_n^R = \begin{bmatrix} 1 & 0 \\ -\omega^2 J & 1 \end{bmatrix}_n \begin{Bmatrix} \theta \\ T \end{Bmatrix}_n^L \quad (9)$$

where $\begin{Bmatrix} \theta \\ T \end{Bmatrix}$ is the state vector and the square matrix is the point inertial matrix.

Next the spring k_n can be examined whose end torques are equal:

$$T_{n-1}^R = T_n^L \tag{10}$$

The spring torques are related to the spring modulus k_n by the:

$$\theta_n^L - \theta_{n-1}^R = \frac{T_{n-1}^R}{k_n} \tag{11}$$

Equations (10) and (11) are now assembled in the matrix form as:

$$\begin{Bmatrix}\theta \\ T\end{Bmatrix}_n^L = \begin{bmatrix}1 & \dfrac{1}{k} \\ 0 & 1\end{bmatrix}\begin{Bmatrix}\theta \\ T\end{Bmatrix}_{n-1}^R \tag{12}$$

where the square matrix above is the field matrix.

The quantities at a station n can now be related in terms of the quantities at the station n-1 by substituting Eq. (12) into Eq. (9) :

$$\begin{Bmatrix}\theta \\ T\end{Bmatrix}_n^R = \begin{bmatrix}1 & 0 \\ -\omega^2 J & 1\end{bmatrix}\begin{bmatrix}1 & \dfrac{1}{k} \\ 0 & 1\end{bmatrix}\begin{Bmatrix}\theta \\ T\end{Bmatrix}_{n-1}^R = \begin{bmatrix}1 & \dfrac{1}{k} \\ -\omega^2 J & (1-\dfrac{\omega^2 J}{k})\end{bmatrix}\begin{Bmatrix}\theta \\ T\end{Bmatrix}_{n-1}^R \tag{13}$$

Since the state vector at n-1 is transferred to the state vector at n through the above square matrix, it is called the *transfer matrix* for section n. With known values of the state vector at station 1 and a chosen value of ω^2, it is possible to progressively compute the state vectors to the last station n. Depending on the boundary conditions, either θ_n or T_n can be plotted as a function of ω^2; the natural frequencies of the system are established when the boundary conditions are satisfied. Problems of this type, where only one displacement is associated with each disk, are called *Holzer-type* problems.

When damping is included, the form of the transfer matrix is not altered, but the mass and stiffness elements become complex quantities. This can be easily shown, see Fig. 4.
The torque equation for disk n is:

$$-\omega^2 J_n \theta_n = T_n^R - T_n^L - i\omega B_n \theta_n$$

or:

$$(i\omega B_n - \omega^2 J_n)\theta_n = T_n^R - T_n^L \tag{14}$$

The elastic equation for the nth shaft is

$$T_n^L = k_n(\theta_n - \theta_{n-1}) + i\omega c_n(\theta_n - \theta_{n-1}) = (k_n + i\omega c_n)(\theta_n - \theta_{n-1}) \tag{15}$$

Thus, the point inertial and the field matrices for the damped system become:

$$\begin{Bmatrix}\theta \\ T\end{Bmatrix}_n^R = \begin{bmatrix}1 & 0 \\ (i\omega B - \omega^2 J) & 1\end{bmatrix}_n\begin{Bmatrix}\theta \\ T\end{Bmatrix}_n^L \tag{16}$$

$$\begin{Bmatrix}\theta \\ T\end{Bmatrix}_n^L = \begin{bmatrix}1 & \dfrac{1}{(k+i\omega c)} \\ 0 & 1\end{bmatrix}_n\begin{Bmatrix}\theta \\ T\end{Bmatrix}_{n-1}^R \tag{17}$$

which are identical to the undamped case, except for the mass and stiffness elements, these are now complex quantities.

A Matlab M-file for the Holzer-type problem was written for the torsional system in Fig. 4. The program was written in such a manner that, by changing the data, it is applicable to any other torsional system.
Equations (16) and (17) can be expressed as:

$$\theta_{I,n+1} = \theta_{I,n+1} + T_{I,n}{}^{R}/(k_{n+1} + i\omega_I c_{n+1}) \tag{18}$$

$$T_{I,n+1}{}^{L} = T_{I,n}{}^{R} \tag{19}$$

$$T_{I,n+1}{}^{R} = T_{I,n+1}{}^{L} + (i\omega_I B_{n+1} - \lambda_I J_{n+1})\,\theta_{I,n+1} \tag{20}$$

$\lambda_I = \omega_I^2$
T^L = torque to the left of the disk
T^R = torque to the right of the disk

The index n defines the position along the structure and index I defines the frequency to be used. The three above stated equations are to be solved for θ and T at each point N of the structure and for various values of ω. At the natural frequencies, θ must be zero at the fixed end. Starting with the boundary conditions:

$$\theta_{I,1} = 1 \tag{21}$$

$$T_{I,1}{}^{R} = T_{I,1}{}^{L} + (i\omega_I B_1 - \lambda_I J_1)\,\theta_{I,1} \tag{22}$$

The equations (18), (19), (20) are computed for each n (position in structure), keeping I (or frequency) fixed.

4 EXPERIMENTAL INVESTIGATION OF DRIVELINE NVH

The rig, shown in Figure 5, simulates a rear wheel drive light truck's drivetrain system, consisting of flywheel, clutch, transmission, a two piece driveshaft assembly, differential and rear axle assembly, comprising rear axle half-shafts, brake assembly and road wheels. The rear wheels were fixed to the ground. A preload torque was applied to the flywheel via a low inertial disc brake system as stored energy, acting through the torsional clutch springs in the system. This pre-load was instantaneously released (the action amounting to the application of a Dirac type function) and was reacted at the fixed rear wheel assemblies. Accelerometer pick-ups were located along the driveline, as shown in Fig. 5, to monitor the response of the system components by subsequent spectral analysis. Details relating to the to the experimental rig and design of experiments are provided by Menday [8] and Menday *et al.* [9,10].

Although the driver requires a responsive vehicle, the impulse may excite an unwanted low frequency longitudinal mode of vibration known as shuffle. Each cycle of the shuffle response may generate clonk noise [11].

Time histories at 9 accelerometer pick-ups along the driveline (as shown in Fig. 5) were recorded for 16 different configurations at the test rig. Figure 6 shows the time response of one of the accelerometer pick-up. Input applied torque was shown in Fig. 7. The transmission flange shows the highest pick-up response in the driveline.

Since the impacting energies at any inertia location will be partly radiated as airborne noise due to the rapid velocity changes of the moving bodies, and partly dissipated into the nearby resonant structure, it was necessary to consider the frequency domain response at a structural pickup point. Figure 8 illustrates the typical spectrum of the measured output response at the transmission. The first dominant frequency was shown in at the frequency of 3.07 Hz, corresponding to the first torsional mode of driveline vibration (first natural frequency of the system).

5 SIMULATION RESULTS AND DISCUSSION

The main objective of this analysis was to study the torsional vibration of the driveline system. Two different methods of analysis were undertaken (integration of the equations of motion by the SMULINK model and the transfer matrix method). The system has been considered as a linear system. The effect of non-linear source in the driveline system such as backlash, non-linear spring stiffness, coulomb friction, constant velocity joint angularity of the propeller shaft on the system torsional vibrations has also been investigated.

The values in Table 1 are substituted into the appropriate simulation model of Fig. 2. Figures 9(a) and 9(b) show the variations of angular velocity at different location. The impulse torque, which has been applied to the flywheel, is shown in Fig. (7). This torque was recorded from a test rig. Figure 9(c) and 9(d) illustrate the corresponding variations in angular acceleration at the gearbox and differential respectively.

A DFT (Discrete Fourier Transform) analysis yields the spectrum of Fig 10 up to the maximum frequency of 100 Hz and with a resolution of 0.5 Hz [12]. The spectrum has a strong component at 3.34 Hz. In fact, the first natural frequency of the driveline system is 3.34 Hz (the shuffle response of the driveline system), as corroborated by the experimental results.

To obtain the spectrum of vibration, using the formulation given by the transfer matrix method, a MATLAB m-file has been developed according to Eqs. (18) to (22). The computer results are shown in Fig. (11). The figure shows the variation of angle θ_4 *with* frequency ω. The natural frequencies of the system correspond to frequencies where θ_4 becomes zero. It can be seen from Fig. (11) that they are at $\omega_1=3.34$ *Hz and* $\omega_2=157$ *Hz*. These results were found to agree with the SIMULINK model results, described earlier.

Although the mathematical model was considered as a linear system, the experimental results were found to agree with mathematical models (SIMULINK model and transfer matrix model). As one can observe in the spectrum of transmission, the first dominant frequency is at 3.07 Hz. Therefore, these models are valid and capable of modification and extension to investigate more complex systems with different loading conditions and characteristics.

6 CONCLUSION

The numerical models developed based upon TMM or Newton-Euler equations of motion predict vehicular driveline shuffle response in an adequate manner. The next stage of research should concentrate upon the inclusion of the FE models to represent component flexibility, thus leading to investigation of structural-acoustic modes, referred to as clonk.

ACKNOWLEDGEMENTS

The authors would like to express their gratitude to Ford Motor Company for the technical support extended to this research paper.

REFERENCES

1. . Rahnejat, H., "Multi-Body Dynamics Vehicle, Machines, and Mechanisms", Published by Society of Automotive Engineering, Inc., 1998.
2. Rooke, G., Chan, E. A., and Crossley, P. R., "Computer Modelling of a Vehicle Powertrain for Driveability Development" , IMechE report C462/31/035, 1995, pp. 1-9.
3. Hawthorn, J., "A Mathematical Investigation of Driveability", IMechE report C420/003, 1995.
4. Fothergil, D. J., Swierstra, N., "The Application of Non-linear Displacements Modelling Techniques to an Automotive Driveline for the Investigation of Shunt", VDI Berichte, Nr. 1007,1992, pp 163-179.
5. Chikamori, S., Yoshikawa, N. "Analysis of drive train noise and vibration", Int. J. of Vehicle Design, Vol. 2, No. 4, 1981, pp 408-427.
6. Tobler, W., "Modelling of ATX Powertrains; Application to Clonk and Suffle", Ford Motor Company Internal Report, 1983.
7. Thomson, W. T., "Theory of Vibration with application", Prentice-Hall, Inc., 1972, p 466.
8. Menday, M. T., "Torsional impact in an automotive vehicle driveline", MSc Thesis, University of Bradford, 1997, UK.
9. Menday, M. T., Ebrahimi, M., "The Application of Dynamic Modelling Using Parametric Design Methods for an Automotive Driveline System", Multi-body Dynamics Monitoring and Simulation Techniques Conference, MEP Publication, 1997, pp 67-79.
10. Menday, M., Rahnejat, H., and Ebrahimi, M., "Clonk: High Frequency Onamatapaeic Response Of vehicular Drivelines, Proc. IMechE, Part D, J. Automobile Eng., 1998.
11. Krenz, R. A., "Vehicle Response to Throttle Tip-in/Tip-out", Proc. of Surface Vehicle Noise and Vibration Conference, Michigan, 15-17 May, 1985, SAE 850967, pp 45-51.
12. Harman, T. L., Dabney, J., Richert, N., "Advanced Engineering Mathematics Using MATLAB V.4", PWS Publishing Company, 1997, p 645.

Stiffness Nm/rad		Moment of inertia $Kg.m^2$		Equivalent system damping Nms/Rad	
Parameter	Value	Parameter	Value	Parameter	Value
Clutch spring $=k_c$	527.12	Flywheel & pressure plate assembly = J_f	0.3076	Flywheel damping = B_1	0.2
Propeller spring = k_p	10714.3	Transmission assembly $=J_g$	0.003	Gearbox damping $=B_2$	2
Axle spring = k_a	13000	Differential assembly = J_d	0.0265	Differential damping = B_3	1
				Clutch damping = c_1	10
				Propeller damping = c_2	0
				Axle damping = c_3	0

Clutch characteristic with predamper

Angle, deg	-9.5	-2.5	0	8.5	27
Torque, Nm	-130	-0.7	0	2.6	290

Clutch characteristic without predamper

Angle, deg	-9.5	0	27
Torque, Nm	-130	0	290

Table 1 Model constant value

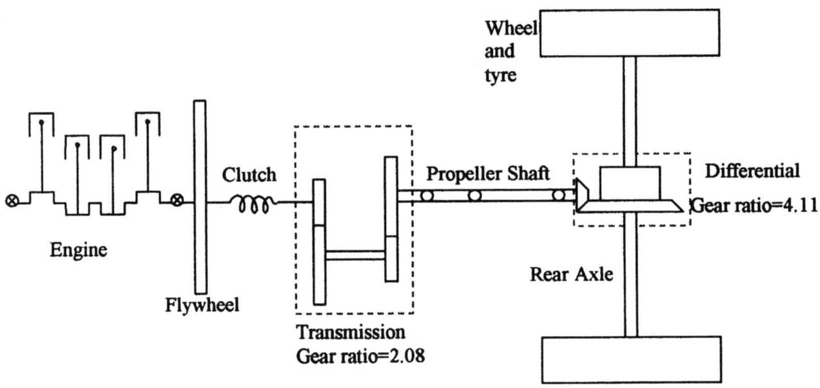

Fig. 1a Schematic layout of a complete driveline for a conventional rear wheel drive vehicle

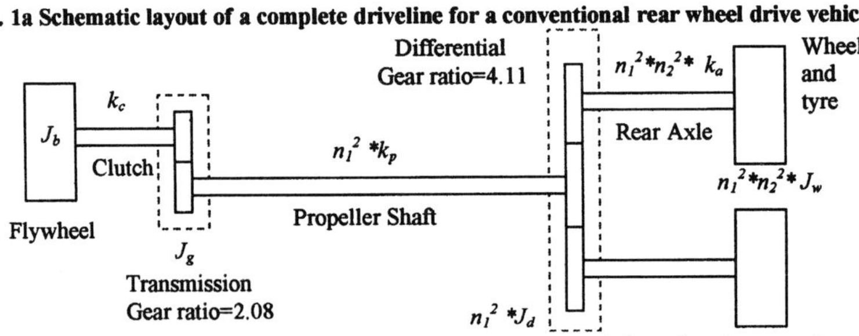

Fig. 1b Branched driveline system reduced to common speeds by a 1 to 1 gear ratio.

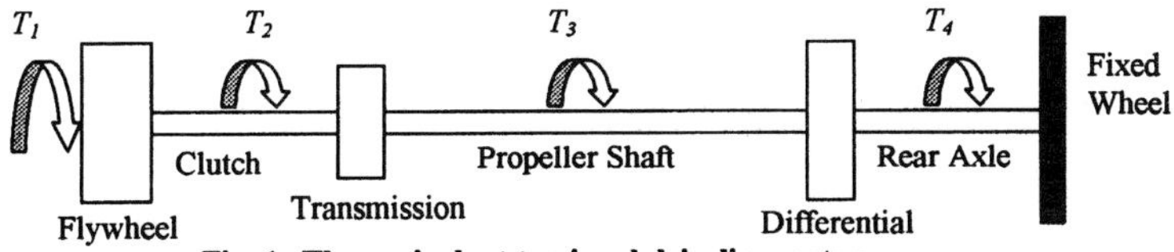

Fig. 1c The equivalent torsional driveline system

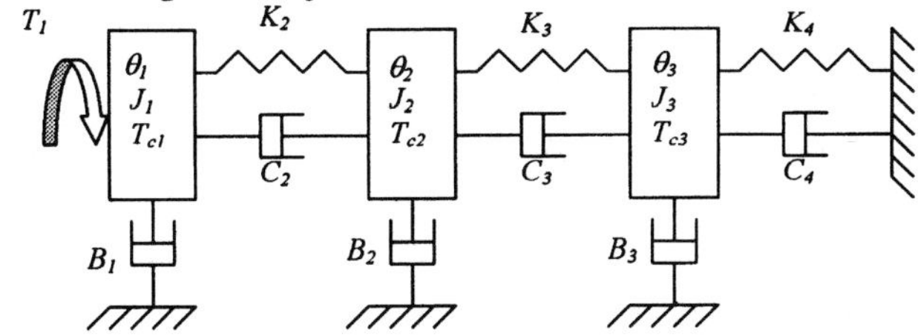

Fig. 1d The driveline system model

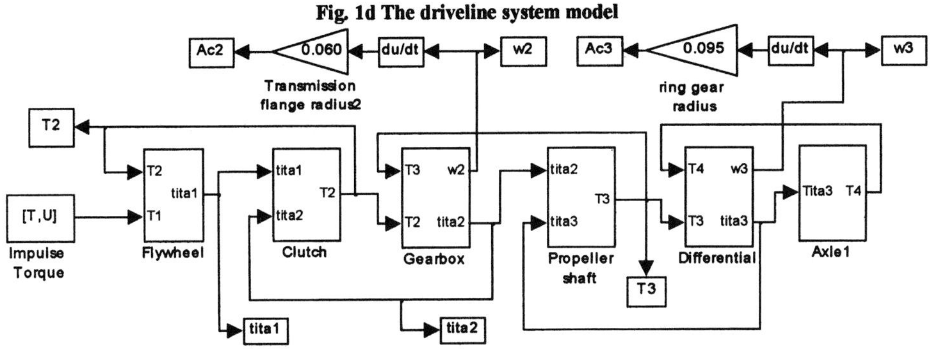

Fig. 2a Simulation model of the driveline system

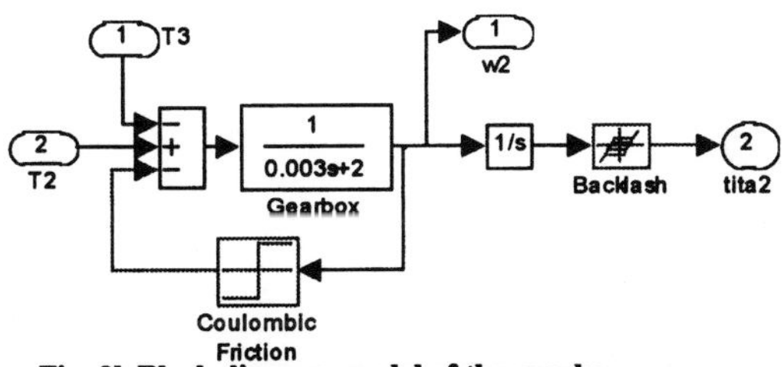

Fig. 2b Block diagram model of the gearbox

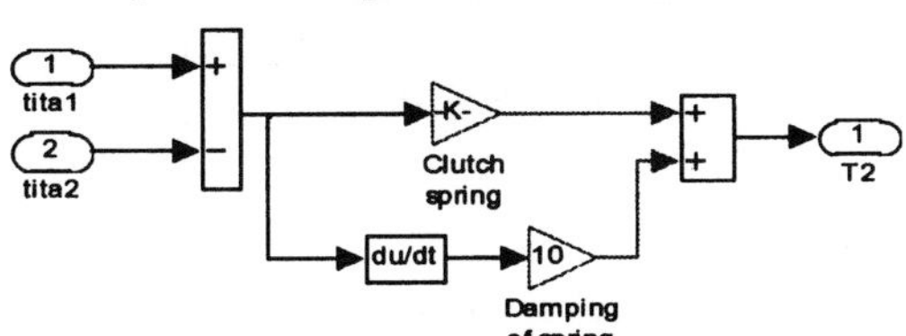

Fig. 2c Block diagram model of the clutch

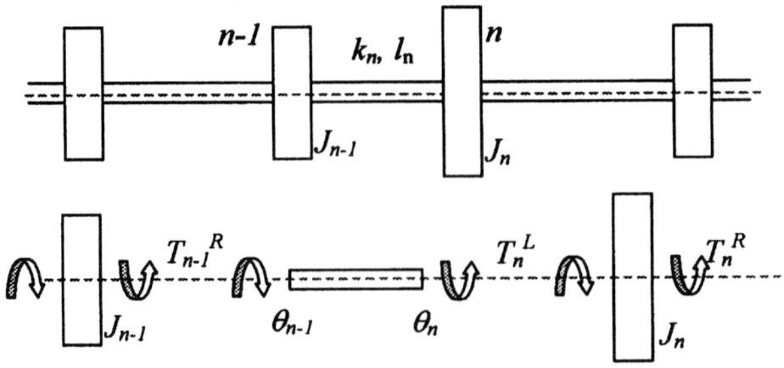

Fig. 3 Torsional system with the subsections

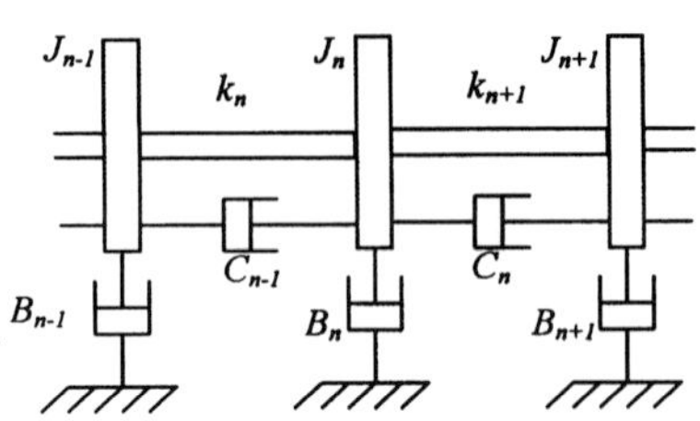

Fig. 4 Torsional system with damping.

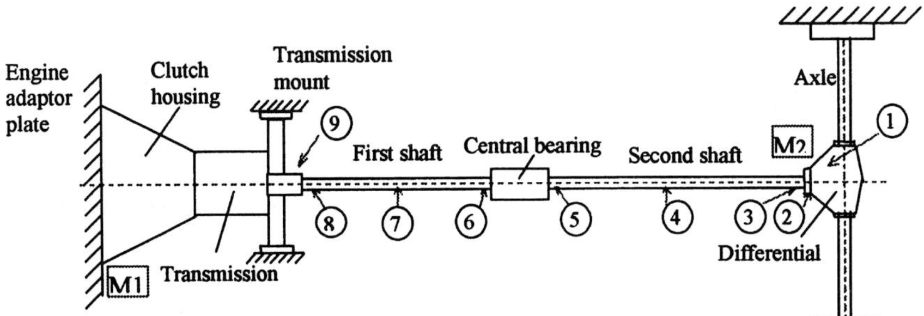

Location info:

#1 Ring gear	#2 first driveshaft, centre bearing location
#2 Axle diff. Flange	#7 first driveshaft, centre location
#3 second driveshaft, diff. End location	#8 first driveshaft, transmission end
#4 second driveshaft, centre location	#9 transmission flange
#5 second driveshaft, centre bearing location	#M1 microphone at transmission bell housing

#M2 microphone at axle differential

Fig. 5 Schematic diagram of the driveline experimental rig.

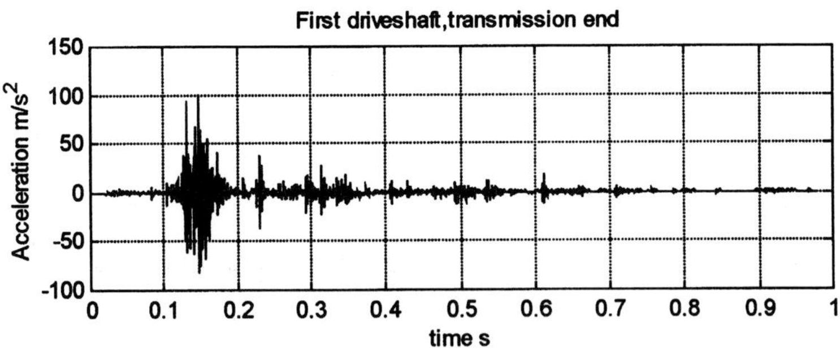

Fig. 6 Measured output from the test rig

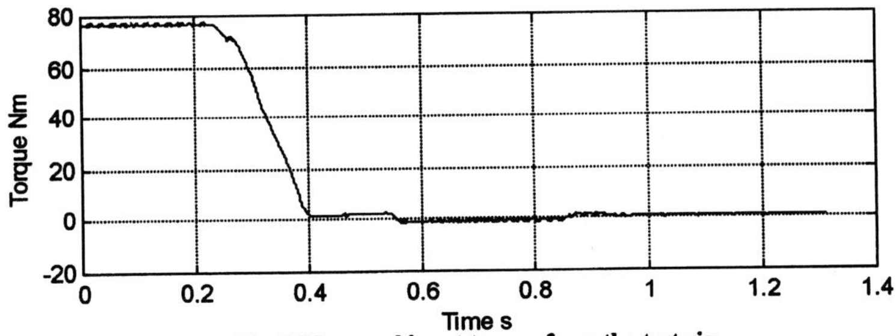

Fig. 7 Measured input torque from the test rig

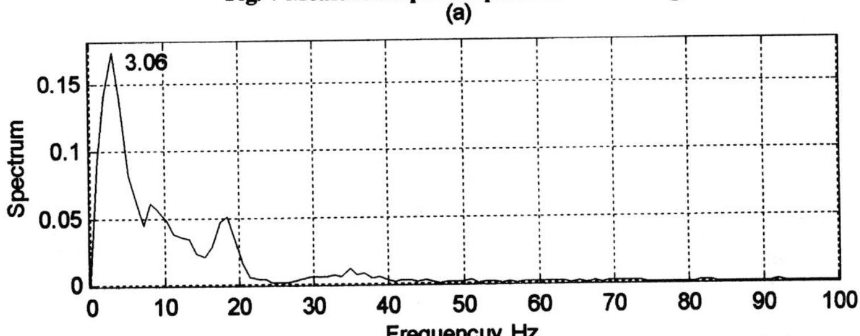

Fig. 8 Typical spectrum of the measured output response at transmission

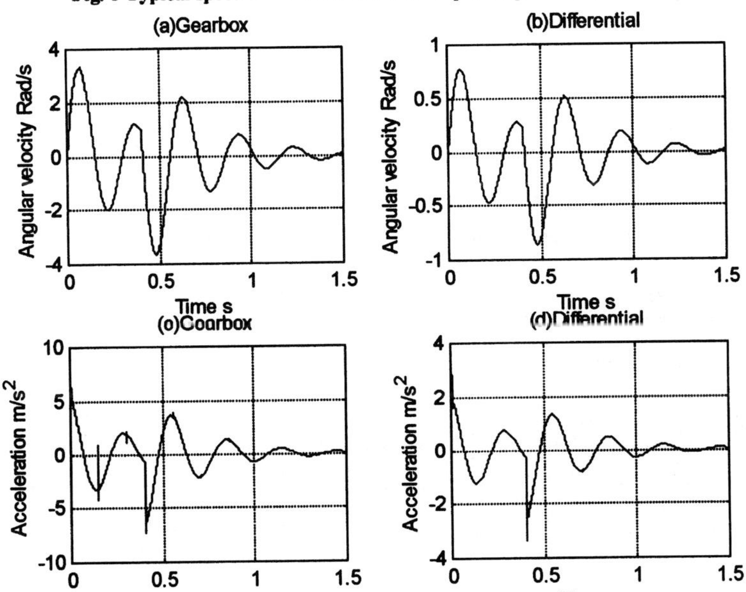

Fig. 9 Simulation results of SIMULINK model of driveline

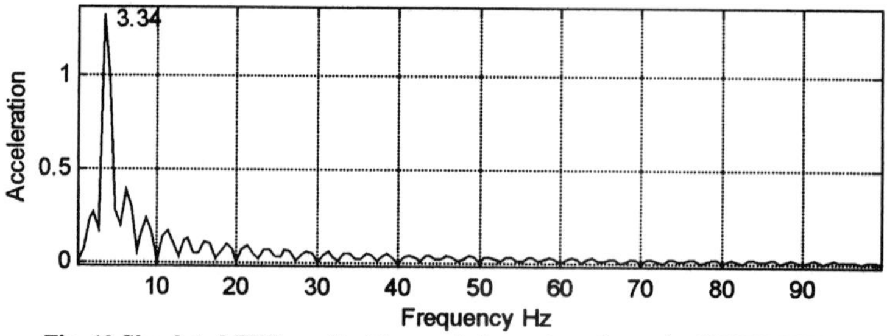

Fig. 10 Simulated FFT result at transmission output from the SIMULINK model

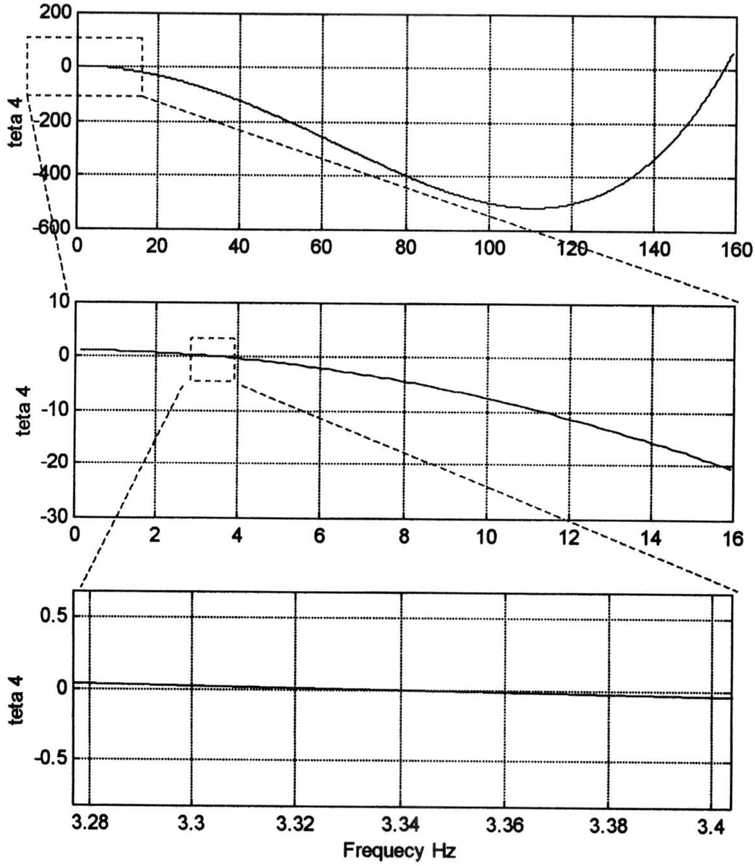

Fig. 11 Natural frequencies correspond to $\theta_4=0$

Simulation of driveline actuation cables to improve cable design

C BREHERET
Dura Automotive Systems Limited, University of Central England, Birmingham, UK
R CORNISH, M DANIELS, and **G A ATKINSON**
Faculty of Engineering and Computer Technology, University of Central England, Birmingham, UK

SYNOPSIS

This paper describes research work relating to the prediction of noise and vibration produced by driveline actuation cables. The cables are used extensively in front wheel drive vehicles. The noise generation is produced by impact between the inner and outer members of the cables which are similar to Bowden cables. The noise and vibration has been studied both practically and via simulations by the authors who report numerical simulation methods in this paper. The paper will be of interest to both NVH engineers and Multi-body Dynamicists.

1. INTRODUCTION

The numerical prediction of cable vibration performance in vehicles has been identified as a beneficial target in automotive actuation cable design. Work is ongoing by the authors combining dynamic test, statistical experimentation and numerical modelling (Ref. 1).

The problem has not been tackled very much in the literature, due to the obvious difficulty in deriving equations of motion and boundary conditions. Most work to date has concentrated on experimental investigations. Work in Ford has been reported in which clutch noise was transmitted by the clutch cable to the clutch pedal and driver, and fixed using an attached inert mass (Ref. 2 and 3). Statistical experimental techniques played a key role in that development.

The authors believe that the time is now appropriate to use automatic dynamic solvers such as ADAMS (Ref. 4) to predict the cable vibration phenomena. There are clear advantages in being able to model the geometry and material properties in cable dynamics from the outset. The model can be used for non-linear dynamic simulation including friction and clearance between components.

The work reported here answers questions about what the cable model should best be like, its constituents and the number of degrees of freedom required. The research is focussed on determining the benefits that can be derived from a simplified discrete model versus a general flexible approach.

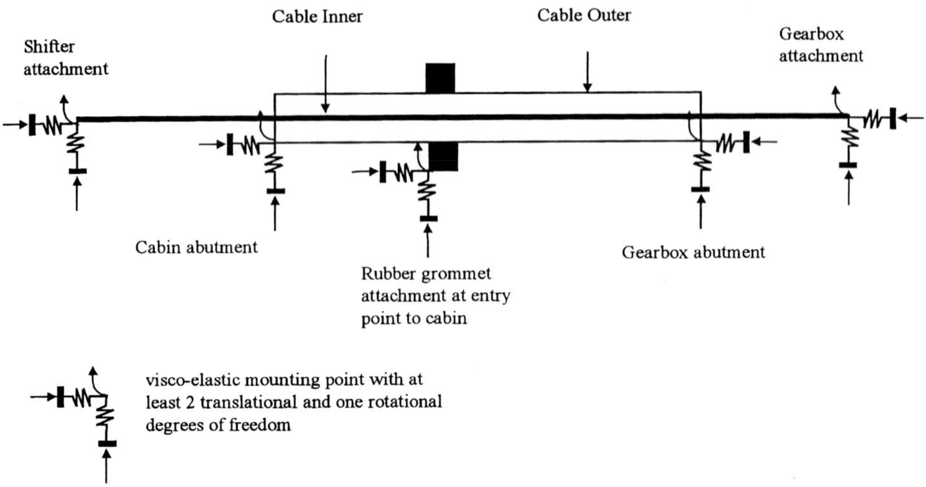

Fig.1: Schematic diagram of vibration transfer mechanisms for a control cable

Figure 1 shows the complexity of the environment of the mechanical actuation cables. Experimental tests have been made using single input excitation. The authors believe that correlation with a single input numerical model is the first step to simulate this complex environment.

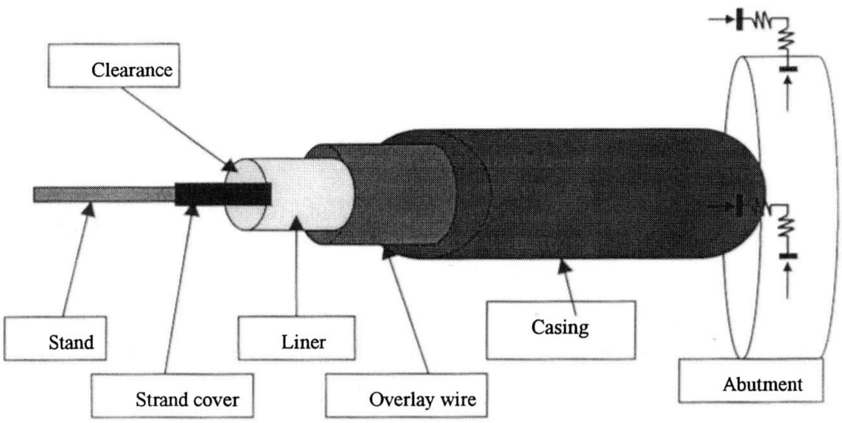

Fig.2: Construction of the actuation cable.

2. METHOD OF MODEL DEVELOPMENT

The method employed was the empirical development of a model of the actuation cable in ADAMS/View. The initial idea was to use a discrete chain of links to model the cable with stiffness and damping provided by the joints between the links. This model is referred to as the Basic Link Model.

The next development was to add contact forces between the elements of two chains representing the clearance of the inner and outer cables which can produce impact noise at certain frequencies.

The third development, for purposes of comparison, was to model the inner and outer members as pure classical flexible bodies. This third model was used to find out what real phenomena had been lost by the simplifications made in the earlier models.

2.1 Basic Link Model

For initial modelling work, the cable was simplified to a two dimensional problem. The inner and the outer cables were modelled as discrete chains of pin-jointed links. Each pin-joint was given a torsion spring and damper. The mass, inertia, stiffness and damping of the real cable components can be built into the model at the relevant discrete elements.

The model simplifies the true complexity of the cable and leaves out the inluence of strand rotation (helical overlay) which is an important parameter in the load/no load push-pull response. The model is currently used at the two extremes of tolerance specified e.g. go no go test.

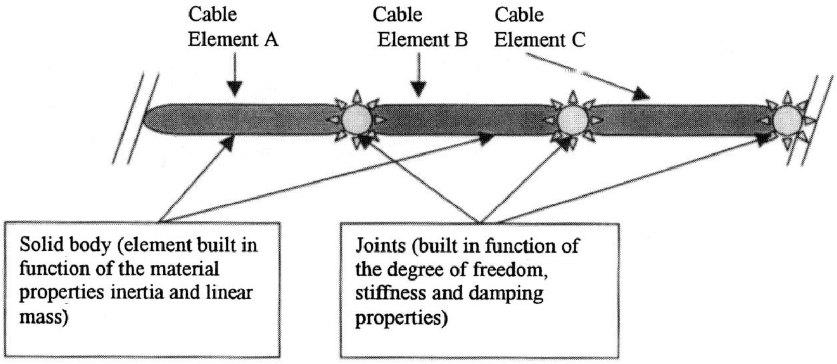

Fig.3: Construction basic link model

2.2 Contact Force Model

The contact forces were applied between two basic link models representing the inner and outer cables. The two chains were limited in their relative displacement by the construction of the ADAMS 'bistop' contact force. This contact force is a built in function of the material restitution contact properties. The force is applied between each of the chain elements to simulate the real life clearance of a mechanical actuation cable. The forces are calculated on the surface of the element centres within a defined penetration and plane surface around the application point. Figure 4 shows the two basic link models assembled together with contact forces.

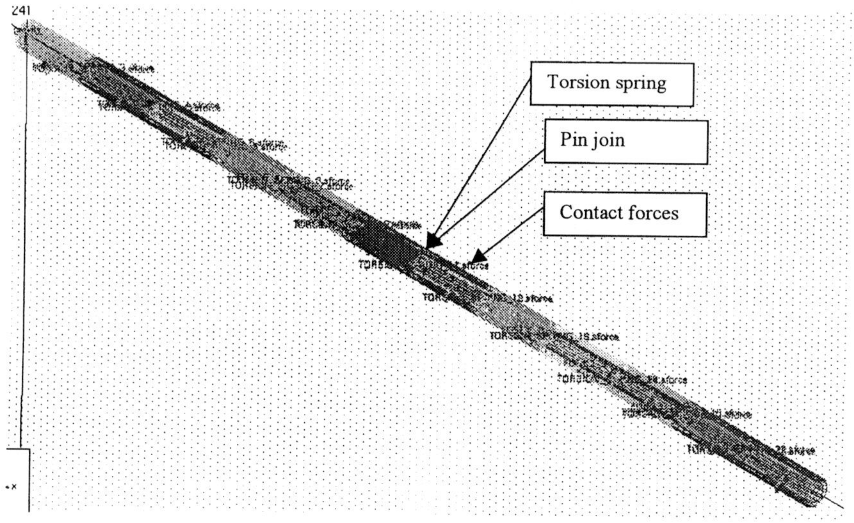

Fig.4: contact force model with 24 degrees of freedom

2.3 Flexible Body Model

The construction of the flexible model is based on the material properties (young modulus, damping ratio at different frequency, inertia matrix e.t.c). The flexible links enable incorporation of components created in Finite Element Analysis (FEA). The flexibility has a considerable impact on the load distribution that influences the mechanical performances such as the push pull losses of the mechanical actuation cable.

The starting flexible model has no tension in it and had pin joints at the end of the flexible element. This allowed correlation with real life transverse mode shape as show with stroboscope.

3. RESULTS

3.1 Basic Link Model

The basic link model figure can be used to confirm the displacement amplitude of the cable. It has been used at different tensions. This model has the advantage of computing rapidly because it uses a limited number of degrees of freedom.

3.2 Contact Force Model

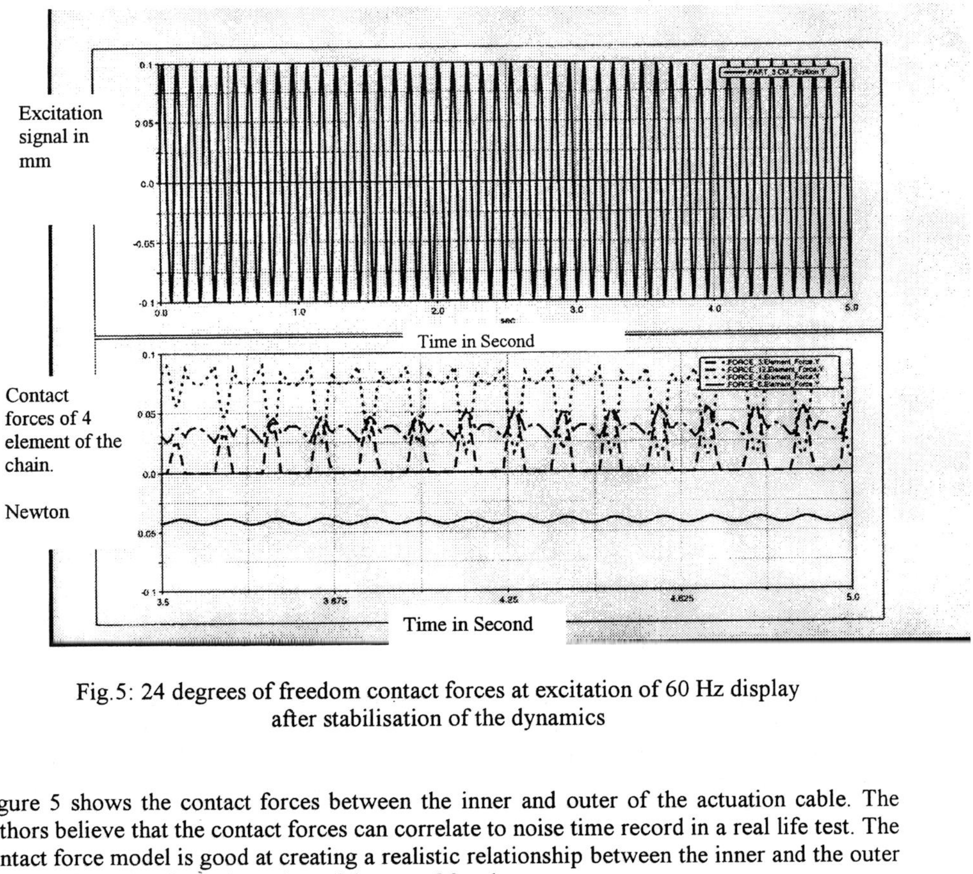

Fig.5: 24 degrees of freedom contact forces at excitation of 60 Hz display after stabilisation of the dynamics

Figure 5 shows the contact forces between the inner and outer of the actuation cable. The authors believe that the contact forces can correlate to noise time record in a real life test. The contact force model is good at creating a realistic relationship between the inner and the outer of the cable with a limited number of degree of freedom.

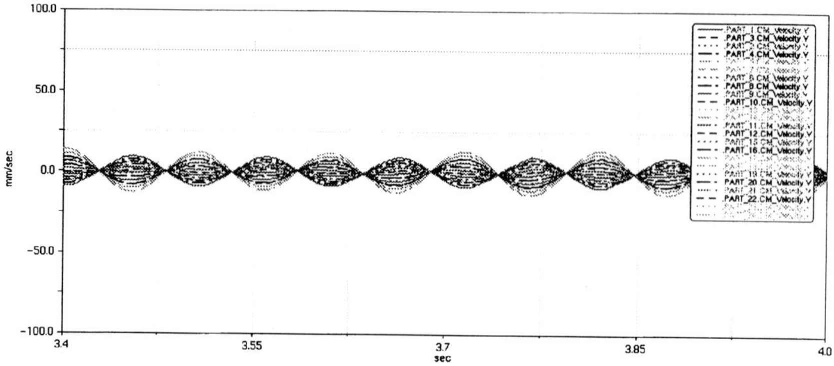

Fig.6: Mode shape predicted for contact force model (24 degree of freedom)

Figure 6 shows the phase relation between the different element of the contact force model.

3.3 Flexible Body Model

This model has been built with pin-joints at the end of the flexible element. It has no tension and only gravity forces are present in order to have an easy correlation with real life test using a sweep sine test of the inner while observing the transverse mode shape using a stroboscope.

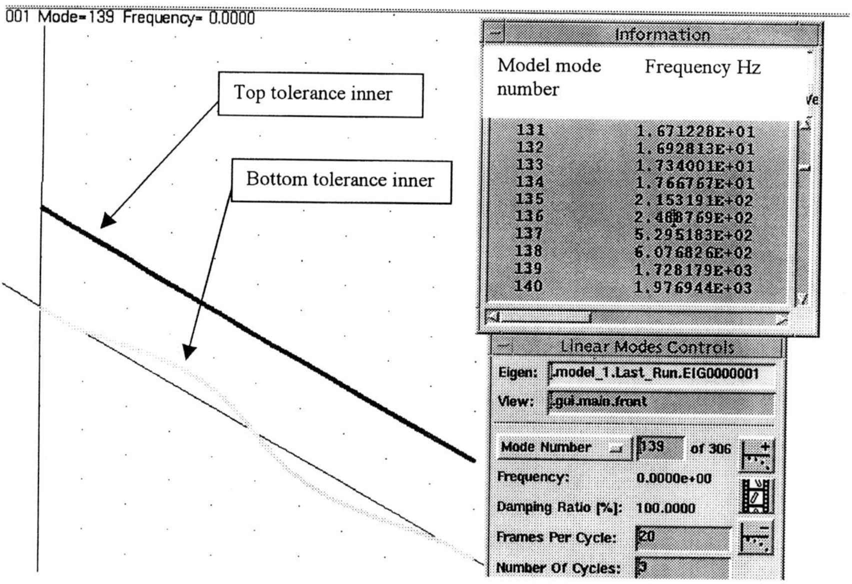

Fig.7: Bottom tolerance inner member using flexible element, natural frequency second mode at 1.728 kHz

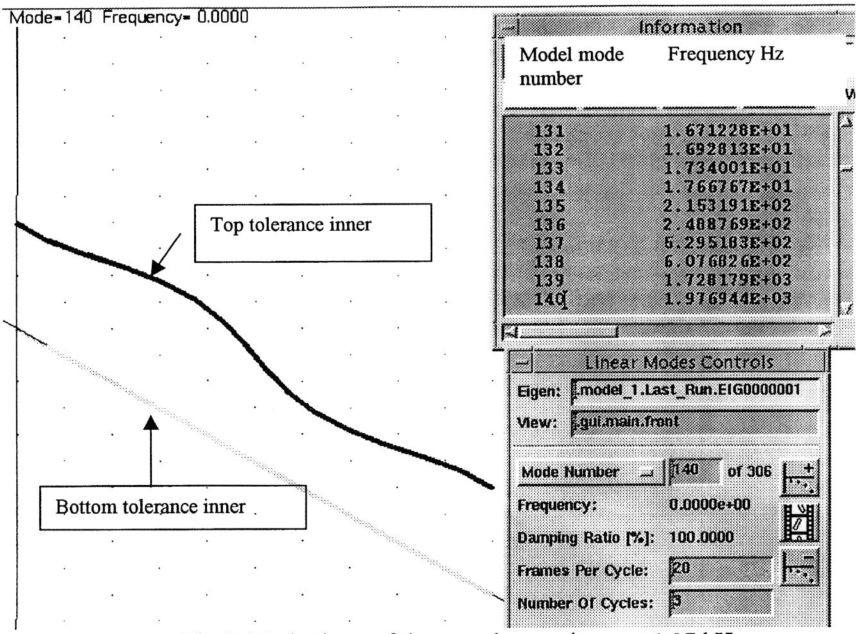

Fig.8: Mode shape of the top tolerance inner at 1.97 kHz.

The figures 7 and 8 show results of the model of a cable at the two-tolerance boundary. Figure 7 and 8 help confirm the model by reference to the pure sine input vibration frequencies which produce cable noise in production cables. The model predict the effect of increasing the gap between the inner and outer cable members.

4. CONCLUSIONS

The complexity of the cable construction such as the strand lay lengths and its influence on the strand rotation during its tension are parameters, which could only be assumed in the model parameter.

The construction of the basic link model allowed a better understanding of the displacement and has the capacity to limit easily the problem into a two dimensional problem. On the contrary the flexible model body will compute all the degrees of freedom of the element increasing their computational time and making interpretation of the results more difficult.

The modelling of the contact forces are an ongoing process: a limitation on the flexibility has been found because of the nature of the construction of the use contact forces. The model currently specified 'line of sight' contact forces between adjacent elements of the inner and outer. However, if actuation of the cable is occurs these elements are no longer adjacent, and the model is no longer valid.

Fig.5 show that the introduction of contact forces distorts the vibration time history sufficiently to produce extra vibration frequencies and harmonics such as are detected in real cables. The authors believe that in the future a correlation between the spectral content of the model forces and the spectral content of experimental noise will be obtained.

The importance of the variances of the production recorded on the sample range during experimental work can be confirmed using the solid body model or the flexible element model.

The flexible element model has the major advantage of being able to compute the mode shape and natural frequency of inner and outer without having to compute a swept sine test as in the basic link model. The flexible element model also has a certain advantage of visualisation of the mode shapes.

In the study of the modelling of the actuation cable the authors believe that the flexible element will have to be used in future modelling work, even if the computation time due to the number of degree of freedom of the flexible element is high. The introduction of the contact forces will be restricted to the non-actuated situation (no translation of the inner in the liner).

5. ACKNOWLEDGEMENTS

The authors would like to thank the Directors of DURA Automotive Systems Ltd. And the Teaching Company Directorate of the DTI for sponsoring this work and allowing its publication.

6. REFERENCES

1. 'Modelling Noise and Vibration of Automotive Manual Transmission Gearshift Cables', C.Breheret, R.H. Cornish, C.Evans, European Conference on vehicle noise and vibration 2000, 10-12 May 2000, IMechE HQ London.
2. Combining design of experiments and modelling techniques to resolve complex clutch pedal noise and vibration problems. I.Mech.E paper C521/012/98 by P.Kelly, H. Rahnejat, J.W. Niermann and B. Hagerodt
3. Clutch Pedal Dynamics Noise and vibration investigation using Taguchi Methods, 6th Aachen Colloqium, Automotive and Vehicle Technology, Germany, 22nd October 1997, by P. Kelly, J. W. Biermann, B. Hagerodt.
4. ADAMS/View Mechanical Dynamics Inc., Ann Abor, Michigan 48105

Vibration Monitoring and Modelling

Vibration and grinding

S J DREW and **B J STONE**
Department of Mechanical and Materials Engineering, The University of Western Australia, Australia
M A MANNAN
Department of Mechanical and Production Engineering, National University of Singapore, Singapore

SYNOPSIS

Vibration in grinding is a serious problem and this is particularly true of the unstable vibration called chatter. In addition vibration resulting from out-of-balance and out-of-round workpieces and grinding wheels can also be significant. This paper describes a model of grinding that includes more effects than have previously been described in the literature. The model indicates that the torsional characteristics of the workpiece drive system may either improve chatter performance or make it worse. It is also possible to investigate how long it takes for out-of-round effects to disappear and what happens on spark out.

1 INTRODUCTION

The problem of vibration in machining and in particular in grinding still exists even though there has been much research conducted in the area. Many solutions have been proposed but publications still abound on methods of suppressing chatter (1-5). An interesting observation is that most of the modelling of vibration in machining has ignored any contribution from torsional vibration. There are historical reasons for this. Until recently it was extremely difficult to measure torsional vibration under working conditions so it was not known if significant torsional vibration was occurring in machining operations. At the same time, the models of chatter that were developed explained most of the effects observed practically. In some recent work Entwistle (6-8) reported an improvement in grinding chatter performance as a result of changing the torsional stiffness of the workpiece drive. He also developed a frequency domain model that included torsional effects. Using a grinding force equation developed by Malkin (9) for steady grinding conditions he showed that the improvements he found experimentally could also be predicted theoretically.

The major question concerning the modelling of Entwistle was whether the steady force equation of Malkin applied under dynamic conditions. A collaborative project between the National University of Singapore and the University of Western Australia has been undertaken to measure grinding forces under oscillating conditions. This project has found that so far Malkin's equation is consistent with the data obtained for vibration conditions (10-11). At the same time, to help with the analysis of the grinding force data it became necessary to develop a time domain model of the grinding process that allowed for additional effects to those included by Entwistle. The initial time domain model allowed for loss of contact (12) and more recent work (13) has included the effects of contact stiffness.

As part of the work into grinding forces it was necessary to include the effects of an out-of-round grinding wheel. As a result it is possible to examine the effects of out of round grinding wheels and in particular what happens during spark out.

2 MODEL

An early version of the grinding model and its simulation on a computer has been described in detail (12), but this was prior to contact stiffness being included. A very brief description of the inclusion of contact stiffness is given in (13). The effects of out of round workpieces and grinding wheels have now been included in the model. To aid with the understanding of this paper a brief description of the model and program will be given.

Figure 1 shows a model of plunge grinding. The model includes transverse modes in the direction joining the centres and in the direction tangential to the contact point. Torsional modes are also included for the workpiece and its drive and also the wheel and its drive. It would be possible to include multiple modes for each direction but single modes are representative and the model is complex enough with them. For the notation shown in figure 1 and assuming the modes are not coupled, the equations of motion are,

$$m_n \ddot{u}_n + c_n \dot{u}_n + k_n u_n = F_n + k_n x_n + c_n \dot{x}_n$$

where $\dot{x}_n$ is the infeed rate (f) and $x_n = f\,t$.

$$m_t \ddot{u}_t + c_t \dot{u}_t + k_t u_t = F_t$$

$$I_w \ddot{\theta}_w + c_w \dot{\theta}_w + k_w \theta_w = T_w$$

$$I_G \ddot{\theta}_G + c_G \dot{\theta}_G + k_G \theta_G = T_G$$

and from geometry considerations,

$$T_w = Radius_w F_t \quad \text{and} \quad T_G = Radius_G F_t$$

Using Malkin's equation for force

$$F_t = \frac{u_{ch} V_w \delta_w b}{V_g} \text{ and } F_n = k_1 F_t$$

where, u_{ch} is assumed a constant; δ_w is the instantaneous depth of cut on the workpiece;

b is the width of cut; V_w is the surface speed of the work and V_g of the wheel.

It is convenient to work in terms of a cutting force coefficient R such that

$$R_w = \frac{u_{ch} V_w}{V_g} \quad \text{and therefore} \quad F_t = R_w \delta_w b$$

A similar coefficient can be defined for material removed from the grinding wheel so that $F_t = R_G \delta_G b$. Using the grinding ratio G (the volume removed from the workpiece divided by the volume removed from the grinding wheel), we obtain,

$$R_G = \frac{G R_w Radius_G \Omega_G}{Radius_w \Omega_w}$$

It is convenient to work in terms of the reduction in radius of the workpiece and wheel. As the simulation is an incremental time approach, an integer number of radial lines are considered for the workpiece and wheel. At each time increment a radial line on the workpiece and wheel will be in contact and be reduced in radius by wear. If the reduction in radius of the workpiece radius is x_w and of the wheel is x_G, then geometric constraint and deflections requires,

$$\delta_w = \frac{u_n - x_w - x_G}{1 + \frac{R_w}{R_G} + \frac{R_w b}{k_c}} \quad \text{and} \quad \delta_G = \frac{\delta_w R_w}{R_G}$$

where kc is the contact stiffness. If the depth of cut is negative then it is made zero and the force becomes zero, as loss of contact has occurred.

If contact has been maintained then,

$$F_t = R_w \delta_w b \frac{\Omega_G}{\Omega_G - \dot{\theta}_G} \frac{\Omega_w - \dot{\theta}_w - \dot{u}_t / Radius_w}{\Omega_w}$$

It should be noted that the force is effectively corrected here for any variation in surface velocities caused by torsional vibration and tangential vibration. In passing it should be noted that if the workpiece rotation is reversed so that it is opposite to the grinding wheel rotation, the $\dot{\theta}_w$ term becomes positive. Finally, the reduction in radius is increased by the depth of cut at any instant.

A computer program was written to simulate the grinding process and this has confirmed the results of Entwistle (6-8) that the torsional characteristics of the workpiece drive can improve chatter performance (13). The program used a fourth order Runge Kutta approach and information was stored for 1000 or 2000 points around the workpiece and a proportionate number around the wheel, so that the time between points was the same.

3 OUT OF ROUND EFFECTS

The simulation described above may easily be extended to include the effects of out-of-round grinding wheels and workpieces. Thus as an initial condition the wheel and/or work may be defined as having a reduction of radius that varies with the circumferential position (θ) as follows,

$$x_G = X_G \sin(6/2\pi) \text{ and } x_w = X_w \sin(6/2\pi)$$

When the model is run with just out-of-round on the grinding wheel it is still possible for chatter to occur. For the example given below of the build up of surface profile on the workpiece conditions have been chosen that are stable. The infeed per workpiece revolution for the simulation was 1.82 μm and the out-of-round amplitude was 2 μm. The structural characteristics were those of the initial research rig developed at NUS (11). The first significant transverse natural frequency was 160 Hz. The excitation frequency caused by the out-of-round wheel was 28.67 Hz as the wheel speed was 1720 rev/min. Thus the excitation is well below any resonant condition. The workpiece rotational speed was 270 rev/min, which is high. This was chosen because of computational difficulties. There was a limit to the number of points around the circumference that could be stored. When there were too few points on the grinding wheel, numerical instability occurred. With a limit of 2000 points on the wheel the workpiece speed needed to be fast enough to prevent the number of points on the wheel being too low.

The method of representing surface profile is to present a Talyrond trace. This is a graph of the variation from the mean radius plotted on a reference base circle. The variation is greatly amplified so that it is more clearly seen. Figure 2 shows such Talyronds for the workpiece after one and three revolutions, with the measured workpiece torsional characteristics included. The profile at the end of the first workpiece revolution shows that the engagement of the wheel with the work is intermittent. There is some evidence of vibration at 160 Hz (the transverse mode of vibration. At the end of the third workpiece revolution the grinding wheel is always in cut and the number of lobes on the workpiece is the ratio of wheel speed to workpiece speed, 1720/270 = 6.37. The scale has been halved but it is clear that since the amplitude of the 160 Hz vibration is not increasing, chatter will not occur.

Figure 3 shows the same two Talyronds for the case when the workpiece torsional characteristics are removed from the model. In many respects they are similar to those shown in figure 2. However it is clear that the vibration around 160 Hz is growing and that the process will go unstable. This in itself confirms the observations of Entwistle (6-8) that the torsional characteristics can be used to suppress chatter in grinding. However it is still conditional on the force equation of Malkin applying under oscillating conditions.

4 FORCE MODEL

Attempts have been made to measure the force under oscillating conditions. As may be imagined there is some difficulty in measuring an oscillating depth of cut when the mean depth is around a few microns. The most successful attempt so far has involved oscillating the workpiece speed and measuring the oscillating force. A description of this work is to be found in (11) and a summary in (13). The results obtained were consistent with Malkin's equation but with force amplitudes that were less than were expected from measurements of mean forces. Also it was found that varying workpiece did cause the force to vary but this in turn caused a variation in depth of cut. Thus two parameters were varying and it was difficult to measure the oscillating depth of cut.

Current work has attempted to vary the depth of cut with all other parameters constant. The easiest way to do this has been to use an out-of-round grinding wheel or an out-of-balance wheel. As a result an extensive investigation has been conducted using the computer model and including out-of-round effects on the grinding wheel. As it has proven difficult to continuously measure the instantaneous depth of cut, it was proposed to rapidly reverse the feed and then to measure the surface profile on a Talyrond machine. From the known frequency of vibration the oscillating depth of cut may be inferred.

The modelling of the out-of-round wheel has led to some interesting observations and these are presented in this paper. First it is necessary to show that the model was accurately representing the grinding process. A standard experimental test has involved truing the wheel and then plunge grinding for a period of around 60 secs with no rapid retraction. During the test the normal and tangential forces are measured and using two Keyence lasers (10) the reduction in radius of the workpiece may be measured. Measurements from a typical test are shown in figure 4.

For the results shown, the experimental rig had been greatly stiffened compared to the simulations shown in figures 2 and 3. The modelled results for the same conditions as for the results of figure 4 are shown in figure 5. It is clear that they are of substantially the same form. The major difference is that the modelled force variation has a plateau whereas the experimental results show a peak and a gradual fall off until the feed is disengaged. This is probably the result of always starting with a trued (and hence 'sharp' wheel). The condition of the wheel changes. In passing it should be noted that the experimental force measurements were always made at the peak.

As the model was seen to be behaving in a realistic way it was decided to investigate the effects of an out-of-round wheel. In particular the profile of the work when in a steady state and also what happens to the profile during spark out occurs, i.e. the feed is stopped.

5 SPARK OUT

It was decided to continue with modelling the 60 second plunge grinding test as the results would prove useful for the proposed tests involving measuring forces and depth of cut. It was anticipated that the workpiece speed could be significant, as it was possible to have an integer number of waves left on the work if the ratio of wheel speed to work speed was an integer. The results shown in figure 4 and simulated in figure 5 were for a wheel speed of 1720 rev/min and a workpiece speed of 300 rev/min. The feed rate was 3.7 μm/sec. The ratio of wheel to work speed was thus 5.733. It was decided to model at ratios of 5.5 and 6.0, work speeds of 312.7272 rev/min and 286.6667 rev/min. It was also possible to produce modelled Talyronds at different stages of the test.

Figure 6(a) shows the results of the model at a ratio of 6.0 and figure 6(b) shows the results at a ratio of 5.5. The oscillating force caused by the out-of-round of the wheel ($\pm$1.0 μm) causes the force trace to be thick. When the feed is engaged the 'snapshot' Talyrond shows a discontinuity. This is where material is currently being removed and represents (for steady conditions) the feed per workpiece revolution. As the profile samples are taken at regular intervals of 20 seconds, the orientation of the profiles varies with respect to the point in cut. An additional profile at 120 seconds is also shown which is 60 seconds after the feed is stopped.

The main area of interest is in the way the profile varies through the simulation and the difference between the two workpiece speeds. It should be noted that the mean maximum force in both cases is the same as for the speed of 300 rev/min. This is because the feed rate is the same for all simulations and the force model gives a force proportional to both depth of cut and workpiece surface speed. Thus though the force increases with an increase in workpiece speed it decreases because the constant feed rate results in a smaller depth of cut.

The variation in profile is significantly different for each workpiece speed. When the ratio of speeds is an integer the wheel eccentricity reinforces any wave left on the workpiece. In the steady state there are 6 lobes on the workpiece and the oscillating force is that necessary for the structure to deflect the ± 1.0 μm of the wheel out-of-round. When the feed is stopped it takes a relatively long time to move the wave on the workpiece as the wheel 'high spot' is always grinding a 'low spot' on the workpiece. However when the speed ratio is 5.5 the high spot on the wheel is engaged with a high spot on the workpiece. As a result the oscillating force is greater under steady conditions than for the case of the integer ratio. The profiles show only a small amplitude wave as the high spots are continuously being removed.

6 CONCLUSIONS

The simulation program described has been useful in the investigation of chatter in plunge grinding. It has confirmed that the torsional characteristics of the workpiece drive may stop chatter from occurring. However this conclusion is dependent on the force model used in the simulation being a realistic one. Work done to check the force model has produced results which are consistent with the model when the workpiece speed is oscillated. Future work to investigate if oscillating depth of cut causes a consistent force variation is dependent on out-of-round grinding wheels.

This paper has presented an investigation of the effects of out-of-round grinding wheels on forces and workpiece profiles. The results are significant for practice in that better roundness of workpieces is achieved if the ratio of wheel to work speed is non-integer. Also the time for spark out is also reduced. In addition larger force variations are obtained for the non-integer case and this is important for future work on measuring the forces.

7 ACKNOWLEDGMENTS

The authors wish to express their thanks to Professor Andrew Nee of the National University of Singapore (NUS) for encouraging the collaboration for this project. Financial support was provided by the National University of Singapore and the Australian Research Council.

8 REFERENCES

(1) Wang M. and Fei, R. Y., 'Improvement of machining stability using a tunable-stiffness boring bar containing an electrorheological fluid', Smart Materials & Structures, 8: (4) 511-514 August 1999.

(2) Pratt, J. R. and Nayfeh, A. H., 'Design and modeling for chatter control', Nonlinear dynamics, 19: (1) 49-69 May 1999.

(3) Altintas, Y., Engin, S. and Budak, E., 'Analytical stability prediction and design of variable pitch cutters', Journal of Manufacturing Science and Engineering-Transactions of the ASME, 121: (2) 173-178 May 1999.

(4) McFarland, D. M., Bailey, G. E. and Howes, T. D., 'The design and analysis of a polypropylene hub CBN wheel to suppress grinding chatter', Journal of Manufacturing Science and Engineering-Transactions of the ASME, 121: (1) 28-31 February 1999.

(5) Yang, F., Zhang, B. and Yu, J., Chatter suppression via an oscillating cutter, Journal of Manufacturing Science and Engineering-Transactions of the ASME, 121: (1) 54-60 February 1999.

(6) Entwistle, R. D., 'Torsional Compliance and Chatter in Grinding'. PhD Thesis, The University of Western Australia Feb 1997.

(7) Entwistle R. D. and Stone, B. J., 'Torsional compliance in grinding chatter', Proc. Eleventh Annual Meeting of The American Society for Precision Engineering, 9-14 420-423, November 1996.

(8) Entwistle, R.D. and Stone, B.J. 1997. 'The Effect of Workpiece Torsional Flexibility on Chatter Performance in Cylindrical Grinding.' Proceedings of the Fifth International Congress on Sound and Vibration, University of Adelaide, South Australia, December 15-18, 1997.

(9) Malkin, S., Grinding Technology: Theory and Application of Machining with Abrasives, Ellis Horwood, Chichester and John AWiley, 1989.

(10) Drew, S. J., Mannan, M. A., Ong, K. L. and Stone, B. J., 'An Investigation of In-process measurement of ground surfaces in the presence of vibration'. International Journal of Machine Tools and Manufacture, (1999), 39 (12), 1841-1861.

(11) K. L. Ong, Mannan, M. A., Drew, S. J., and Stone, B. J., 'The Effects of Workpiece Torsional Vibration on Forces in Grinding'. International Journal of Machine Tools and Manufacture, Submitted Dec 1999.

(12) Mannan, M. A., Fan, W. T. and Stone, B. J., 'The effects of torsional vibration on chatter in grinding.' Journal of Materials Processing Technology 89-90 (1999), 300-309.

(13) Drew, S. J., Mannan, M. A. and Stone, B. J., 'Torsional Vibration Effects in Grinding?' Accepted for Annals of CIRP, Sydney, 2000.

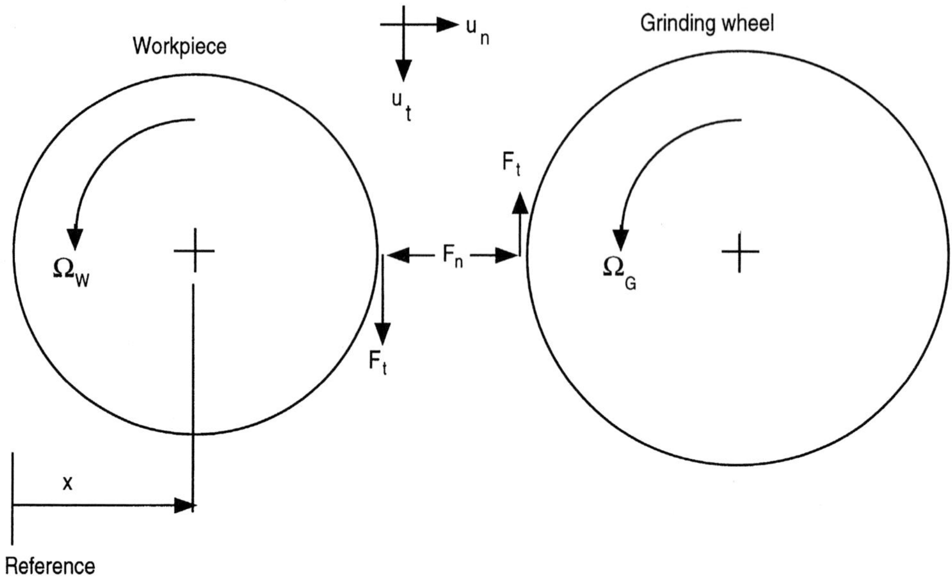

Fig. 1 Schematic of plunge grinding (showing notation).

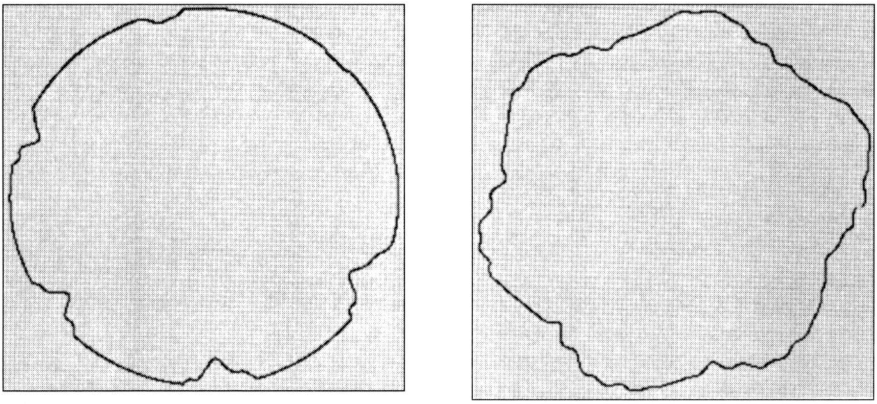

(a) First revolution

(b) Third revolution and half scale of (a)

Fig. 2 Talyronds of workpiece on start up with out-of-round wheel and workpiece torsion included in the model

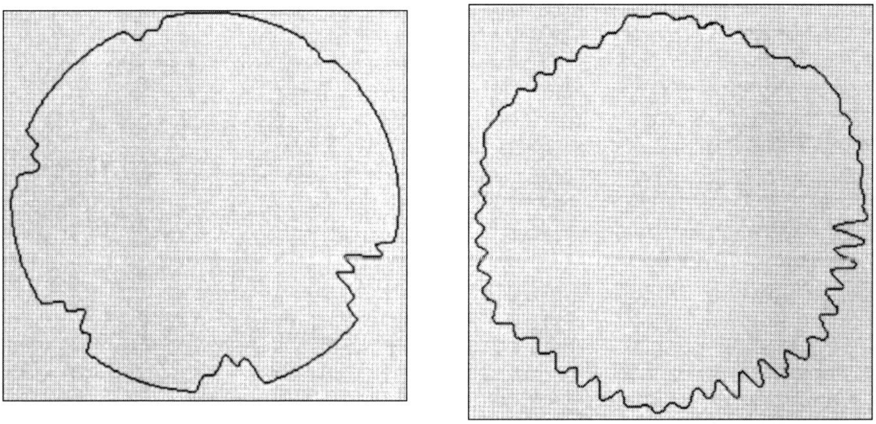

(a) First revolution

(b) Third revolution and half scale of (a)

Fig. 3 Talyronds of workpiece on start up with out-of-round wheel and no workpiece torsion.

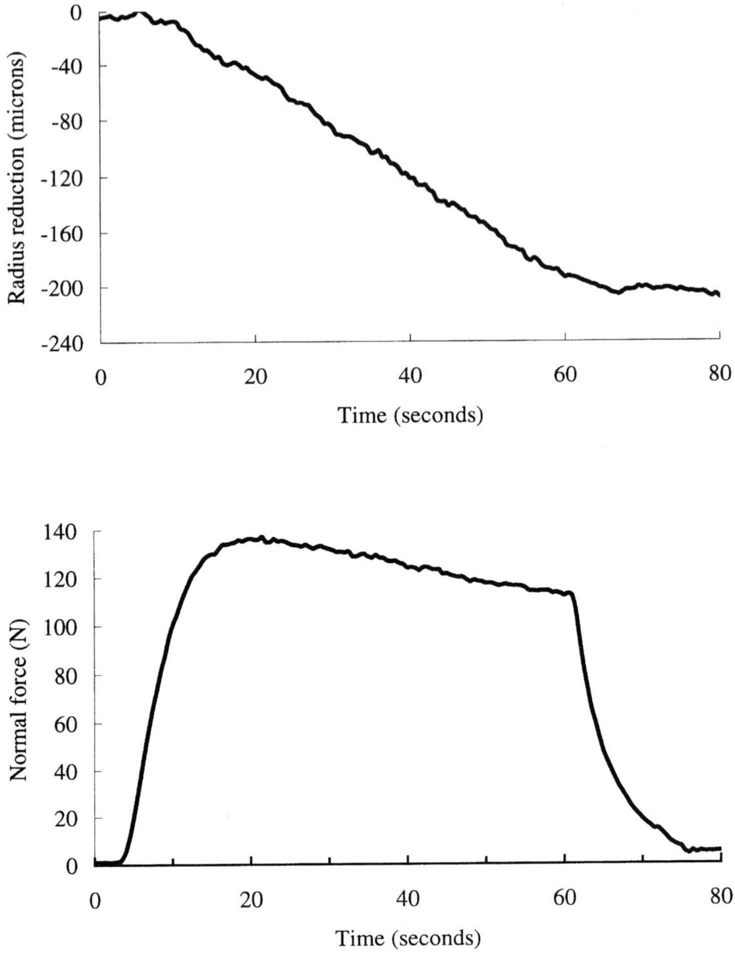

Fig. 4 Experimental force and radius variation with time.

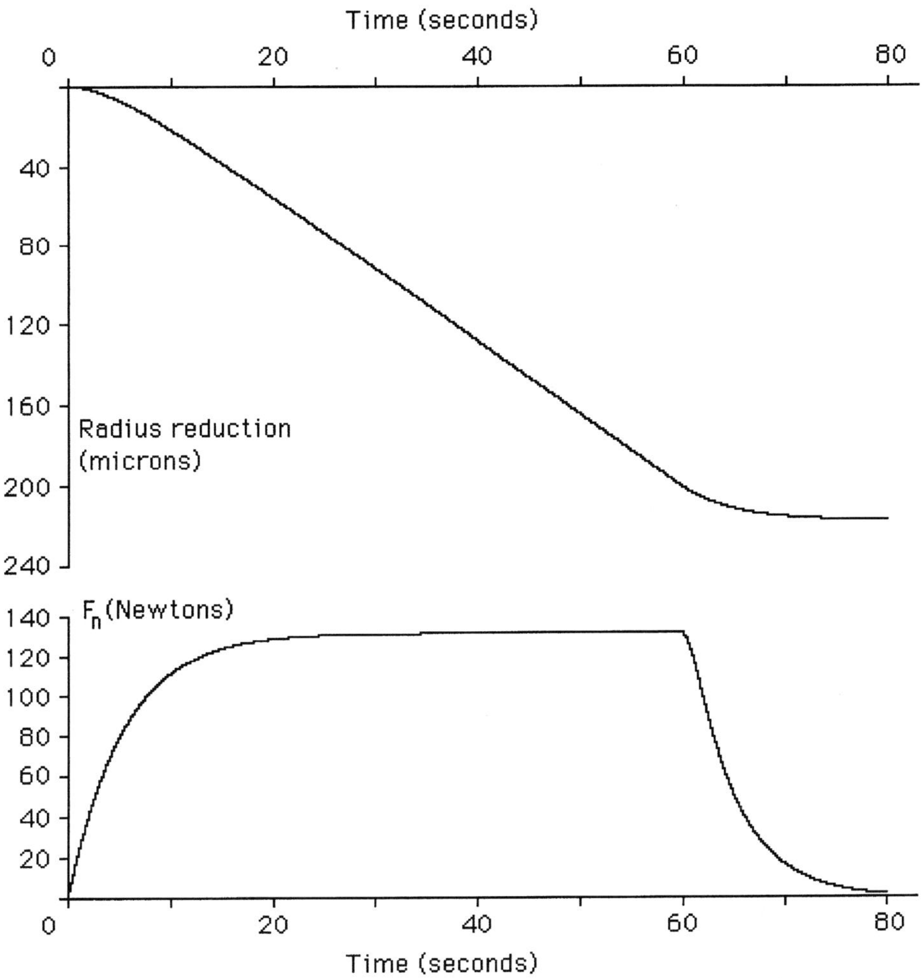

Fig. 5 Theoretical force and radius variation with time.

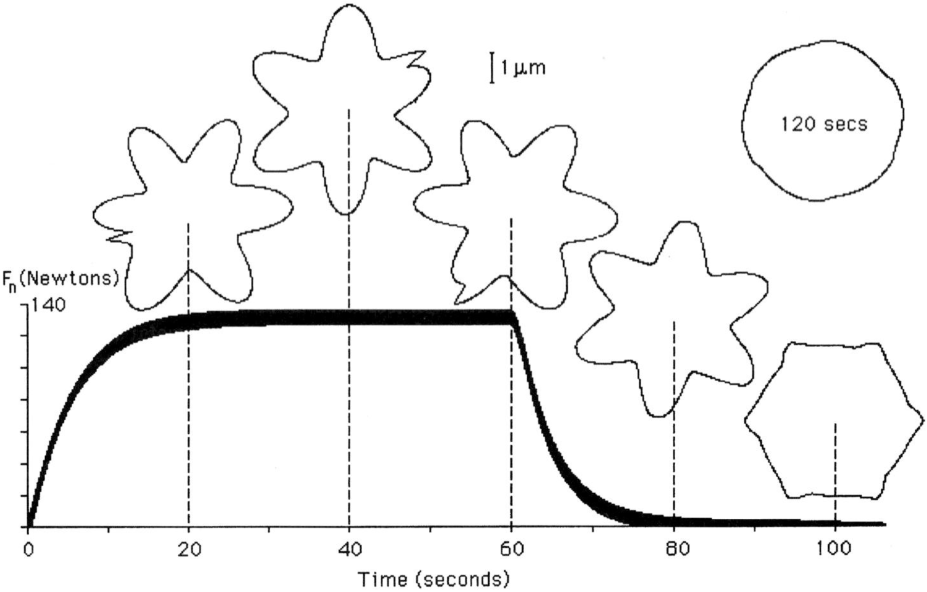

(a)

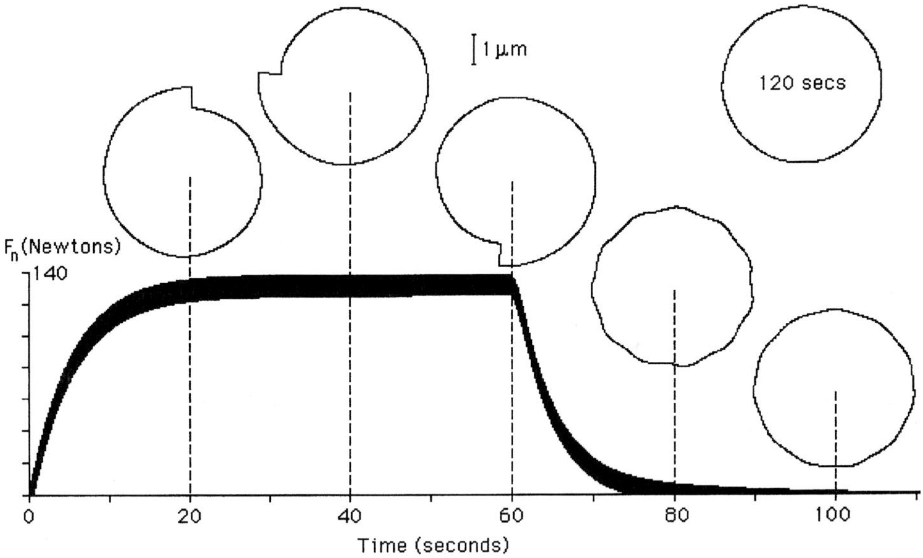

(b)

Fig. 6 Simulation for: (a) wheel to work speed ratio of 6.0, and (b) wheel to work speed ratio of 5.5

Multivariable control of AMB spindles

M ALEYAASIN, M EBRAHIMI, and **R WHALLEY**
Department of Mechanical and Medical Engineering, University of Bradford, UK

ABSTRACT

In this paper an Active Magnetic Bearing (AMB) spindle is considered as a rigid rotor mounted on two bearings with stiffness and damping characteristics. The equations for small perturbations of the rotor are derived in the form of a multi-input, multi-output transfer function matrix description, enabling the multivariable frequency response matrix for the system to be computed. Thereafter a control system configuration comprising four vibration sensors and two actuators is proposed. Using the multivariable frequency response matrix of the system, a model based feedback controller is derived. A numerical example is used to illustrate that the effects of disturbances can be rapidly diminished reducing the net radial displacement of the spindle to zero. The advantages of this scheme for control of AMB spindle is discussed.

1-Introduction

One of the major factors which affects the accuracy of machining processes is the type of the spindle used for machine tools. Manufacturers of high precision grinding machines have in fact implemented various types of spindles in order to improve the accuracy of grinding processes. An experimental study was carried out by Rowe [1] which reports the performance of different types of spindles used in grinding machines. Hydrostatic spindles are recognised as being more accurate owing to employment of bearings with error correction capabilities [2]. This property is limited to specific load and speed ranges [1]. Recently the authors have formulated an alternative mechanism for the error correction of hydrostatic spindles. It has been shown that, in order to use these spindles at any load and speed, a complementary bearing tuning system is required for regulating the dynamic characteristics of the externally pressurised bearings [3]. However using this scheme requires additional machine tool actuators.

This new generation of spindles are supported by Active Magnetic Bearings and are known as AMB Spindles. These kind of spindles also have advantages over fluid film bearings due to the air gap which is required between rotor and bearing. This is sustained by a magnetically elevating rotor system [4]. Moreover the position of the rotor is controlled through magnetic actuators so that, the accuracy of the system is strongly dependent the response of the relevant control system which is used to track the error free position of the spindle. Figure 1 describes a conventional AMB spindle system, which employs two sensors and two actuators to locate the rotor in the zero error position. The advantages of this system would be realised in higher machining speeds, giving greater accuracy and higher material removal rates.

The control scheme used in industrial AMB spindles is an adaptive multi-channel DSP control, which is implemented for a 40000 rpm, high speed milling spindle [5] . The control procedure includes on line identification and signal processing. However the system has only been successful for a specified range of loads. It appears from the analysis herein that a controller based on an accurate multivariable model of the system, would enable the magnetic actuators to apply appropriate forces to eliminate the vibration of the spindle at any machining load.

2. Multivariable model of a rotating spindle

The rotating spindle considered herein is modelled as a single rotor on a radial, compliant, damped suspension, as shown in Figure 2. The axis of rotation is in line with the X axis co-ordinate. Small amplitude vibrations in the Y and Z directions are considered. The model stiffness and damping are lumped at the bearings and the rotor mass is lumped at the centre of gravity, as shown in figure 2.

The general equations for rigid body motion for the centre of gravity in the X ,Y and Z directions and the motion about the centre of gravity can be formulated as:

$$\sum F_x = m.\ddot{x}_g \quad (1) \qquad \sum F_y = m.\ddot{y}_g \quad (2) \qquad \sum F_z = m.\ddot{z}_g \quad (3)$$

$$I_{yy}\dot{\omega}_y + \Omega_z I_{xx}\omega_x - \Omega_x I_{zz}\omega_z = M_y \quad (4)$$

$$I_{zz}\dot{\omega}_z + \Omega_x I_{yy}\omega_y - \Omega_y I_{xx}\omega_x = M_z \quad (5)$$

$$I_{xx}\dot{\omega}_x + \Omega_y I_{zz}\omega_z - \Omega_z I_{yy} = M_x \quad (6)$$

For the above problem the equations are for a reference frame that remains the principal axis of the rotor [6], $\omega_x, \omega_y, \omega_z$ are components of the angular velocity of the body and $\Omega_x, \Omega_y, \Omega_z$ are components of the angular velocity of the frame. When the frame is considered for small precessions of the rotor only, then $\Omega_x = 0$, $\Omega_z = \omega_z$, $\Omega_y = \omega_y$ for the frame, and $\omega_x = \Omega$ is the constant rotational speed of the rotor. Due to symmetry $I_{yy} = I_{zz} = I_t$ so that equation (6) vanishes. Equation (1) is not considered due to the absence of motion in that direction. Therefore the governing equations are (3), (2), (4) and (5). In this application dimensionless parameters are also introduced.

$$i_t = \frac{I_t}{L^2} \quad (7) \qquad i_p = \frac{I_p}{L^2} \quad (8)$$

$$l_1 = \frac{L_1}{L} \quad (9) \qquad l_2 = \frac{L_2}{L} \quad (10)$$

According to (9) and (10) the following relations between the co-ordinates of the centre of gravity and the centre of bearings 1 and 2 exist.

$$\ddot{z}_g = \ddot{z}_1 l_2 + \ddot{z}_2 l_1 \tag{11}$$

$$\ddot{y}_g = \ddot{y}_1 l_2 + \ddot{y}_2 l_1 \tag{12}$$

For small displacements at the centres of bearings 1 and 2 we can write the following relations

$$\omega_z = \frac{\dot{y}_2 - \dot{y}_1}{L} \tag{13}$$

$$\omega_y = \frac{\dot{z}_2 - \dot{z}_1}{L} \tag{14}$$

$$\dot{\omega}_z = \frac{\ddot{y}_2 - \ddot{y}_1}{L} \tag{15}$$

$$\dot{\omega}_y = \frac{\ddot{z}_2 - \ddot{z}_1}{L} \tag{16}$$

Equations (2) and (3) can be further expanded to include internal stiffness and damping forces due to the oil film and suspension. Using equation (11) and (12) results in the following differential equation .

$$\mathbf{M}\ddot{\boldsymbol{q}} + \mathbf{C}\dot{\boldsymbol{q}} + \mathbf{K}\boldsymbol{q} = \boldsymbol{F}(\mathrm{t}) \tag{17}$$

The 4 x 4 matrices M, C and K and the column vectors q , F are :

$$\mathbf{M} = \begin{bmatrix} ml_2^{\,2} + i_t & ml_1 l_2 - i_t & 0 & 0 \\ ml_1 l_2 - i_t & ml_2^{\,2} - i_t & 0 & 0 \\ 0 & 0 & ml_2^{\,2} + i_t & ml_1 l_2 - i_t \\ 0 & 0 & ml_1 l_2 - i_t & ml_1^{\,2} + i_t \end{bmatrix},$$

$$\mathbf{C} = \Omega \begin{bmatrix} 0 & 0 & i_p & -i_p \\ 0 & 0 & -i_p & i_p \\ -i_p & i_p & 0 & 0 \\ i_p & -i_p & 0 & 0 \end{bmatrix} + \begin{bmatrix} C_{y1y1} & 0 & C_{y1z1} & 0 \\ 0 & C_{y2y2} & 0 & C_{y2z2} \\ C_{z1y1} & 0 & C_{z1z1} & 0 \\ 0 & C_{z2y2} & 0 & C_{z2z2} \end{bmatrix},$$

$$\mathbf{K} = \begin{bmatrix} K_{y1y1} & 0 & K_{y1z1} & 0 \\ 0 & K_{y2y2} & 0 & K_{y2z2} \\ K_{z1y1} & 0 & K_{z1z1} & 0 \\ 0 & K_{z2y2} & 0 & K_{z2z2} \end{bmatrix}, \quad \boldsymbol{q}(t) = \begin{bmatrix} y_1(t) \\ y_2(t) \\ z_1(t) \\ z_2(t) \end{bmatrix} \text{ and } \boldsymbol{F}(t) = \begin{bmatrix} f_{y1}(t) \\ f_{y2}(t) \\ f_{z1}(t) \\ f_{z2}(t) \end{bmatrix}$$

Before the computation of the control forces, the characteristic determinant of the system, which determines the stability regions of the rotor , should be evaluated. This ensures that the computed, optimal, dynamic characteristic of the bearings executes a stable whirl motion .

Referring to equation (17) the characteristic determinant of the system could be defined as follows.

$$\Delta(s) = \left|(\mathbf{M}s^2 + \mathbf{C}s + \mathbf{K})\right| \tag{18}$$

The roots of the polynomial $\Delta(s)$ defined in (18) are called the system poles [7] and are generally expressed in the form of complex numbers. If all of the poles are located in the left hand plane the rotor motion is stable.

$$\lambda_i = \sigma_i + j\omega_I \tag{19}$$

where

$$\Delta(\lambda_i) = 0 \tag{20}$$

3. Control force computation

The frequency dependent stiffness matrix of the system is required, for the computation of the control forces. This could be followed by defining the parameters for the **M**, **C** and **K** matrices with:

$$\mathbf{M} = \begin{bmatrix} m_1 & m_2 & 0 & 0 \\ m_3 & m_4 & 0 & 0 \\ 0 & 0 & m_5 & m_6 \\ 0 & 0 & m_7 & m_8 \end{bmatrix} \quad \begin{aligned} m_1 &= m_5 = ml_2^{\,2} + i_t \\ m_2 &= m_3 = m_6 = m_7 = ml_1 l_2 - i_t \\ m_4 &= m_8 = ml_1^{\,2} + i_t \end{aligned} \tag{19}$$

$$\mathbf{K} = \begin{bmatrix} k_1 & 0 & k_2 & 0 \\ 0 & k_3 & 0 & k_4 \\ k_5 & 0 & k_6 & 0 \\ 0 & k_7 & 0 & k_8 \end{bmatrix} \quad \begin{aligned} k_1 &= K_{y1y1} & k_5 &= K_{y1z1} \\ k_2 &= K_{y1z1} & k_6 &= K_{z1z1} \\ k_3 &= K_{y2y2} & k_7 &= K_{z2y2} \\ k_4 &= K_{y2z2} & k_8 &= K_{z2z2} \end{aligned} \tag{20}$$

$$\mathbf{C} = \begin{bmatrix} c_1 & 0 & c_2 & c_3 \\ 0 & c_4 & c_5 & c_6 \\ c_7 & c_8 & c_9 & 0 \\ c_{10} & c_{11} & 0 & c_{12} \end{bmatrix} \quad \begin{aligned} c_1 &= C_{y1y1} & c_5 &= -\Omega i_p & c_9 &= C_{z1z1} \\ c_2 &= \Omega i_p + C_{y1z1} & c_6 &= \Omega i_p + C_{y2z2} & c_{10} &= \Omega i_p \\ c_3 &= -\Omega i_p & c_7 &= C_{y1z1} - i_p \Omega & c_{11} &= C_{z2y2} - \Omega i_p \\ c_4 &= C_{y2y2} & c_8 &= \Omega i_p & c_{12} &= C_{z2z2} \end{aligned} \tag{21}$$

Then thefrequency dependent stiffness matrix of the system could be computed from equations (17) (19) and (21) yielding:

$$\boldsymbol{F}(s) = \boldsymbol{H}.\boldsymbol{q}(s) \tag{22}$$

where the elements of equation (22) are given by:

$$\begin{bmatrix} f_{y1}(s) \\ f_{y2}(s) \\ f_{z1}(s) \\ f_{z2}(s) \end{bmatrix} = \begin{bmatrix} H_{11} & H_{12} & H_{13} & H_{14} \\ H_{21} & H_{22} & H_{23} & H_{24} \\ H_{31} & H_{32} & H_{33} & H_{34} \\ H_{41} & H_{42} & H_{43} & H_{43} \end{bmatrix} \cdot \begin{bmatrix} y_1(s) \\ y_2(s) \\ z_1(s) \\ z_2(s) \end{bmatrix} \tag{23}$$

Where the elements of the H would be:

$$\begin{aligned}
&H_{11}(s)=m_1s^2+c_1s+k_1,\quad H_{12}(s)=m_2s^2,\quad H_{13}(s)=c_2s+k_2,\ H_{14}(s)=c_3s\\
&H_{21}(s)=m_3s^2,\quad H_{22}(s)=m_4s^2+c_4s+k_3,\quad H_{23}(s)=c_5s,\ H_{24}(s)=c_6s+k_4\\
&H_{31}(s)=c_7s+k_5,\quad H_{32}(s)=c_8s,\quad H_{33}(s)=m_5s^2+c_9s+k_6,\ H_{34}(s)=m_6s^2\\
&H_{41}(s)=c_{10}s,\quad H_{42}(s)=c_{11}s+k_7,\quad H_{43}(s)=m_7s^2,\ H_{44}(s)=m_8s^2+c_{12}s+k_8
\end{aligned}\tag{24}$$

The configuration proposed herein comprise four vibration sensors for measuring y_1, y_2, z_1 and z_2 and two actuators as indicated in figure 3, The control force in bearing 1, could then be expressed as:

$$\begin{aligned}
f_{cy_1}&=-\left(H_{11}(j\omega)y_1+H_{12}(j\omega)y_2+H_{13}(j\omega)z_1+H_{14}(j\omega)z_2\right)\\
f_{cz1}&=-\left(H_{31}(j\omega)y_1+H_{32}(j\omega)y_2+H_{33}(j\omega)z_1+H_{34}(j\omega)z_2\right)\\
\vec{f}_{c1}&=\vec{f}_{cy_1}+\vec{f}_{cz1}
\end{aligned}\tag{25}$$

The control force required at bearing 2 is:

$$\begin{aligned}
f_{cy_2}&=-\left(H_{21}(j\omega)y_1+H_{22}(j\omega)y_2+H_{23}(j\omega)z_1+H_{24}(j\omega)z_2\right)\\
f_{cz2}&=-\left(H_{41}(j\omega)y_1+H_{42}(j\omega)y_2+H_{43}(j\omega)z_1+H_{44}(j\omega)z_2\right)\\
\vec{f}_{c2}&=\vec{f}_{cy_2}+\vec{f}_{cz2}
\end{aligned}\tag{26}$$

Upon applying the above control forces, the vibration of the spindle would approach zero, in accordance with the following decay rates.

$$\begin{aligned}
&y_1(t)=\sum_{i=1}^{4}a_{1i}e^{\sigma_i t}\cos(\omega_i t) &\quad& y_2(t)=\sum_{i=1}^{4}a_{2i}e^{\sigma_i t}\cos(\omega_i t)\\
&z_1(t)=\sum_{i=1}^{4}a_{3i}e^{\sigma_i t}\cos(\omega_i t) && z_2(t)=\sum_{i=1}^{4}a_{4i}e^{\sigma_i t}\cos(\omega_i t)
\end{aligned}\tag{27}$$

4. The numerical example

The numerical values of the system parameters, shown in figure 2, are m = 8 kg, $l_1 = l_2 = 0.5$, $i_t = 0.45\,kg$, $i_p = 0.25\,kg$ Ω = 10,000 *rpm*. At a constant rotational speed, the elements of the matrices K and C in equation (17), which are the stiffness and damping of the bearings can be computed by obtaining the values in the y direction for K_{yy} and C_{yy} . The coefficients K_{zz} ,K_{yz},C_{zz} , C_{yz} and K_{zy} ,C_{zy} are functions of the K_{yy} , C_{yy} etc. as shown in [8] . The numerical values from [9] are:

$C_{yy} = C_{zz}$ =300 *N.sec/m*, C_{yz} = -C_{zy} = *20 Nsec/m*, K_{yy} =4×10^6 *N/m*,
K_{zz} = 4.2051×10^6 *N/m* and K_{zy} = -K_{yz} = 8×10^4 *N/m*.

In the configuration described by figure 3, if the steady state vibration signals, measured by the four sensors are represented by:

$$y_1(t) = y_2(t) = 1^{mm} \cos\left(\frac{2\pi\Omega}{60} t\right), z_1(t) = z_2(t) = 1.5^{mm} \cos\left(\frac{2\pi\Omega}{60} t\right)$$

then the numerical values of the control forces calculated from (25) and (26) in Newtons would be:

$$f_{c1} = 2257^N - 628^N j \qquad f_{c2} = -859.6^N - 942^N j$$

$$f_{c1}(t) = 2432.7^N \sin(1047.2t - 15.54^0) \; f_{c2}(t) = 1275.3^N \sin(1047.2t - 132.3^0)$$

Upon applying these control forces, the vibrations of the spindle rapidly reduce to zero rapidly. By considering the equations 27 and 35; Figure 4 can be computed indicating the decay rate of the vibration at the right side of the spindle from 1^{mm} amplitude to zero. This takes $0.15\ sec$. Similarly figure 5 shows the vibration decay at the left side of the spindle which decays from 1.5^{mm} to zero in the same time interval.

Conclusion

In this paper a control system comprising of four vibration sensors and two actuators is proposed for AMB spindles. It is shown that a model based controller enables the magnetic actuators to reduce the vibration of the spindle to zero, increasing the accuracy of the machining by this kind of spindle thereby.

References:

1- Rowe, W.B. "Experience with Four Types of Grinding Machine Spindle" Proceedings of eighth international conference on Machine Tool Design & Research (MTDR), Manchester University 1967, pp. 453-476.

2- Levesque, G.N. "Error Correcting Action of Hydrostatic Bearings" ASME, paper 65-LUBS-12 for meeting on June 6-9, 1965.

3- Aleyaasin, M. Whalley, R. and Ebrahimi, M. "Error Correction in Hydrostatic Spindles by Optimal Bearing Tuning" International Journal of Machine Tools & Manufacture, Vol. 40, pp, 809-822, 2000.

4- Habermann, H. and Liard, G. "An Active Magnetic Bearings System" Tribology International, Vol. 13 pp. 85-89, 1980.

5- Bleuler, H. Gahler, C. Herzog, R. Larsonneur, R. Mizuno, T. Siegwart, R. and Woo, S. J. "Application of Digital Signal Processors for Industrial Magnetic Bearings", IEEE Transactions on Control System Technology, Vol. 2 No. 4, pp, 280-289, December 1994.

6- Huang, T.C. "Engineering Mechanics, Vol. 2, Dynamics" Addison-Wesley 1967, pp. 745-750.

7- Rosenbrock H. H. "State Space and Multivariable Theory", T Nelson and Son, London, 1970.

8- Zhicheng, P., Jingwu S., Wenjie, Z., Qingming, L. and Wei, C., "Dynamic Characteristics of Hydrostatic Bearings " Wear, Vol. 166, No. 2, pp. 215-220, July 1993 .

9- Lee, C-W., Joh, Y-D., "Theory of Excitation Methods and Estimation of Frequency Response Functions in Complex Modal Testing of Rotating Machinery", Mechanical System and Signal Processing Vol. (1), p57-74, Jan 1993.

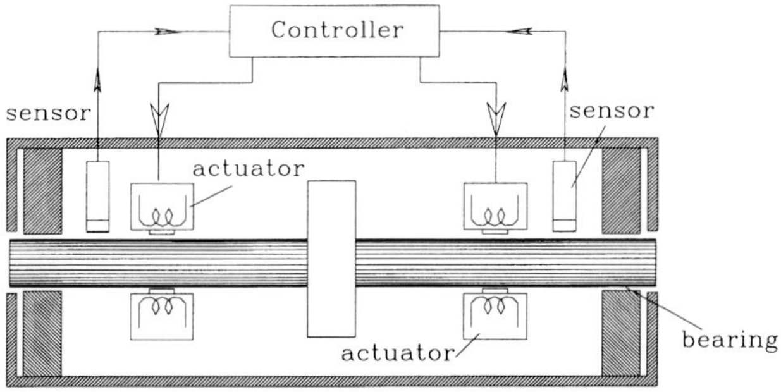

Figure 1. Conventional AMB spindle system.

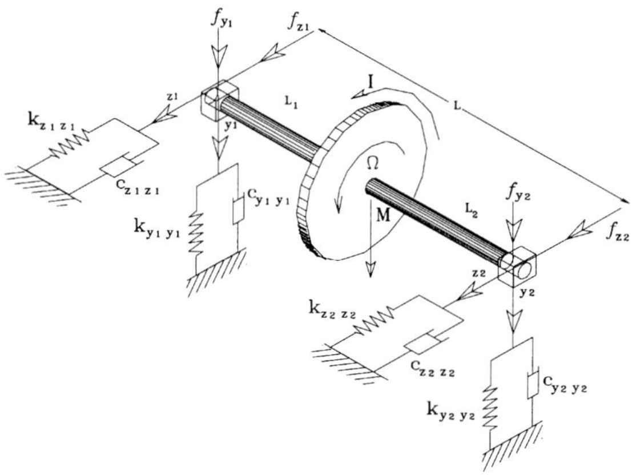

Figure 2. Model of stiffness and damping at bearings in a lumped mass rotor.

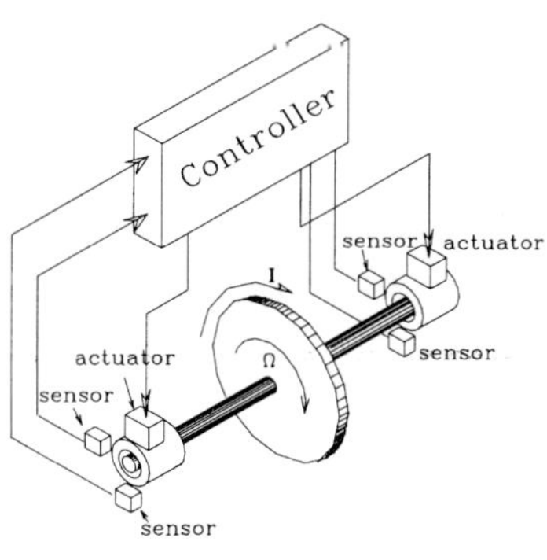

Figure 3. Proposed controller configuration for the AMB spindle.

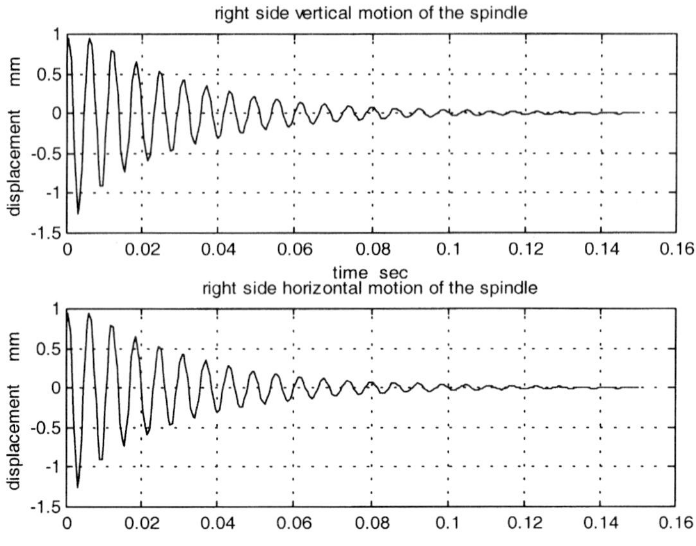

Figure 4. Horizontal and vertical displacement in right hand side of the spindle.

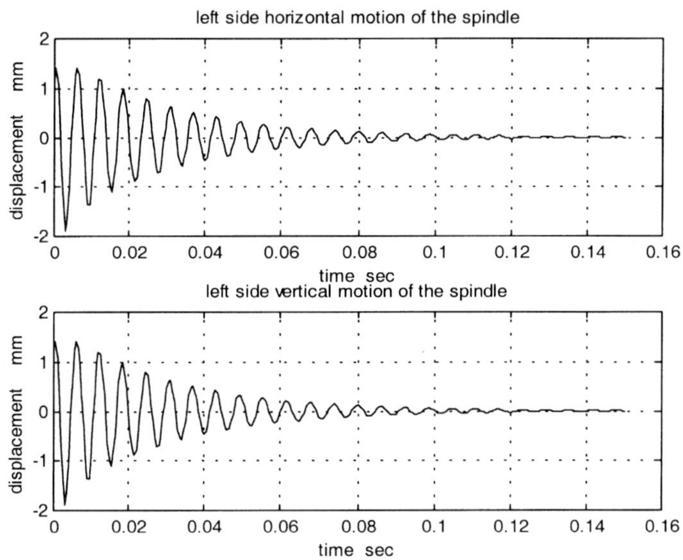

Figure 5. Horizontal and vertical displacement on left hand side of the spindle.

Vibration modelling and identification using Fourier transform, wavelet analysis, and least-square algorithm

G Y LUO, D OSYPIW and **M IRLE**
Buckinghamshire College of Technology, UK

ABSTRACT

Vibration signals are considered as nonstationary signals whose spectral character changes with time. Conventional harmonic Fourier analysis finds it difficult to completely model the vibration signals. In this paper, a novel approach using the global Fourier transforms and local Wavelet analysis for vibration signal analysis and modelling is presented. For the wavelet analysis both continuous and discrete wavelets are discussed. Continuous wavelet transform is a remarkable tool for signal and feature extraction, while discrete wavelet transform is computationally efficient. The effective and accurate analysis of power spectrum, which influences the signal significantly and reflects the inherent property of the systems' nature, is the basis of vibration modelling. The coefficients of the model are identified by a least-squares algorithm, which ensures that the error is minimised. To demonstrate this approach a machine spindle vibration signal is analysed, and the main features of the vibration signal are extracted, which are useful for system monitoring and further analysis.

1 INTRODUCTION

Vibration signals of any machine, engine, or structure contain a great deal of dynamic information related to the exciting forces applied to them and the condition of the system. Therefore, changes in these response signals could be used to identify undesirable external loads or the onset of system faults before drastic failure occurs. Vibration measured at appropriate locations has been shown to present a reliable diagnostic tool by means of observing the overall vibration signal or by processing the vibration signal using certain techniques [Dimarogonas 1992].

The mechanism of vibration generation in machines, engines, and common structures can be modelled and quantified in relation to changing operating or excitation conditions. The models could be used to describe the behaviour of the system, or to predict the trend of changing vibration corresponding to changing excitations or operating conditions, thus providing us with prior knowledge that can be used for monitoring purpose. Naturally these models become useful tools for system simulation, analysis and diagnosis.

2 The underlying principle of Fourier harmonic analysis

The conventional harmonic Fourier analysis states that any function *x(t)* that is periodic over the interval 0 < t < T can be represented with arbitrary accuracy by a Fourier series as an infinite sum of sine and cosine terms [Rao 1990]:

$$x(t) = \frac{a_0}{2} + \sum_{n=1}^{\infty}\left(a_n \cos\frac{2\pi nt}{T} + b_n \sin\frac{2\pi nt}{T}\right) \qquad (1)$$

where
$$a_0 = \frac{2}{T}\int_0^T x(t)dt$$

$$a_n = \frac{2}{T}\int_0^T x(t)\cos\frac{2\pi nt}{T}dt\,,\; b_n = \frac{2}{T}\int_0^T x(t)\sin\frac{2\pi nt}{T}dt$$

The physical interpretation of equation (1) is that any periodic function can be represented as a sum of harmonic function. Although the series in equation (1) is an infinite sum, we can approximate most periodic function with the help of only a few harmonic functions. This formula plays an important role in harmonic analysis such as electrical signal, which globally is periodic. In equation (1), when T approaches infinity, it leads to an infinite number of frequency components [Kraniauskas 1992]. This is Fourier transform.

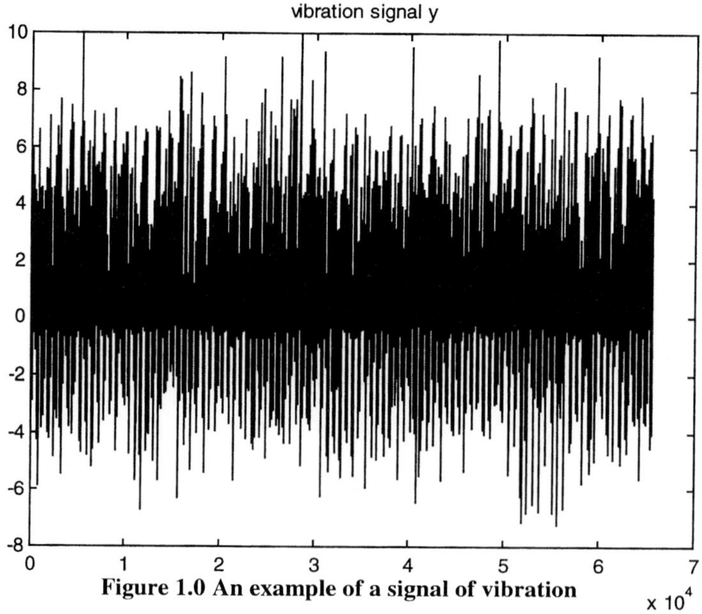

Figure 1.0 An example of a signal of vibration

Normally the readings of signal y at a different time t are available with certain sampling rate. Direct integration to determine the Fourier coefficients is then not possible and numerical methods of integration have to be used, such as trapezium rule or twelve-point analysis [Bolton 1995]. Function $x(t)$ is given by a number of points $y_1, y_2, \ldots, y_N$ and at time $t_1, t_2, \ldots, t_N$ respectively, y is measured by means of a measuring instrument.

Therefore $$a_0 \approx \frac{2}{N}\sum_{i=1}^{N} y_i$$

$$a_n \approx \frac{2}{N}\sum_{i=1}^{N} y_i \cos\frac{2\pi n t_i}{T}, \quad b_n \approx \frac{2}{N}\sum_{i=1}^{N} y_i \sin\frac{2\pi n t_i}{T} \tag{2}$$

The above harmonic Fourier analysis is under the condition that the function (signal) is true periodic. However, engineering systems are highly complex, and the vibration record obtained from measurements on a vibrating machine or structure is an example of such a system. In practice, most vibration signals are actually considered as nonstationary and Figure 1.0 illustrates an example of this. Generally speaking, these signals are normally not periodic or very difficult to determine the periods. Therefore, when analysing nonstationary signals, using harmonic Fourier analysis alone is not sufficient. Huang [1996] recently investigated vibration modelling, which is formulated as a cosine function with a constant frequency and a random walk phase. This model is said to be 2nd order stationary and can be rewritten as an Auto Regression (AR) model as well as an Auto Regression Moving Average (ARMA) model. However, this model may not be satisfied since the natural frequencies are not indicated in the model. McCormick [1998] discussed a periodic time-variant AR model, which can be represented by a Fourier series and emphasised the difficulties in determining the coefficients and frequencies. Although AR and ARMA estimation have proven successful in analysing signals of an evolutionary harmonic or broad band nature, the problem of transient signal analysis can still not be adequately addressed.

3 FOURIER TRANSFORM AND WAVELET TRANSFORM

The Fourier transform (FT) analysis concept is widely used for signal processing, and gives the frequency content, but no time localisation. FT is particularly suited for signals global analysis where the spectral characteristics do not change with time. Windowed FT multiplies the signal by a windowing function, which makes it possible to look at features of interest at different times. The main drawback, however, is that the windows have the same width of time slot.

Wavelet transform (WT) developed during the last decade, overcome these limitations and is known to be more suitable for nonstationary signals where the description of the signal involves both time and frequency. The values of the time frequency representation of the signal provide an indication of the specific times at which certain spectral components of the signal can be observed. WT provides a mapping that has the ability to trade off time resolution for frequency resolution and vice versa. It is effectively a mathematical microscope, which allows the user to zoom on features of interest at different scales and locations.

The WT is defined as the inner product of the signal *y(t)* with a two-parameter family with the basis function:

$$WT(\tau,a)=|a|^{-\frac{1}{2}}\int_{-\infty}^{\infty} y(t)\Psi\left(\frac{t-\tau}{a}\right)dt \tag{3}$$

where $\Psi\tau, a$ is an oscillatory function.
τ is the time delay (translate parameter) which gives the position of the wavelet.
a is the scale factor (dilation parameter) which determines the frequency content.

The value $WT(\tau, a)$ measures the frequency content of *y(t)* in a certain frequency band within a certain time interval. The time-frequency localisation property of the WT and the existence of fast algorithms make it a tool of choice for analysing nonstationary signals [Strang 1996, Aldroubi 1996]. The technique of WT has been applied in such diverse fields as digital communications, remote sensing, biomedical signal processing, medical imaging, astronomy and numerical analysis.

4 VIBRATION MODEL ESTIMATION WITH FT AND HARMONIC WT

To try to analyse the complex vibration system in its finest detail, even more to try to predict its exact performance is an impossible task. However, we are often interested only in the main features of the vibration record. The analysis extracts the inherent properties of the system from the vibration signal. Therefore we measure the vibration signal with free running, with the assumption that the system is in good condition. We also assume:

i) Transient signals with short time duration do not reflect a system inherent property.
ii) The significant frequencies whose amplitudes change irregularly during the measuring do not reflect a system inherent property.

In equation (3), if we choose Morlet wavelet $\Psi(t)=e^{i\omega_0 t}e^{-t^2/2}$, which has been shown to yield the best time-frequency localisation [Aldroubi 1996], where $\omega_0 \geq 5.5$, let $a=\omega_0/\omega$, then from equation (3) we can observe any significant frequency (around it) whose amplitude changes with time by setting a,τ. (Note that the scale parameter appears proportional to the inverse of a frequency). Figure 2.0 is an example with a=0.9371, τ = 0~1000/32768 second, $f=9341Hz$, the original signal is shown in Figure 1.0.

The selection of a is very important [Aldroubi 1996]. When a>1 and increases, the width of the wavelet WT (τ, a) increases such that the frequency resolution increases. When a <1 and decreases, the interval over which the wavelet WT (τ, a) is non-zero is more closely tightened around the point τ, thus increasing temporal resolution, at the expense of frequency resolution.

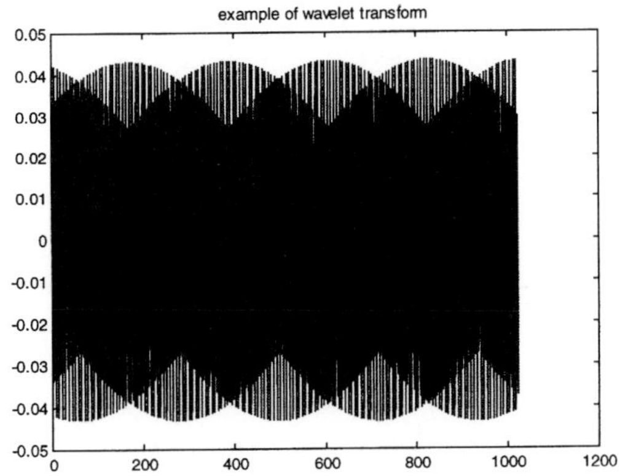

Figure 2.0 An example of a wavelet transform

The uncertainty principle which rules the limitation of localisation at any frequency in space time is expressed by:

$$\Delta t \cdot \Delta f \geq \frac{2}{\pi},$$

where Δt and Δf represent temporal and frequency deviations from a base-line. In the case of a Gaussian function, Δt is chosen equal to $2a$, so

$$\Delta f \geq \frac{2}{2\pi a} = \frac{2f}{\omega_0}.$$

In practice, variation range of the parameter a is chosen by the user according to the frequency domain of interest in the signal. This is continuous wavelet transform (CWT). It provides a natural tool for time-frequency signal analysis since each template $\Psi_{\tau,a}$ is predominantly localised in a certain region of the time-frequency plane with a central frequency that is inversely proportional to a. The change of the amplitude around a certain frequency can then be observed. What distinguishes it from the windowed FT is the multi-resolution nature of the analysis. However, CWT is time consuming, while discrete WT is computationally efficient. As discrete WT, harmonic wavelet transform (HWT) with fast algorithm can be used for vibration (with transient signal) analysis.

The advantage of using HWT in this case, over wavelets formed from dilation equations, is that each level of the HWT represents a distinct frequency band [Newland 1993]. Thus, by choosing an appropriate series length and sampling rate, the spectral frequency bands recommended by global FT analysis can be analysed more detailed.

Newland [1993] describes a form of wavelet analysis, the discrete harmonic wavelet transform (DHWT), which has the desirable property that the wavelet levels represent nonoverlapping frequency bands and can therefore be used to quantify specific frequency bands identified as having significant influence on the signal. In general a harmonic wavelet at level j (j an integer ≥ -1) translated by k steps of size $1/2^j$ is defined by:

$$W(\omega) = \begin{cases} \frac{1}{2^{j+1}\pi} e^{-i2\omega k/2^j} & for 2\pi 2^j \leq \omega \leq 4\pi 2^j \\ 0 & elsewhere \end{cases}$$

The elements of this wavelet family are mutually orthogonal and their coefficients can be computed using the fast Fourier transform. The HWT coefficients are used to construct a time-frequency map, as described in [Newland 1993], that shows how the frequency spectrum evolves with time. This is helpful to observe the duration of the significant frequencies that affect the signal. For a signal y of length $N=2^m$ there are m+1 wavelet levels (0,1,..., m) with a signal coefficient at level 0 and m, and 2^{j-1} coefficients at each subsequent level j = 1,2,..., m-1. The band of frequencies f_j represented by level j is given by

$$\frac{f_s}{2^{(m-j)+1}} \leq f_j \leq \frac{f_s}{2^{(m-j)}}$$

where f_s is the sampling frequency.

In general, DHWT is suitable for the detection of vibration transients, but the disposition in scale (as shown in Figure 3.0) is usually too coarse to correctly track the evolution of the WT coefficients through the different scale levels. In other words, DWT offers limited detailed information. Therefore, it would be better to combine the CWT and DHWT for vibration analysis.

From global Fourier analysis, we can calculate the significant frequencies whose amplitudes in the power spectrum are larger than a certain value. This value is selected with acceptable accuracy. Other frequencies may be considered as noises or negligible harmonics. Then CWT and DHWT analysis can distinguish the transient signals and the signals whose amplitude frequently changes irregularly from the significant frequencies. Only the significant frequencies, which reflect the system inherent properties, are kept.

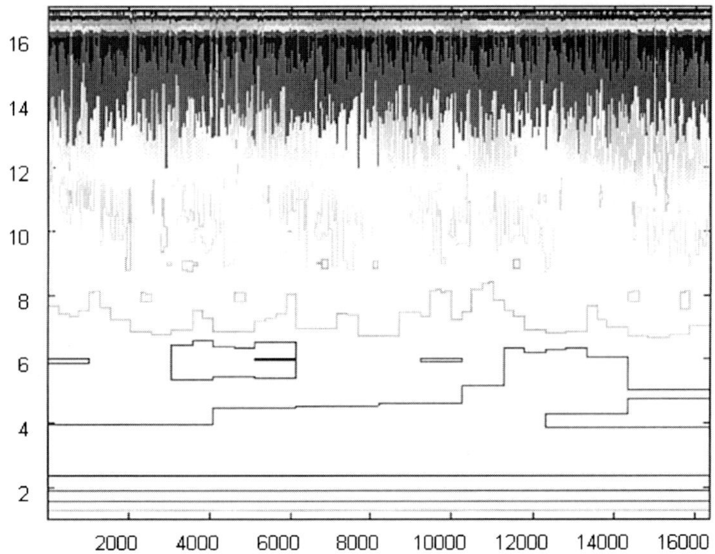

Figure 3.0 Map of DHWT (vibration signal *y* in Figure 1.0)

Finally, we can summarise these frequencies as $f_1, f_2, \ldots f_M$ with finite number M. Therefore, from equation (1), we have the estimated vibration model which can be written as

$$y(t) \approx \frac{a_0}{2} + \sum_{n=1}^{M} \left(a_n \cos 2\pi f_n t + b_n \sin 2\pi f_n t\right) \tag{4}$$

How to identify the values of coefficients $a_0, a_1, \ldots, a_M, b_1, \ldots, b_M$ is a problem that is described in the next section.

5 COEFFICIENTS IDENTIFICATION WITH LEAST-SQUARES ALGORITHM

Since T is unknown or does not exist, equation (2) can not be used. Alternatively a least-squares algorithm may be chosen. The least-squares principle is widely applicable to the design of signal processing system [Stearns 1990] and control system identification [Ogata 1987]. In this case, the signal *y* (*y*(1), *y*(2),..,*y*(*N*)) is given. Thus according to equation (4), we have

$$
\begin{bmatrix} y(1) \\ y(2) \\ . \\ . \\ . \\ y(N) \end{bmatrix} = \begin{bmatrix} \frac{1}{2} & \cos 2\pi f_1 t_1 \ldots \cos 2\pi f_M t_1 & \sin 2\pi f_1 t_1 \ldots \sin 2\pi f_M t_1 \\ \frac{1}{2} & \cos 2\pi f_1 t_2 \ldots \cos 2\pi f_M t_2 & \sin 2\pi f_1 t_2 \ldots \sin 2\pi f_M t_2 \\ . & . \quad . & . \quad . \\ . & . \quad . & . \quad . \\ . & . \quad . & . \quad . \\ \frac{1}{2} & \cos 2\pi f_1 t_N \ldots \cos 2\pi f_M t_N & \sin 2\pi f_1 t_N \ldots \sin 2\pi f_M t_N \end{bmatrix} \begin{bmatrix} a_0 \\ a_1 \\ . \\ . \\ . \\ a_M \\ b_1 \\ . \\ . \\ . \\ b_M \end{bmatrix} + \begin{bmatrix} \varepsilon(1) \\ \varepsilon(2) \\ . \\ . \\ . \\ \varepsilon(N) \end{bmatrix} \tag{5}
$$

where $\varepsilon(n)$ is the error. Let

$$
\mathbf{y} = \begin{bmatrix} y(1) \\ y(2) \\ . \\ . \\ . \\ y(N) \end{bmatrix}, \mathbf{C} = \begin{bmatrix} \frac{1}{2} & \cos 2\pi f_1 t_1 \ldots \cos 2\pi f_M t_1 & \sin 2\pi f_1 t_1 \ldots \cos 2\pi f_M t_1 \\ \frac{1}{2} & \cos 2\pi f_1 t_2 \ldots \cos 2\pi f_M t_2 & \sin 2\pi f_1 t_2 \ldots \cos 2\pi f_M t_2 \\ . & . \quad . & . \quad . \\ . & . \quad . & . \quad . \\ . & . \quad . & . \quad . \\ \frac{1}{2} & \cos 2\pi f_1 t_N \ldots \cos 2\pi f_M t_N & \sin 2\pi f_1 t_N \ldots \cos 2\pi f_M t_N \end{bmatrix}, \mathbf{x} = \begin{bmatrix} a_0 \\ a_1 \\ . \\ . \\ . \\ a_M \\ b_1 \\ . \\ . \\ . \\ b_M \end{bmatrix}, \mathbf{e} = \begin{bmatrix} \varepsilon(1) \\ \varepsilon(2) \\ . \\ . \\ . \\ \varepsilon(N) \end{bmatrix}
$$

From equation (5), we have **y=Cx+e**, therefore **e=y-Cx** (6)
Define

$$
J_N = \frac{1}{2}\sum_{n=1}^{N} \varepsilon^2(n) = \frac{1}{2}(\mathbf{e}^*\mathbf{e})
$$

Therefore

$$
J_N = \frac{1}{2}(\mathbf{y} - \mathbf{Cx})^*(\mathbf{y} - \mathbf{Cx}) = \frac{1}{2}(-\mathbf{x}^*\mathbf{C}^* + \mathbf{y}^*)(\mathbf{y} - \mathbf{Cx})
$$

$$
= \frac{1}{2}(-\mathbf{x}^*\mathbf{C}^*\mathbf{y} + \mathbf{y}^*\mathbf{y} + \mathbf{x}^*\mathbf{C}^*\mathbf{Cx} - \mathbf{y}^*\mathbf{Cx})
$$

To minimise J_N with respect to vector **x,** let

$$
\frac{\partial J_N}{\partial \tilde{\mathbf{x}}} = \frac{1}{2}(-\mathbf{C}^*\mathbf{y} + \mathbf{C}^*\mathbf{Cx}) = 0 \tag{7}
$$

here $\tilde{\mathbf{x}}$ is the estimation of coefficient vector **x**, from equation (7), we have

$$
\tilde{\mathbf{x}} = (\mathbf{C}^*\mathbf{C})^{-1}\mathbf{C}^*\mathbf{y} \tag{8}
$$

where $\mathbf{C}^*\mathbf{C}$ is not singular. That means $\mathbf{C}$ is not a singular matrix that is naturally satisfied with different time t and frequency f. Thus equation (8) is the optimal estimation of coefficients $a_0, a_1, \ldots, a_M, b_1, \ldots, b_M$.

The coefficients $a_0, a_1, \ldots, a_M, b_1, \ldots, b_M$ calculated are one kind of mean values which are the amplitudes of oscillation related to the significant frequencies.

6 SIMULATION

In order to simulate the vibration signal, we define a test signal with frequency and amplitude modulation as an approximation:

$$y = 2 + 30\sin(2\pi 100t + \sin 2\pi 20t) + 15\sin 2\pi 300t$$
$$+(30 + 8\sin 2\pi 10t + 20\sin 2\pi 20t)\sin 2\pi 200t + 25\sin 2\pi 30t + 28\sin 2\pi 400t$$

We choose the sampling frequency f_s =1000Hz, by Fourier and wavelet analysis, we may recognise $M = 13$ significant frequencies as f =30, 60, 80, 100, 120, 140, 180, 190, 200, 210, 220, 300, 400Hz. Therefore, equation (4), becomes:

$$y(t) \approx \frac{a_0}{2} + \sum_{n=1}^{13}\left(a_n \cos 2\pi f_n t + b_n \sin 2\pi f_n t\right)$$

From equation (5), and (8), the optimal coefficient estimation is calculated and given as follows:

$$a_0 = 4, a_1 = a_2 = a_3 = a_4 = a_5 = a_6 = a_9 = a_{12} = a_{13} = b_8 = b_{10} = 0, a_7 = 10, a_8 = 4,$$
$$a_{10} = -4, a_{11} = -10, b_1 = 25, b_2 = 3.4471, b_3 = -13.2015, b_4 = 22.9559, b_5 = 13.2015,$$
$$b_6 = 3.4471, b_7 = 0.0743, b_9 = 30.0075, b_{11} = 0.0006, b_{12} = 15, b_{13} = 28$$

Then the estimated model is achieved.

Comparing the Fourier transform of the model with that of the test signal y, it is evident that both are quite similar (Figure 4.0). That means the model approximates the test signal since the signal and its Fourier transform are invertible.

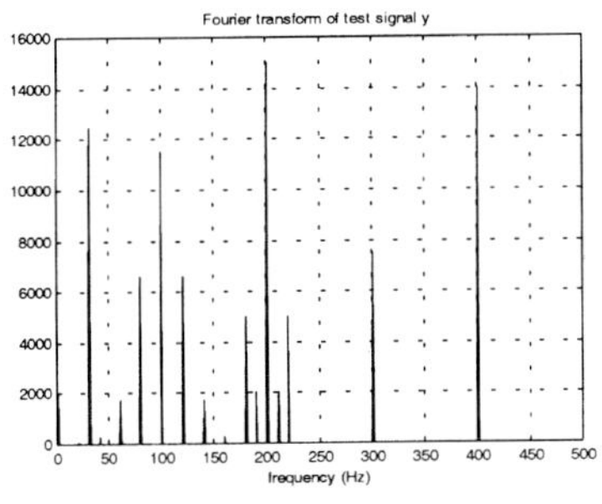

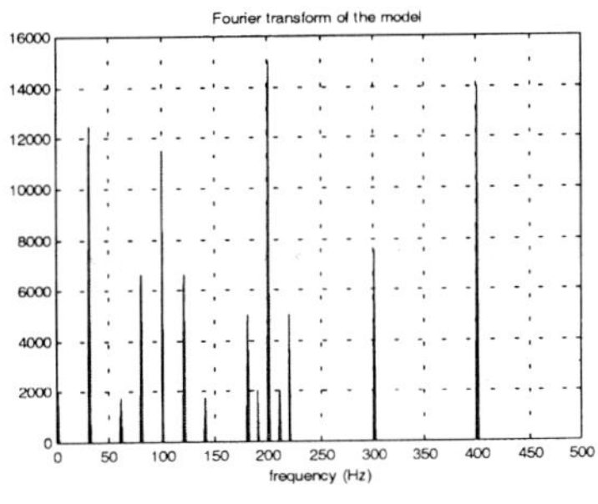

Figure 4.0 Fourier transform of test signal *y* and the model

7 RESULTS AND DISCUSSION

The example vibration signal *y* shown in Figure 1.0, was measured with the machine spindle nose running in the idle condition. A sampling frequency of f_s = 32768 Hz was used with a measuring time of 2 seconds. The FT of this signal is shown in Figure 5.0. There are a number of significant frequencies identified from the analysis, these are as follows:

f = 100, 150, 200, 2374, 2402, 2497, 4292, 4443, 4746, 4804, 4898, 6541, 9038, 9190, 9291, 9341, 9391, 11686, 11714, 11742, 11838, 14088, 14144, 14239, 14391, 15882, and 16034 Hz.

In order to analyse the signal in more detail, the method of continuous and discrete wavelet analysis can be used. The signal length is $N = 2^{16}$, therefore m = 16. Figure 3.0 is the frequency-time map calculated by DHWT, from which we can see that frequencies under level 9 are all transient signals.

By CWT and DHWT analysis the transient signals and the signals whose amplitude frequently changes irregularly, can be distinguished from the significant frequencies. Figure 6.0 is an illustration of a signal whose amplitude

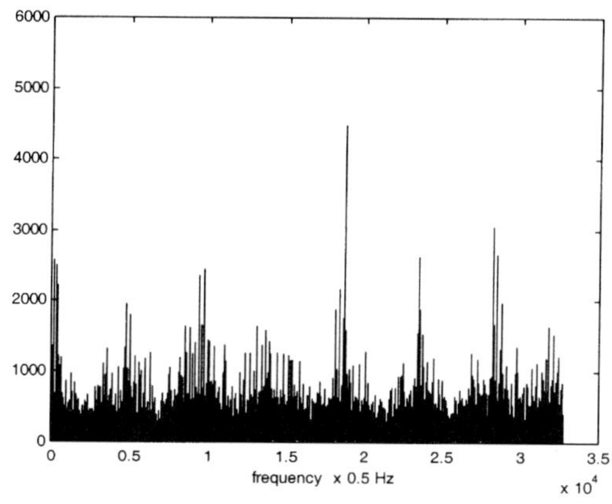

Figure 5.0 FT analysis of signal shown in Figure 1.0

frequently changes irregularly, (f = 11838 Hz, τ = 1000/32786 ~ 1500/32786 second).

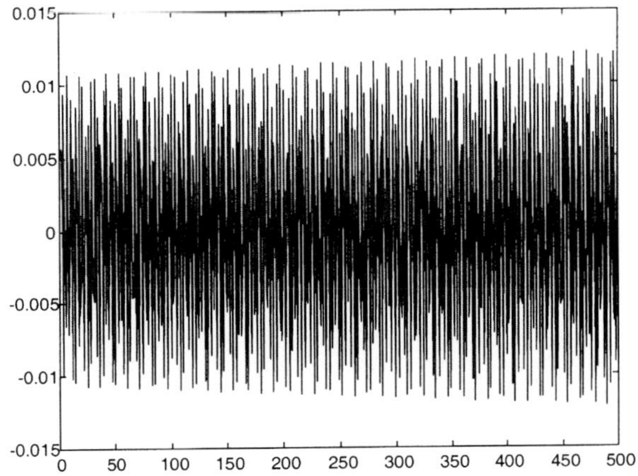

Figure 6.0 A signal with irregular changing amplitudes.

The regular frequencies are given in Table 1.

Table 1 Regular frequencies identified from the signal illustrated in Figure 6.0.

Level j	Frequency band (Hz)	Significant frequencies (Hz)
0		
1		
. . .		
9	128-256	150, 200
10	256-512	
11	512-1024	
12	1024-2048	
13	2048-4096	2374, 2402
14	4096-8192	4746, 4898, 6844
15	8192-16384	9038, 9190, 9341, 11686, 11714, 11742, 14088, 14239, 14391, 15882

Then we get the significant frequencies of the signal as shown in Table 1, where M = 17. It is these frequencies (with periodic nature) that reflect the systems' inherent properties. The estimated model is obtained according to equation (4).

$$y(t) \approx \frac{a_0}{2} + \sum_{n=1}^{17} \left(a_n \cos 2\pi f_n t + b_n \sin 2\pi f_n t \right)$$

We need to identify the coefficients, therefore using equation (5) we have

$$
\mathbf{y}=\begin{bmatrix} -0.2440 \\ 0.7420 \\ \cdot \\ \cdot \\ \cdot \\ 0.0980 \end{bmatrix},\quad
\mathbf{C}=\begin{bmatrix}
\frac{1}{2} & \cos 2\pi 150 t_1 \; \dots \cos 2\pi 15882 t_1 & \sin 2\pi 150 t_1 \; \dots \cos 2\pi 15882 t_1 \\
\frac{1}{2} & \cos 2\pi 150 t_2 \; \dots \cos 2\pi 15882 t_2 & \sin 2\pi 150 t_2 \; \dots \cos 2\pi 15882 t_2 \\
\cdot & \cdot \quad\quad \cdot & \cdot \quad\quad \cdot \\
\cdot & \cdot \quad\quad \cdot & \cdot \quad\quad \cdot \\
\cdot & \cdot \quad\quad \cdot & \cdot \quad\quad \cdot \\
\frac{1}{2} & \cos 2\pi 150 t_N \; \dots \cos 2\pi 15882 t_N & \sin 2\pi 150 t_N \; \dots \cos 2\pi 15882 t_N
\end{bmatrix}
$$

From equation (8), we obtain: $a_0 = 0.1808$

$a_1 = 0.0031$	a_2 =-0.0328	$a_3 = 0.0096$	a_4 =-0.0026	$a_5 = 0.0443$	$a_6 = 0.0682$
a_7 =-0.0149	a_8 =0.0009	a_9 =0.0072	a_{10} =0.0890	a_{11} =0.0045	a_{12} =-0.0388
a_{13}=0.0053	a_{14} =0.0109	a_{15} =0.0125	a_{16} =0.0000	a_{17} =-0.0081	
b_1 =-0.0763	b_2 =-0.0594	b_3 =0.0019	b_4=-0.0032	b_5 =0.0565	b_6 =-0.0310
b_7 =-0.0458	b_8 =-0.0044	b_9 =0.0055	b_{10} =-0.0258	b_{11}=-0.0066	b_{12} =-0.0444
b_{13}=-0.0571	b_{14} =0.0076	b_{15} =-0.0106	b_{16} =0.0128	b_{17} =0.0023	

To finalise the modelling.

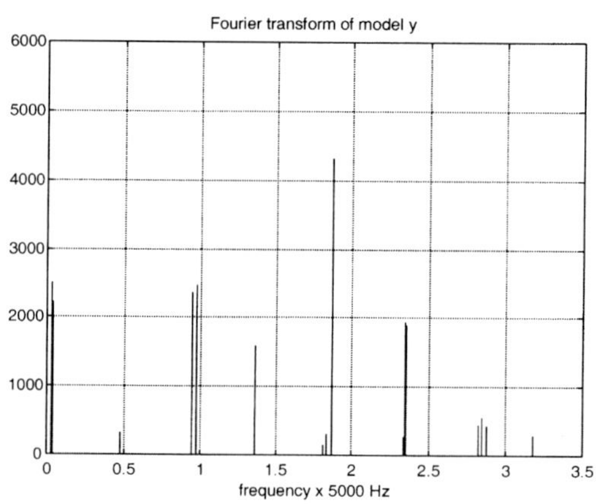

Figure 7.0 The FT of the model.

Most of the natural frequencies, which normally are periodic and have significant amplitude peak in power spectrum, can be identified in the model. Since the model extracts the system inherent property, a change in the signal, possibly due to a change in machine condition, will result in parameter change in the model. On the other hand, it is possible to avoid the resonant frequencies, which may appear in self-excited vibration (cutting operation) and damage the machine and workpiece unacceptably.

The excellent frequency resolution of WT allows us to analyse the signal in more detailed. Then the vibration signal can be decomposed and it becomes possible to obtain more information, allowing a more accurate model to be achieved. Least-squares algorithms seems to be a good solution for the purpose of coefficient identification, whilst the harmonic Fourier method for coefficient calculation involves the use of period T, which is unknown in the vibration signal.

Vibration is essentially lost energy caused by rotational problems of component imbalance, misalignment, bearing faults, and mechanical fits, *etc.*, which are common to all plants. A change of vibration signal or signature highlights a deviation from normal operation, whilst our vibration model is based on the normal condition and stable-state operation. Therefore, monitoring for the changes of the vibration signal can provide us with diagnosis, including the detection of faults in structures, rotor systems, bearings, gearboxes, engines, and so on. This can be achieved by analysing the change of natural frequencies and their amplitudes in the model.

8 CONCLUSIONS

A modelling approach for complicated vibration behaviour has been proposed. The model approximately represents the vibration signal in a pure mathematical function. Two methods of signal processing, Fourier transform and wavelet transform were investigated for this purpose. The high performance in time-frequency domain of wavelet analysis demonstrates its potential for signal analysis of this area.

Vibration analysis provides a quick and relatively easy way to detect and identify minor mechanical problems before they become serious and force costly unscheduled shutdown. Worn bearings, loose belts, improperly meshed gears, unbalanced shafts, misalign coupling, tool wear, *etc.*, are accompanied by specific change in signature. Monitoring of these changes permits maintenance to be planned in advance of a major breakdown. In addition, it may be possible to identify workpiece material property and finish surface quality during machining by vibration analysis. This could provide the opportunity for the setting of optimal control parameters in machining, determined on-line by wavelet analysis of the vibration signal. This is the topic for future work.

9 ACKNOWLEDGEMENT

The authors wish to acknowledge Dr Homer Rahnejat and his team in the Department of Mechanical and Manufacturing Engineering, University of Bradford, for their assistance.

10 REFERENCES

ALDROUBI, A., UNSER, M. 'Wavelets in medicine and biology'. CRC Press, Inc., Florida, 1996.

BOLTON, W. 'Fourier series' Longman Scientific and Technical, 1995.

DIMAROGONAS, A.D. 'Vibration for engineers'. Prentice-Hall, Inc., New Jersey, 1992.

HUANG, D.W., and SPENCER, N.M. 'On a random vibration model', Journal of Applied Probability, Sheffield, Dec 1996. 1141-1158.

KRANIAUSKAS, P. 'Transforms in signals and systems' (Addison-Wesley Publishing Company Inc., 1992.

McCORMICK, A.C., NANDI, A.K., and JACK, L.B. 'Application of periodic time-varying autoregressive models to the detection of bearing faults', Proc Instn Mech Engrs 1998. **212**:Part C, 417-428.

NEWLAND, D.E. 'An introduction to random vibration, spectral and wavelet analysis'. Longman Scientific and Technical, 1993.

OGATA, K. 'Discrete-time control systems' Prentice-Hall, Inc., 1987.

RAO, S.S. 'Mechanical vibration'. Addison-Nesley Publishing Company, Inc., 1990.

STRANG, G., and NGUYEN, T. 'Wavelet and filter banks'. Wellesley-Cambridge Press, 1996.

STEARNS, S.D., and HUSH, D.R. 'Digital signal analysis'. Prentice-Hall, Inc., New Jersey, 1990.

Modelling and simulation of a vehicle dynamometer using hybrid modelling techniques

A A ABDUL-AMEER, H BARTLETT, and **A S WOOD**
Department of Mechanical and Medical Engineering, University of Bradford, UK

ABSTRACT

This paper details the modelling, simulation and control of a vehicle dynamometer using distributed-lumped parameter techniques. This technique has been successfully applied to a variety of engineering applications in which the system parameters are partially distributed and partially lumped. The vehicle dynamometer lends itself to this method since the inertia of the rolls and shaft are distributed whilst the damping due to the bearings and the motor can be considered as lumped systems/elements. This method enables the analysis of the dynamic performance of the dynamometer which is not possible using lumped or point wise techniques.

The dynamic performance of the dynamometer is investigated in this paper, using a conventional three-term controller for maintaining the roll speed under step, ramp and impulse disturbances.

NOTATION

c_d	damping coefficient (bearing friction)
d_0, d_2	diameter of the shafts
d_{1i}, d_{3i}	inside diameters of the rolls
d_{1o}, d_{3o}	outside diameters of the rolls

E_a	armature voltage
G	modulus of rigidity (shear modulus)
J_j	shaft or roll polar moment of inertia (j = ith element –1)
$l_o, l_1, l_2, l_3,$	lengths
K	gain constant
K_m	motor gain constant
s	Laplace variable
T_j,	shaft torque (j = ith element –1)
T_o	load torque
t	time
x	distance along distributed section
z	finite time delay
ω_j	angular velocities (j = ith element –1)
ξ_j	characteristic impedances (j = ith element –1)
Γ_j	propagation operator (j = ith element –1)
τ_m	motor time constant

1 INTRODUCTION

Rolling-Road dynamometers are used to simulate the moving road. Consequently, the rolling road dynamometer can be considered as an instrument used for measuring power, torque and represents the dynamic performance of the vehicle engine. In recent years, almost all vehicle manufactures use dynamometers to avoid the inaccuracy and non-repeatability of driving a car on the open road. Fuel economy, dynamic performance, brake testing, NVH (noise, vibration, harshness), mileage accumulation, emissions and EMC (electromagnetic compatibility chambers) are a few of the areas which vehicle dynamometers may be used for research and future development [1].

In the early 1980's most of the Rolling-Road dynamometers utilised eddy current or hydraulic brakes to absorb the generated power. These systems suffered from two main problems. Firstly, their operating characteristics could not accurately match the operating characteristics of a car on the road, which is affected by inertial, frictional and aerodynamic forces. Secondly, at low speed the frictional forces of such brakes are greater than the frictional forces that must be absorbed from the modern light-weight small cars [2]. Therefore, testing using dynamometers requires the use of large diameter rollers to minimise the tyre deflection.

With the consecutive changes in machine and control design, there is a large demand to improve the performance of Rolling-Road dynamometers because of their wide use in the automotive industry. Therefore, in order to achieve a good dynamometer performance, a proper simulation model for the dynamometer is required in order to study all variable effects on the dynamometer's performance.

This paper describes the development and application of a simulation model for a vehicle Rolling-Road dynamometer. Due to the topology of such a system it can not be considered as a purely lumped system nor would it be practical to assume it to be purely distributed. Consequently a partially distributed and partially lumped assumption is adapted. The drive shaft, rolls and interconnecting shaft are considered to be distributed elements whilst the motor and bearing characteristics are taken to be lumped elements. This forms the basis of the hybrid modelling technique developed by Whalley [3,4] and Bartlett [5] and is used in this paper to model the vehicle Rolling-Road dynamometer. Results from the simulation model following step and ramp speed changes, together with impulse, disturbances are presented.

2 VEHICLE ROLLING-ROAD DYNAMOMETER

By definition a dynamometer is an instrument used for measuring power. In the case of the vehicle dynamometer, the power is produced by a test vehicle. A common phrase used to describe a vehicle dynamometer is 'Rolling-Road' [1]. Generally dynamometers enable vehicle manufacturers to carry out full dynamic testing on a vehicle under laboratory conditions without the inaccuracy and poor repeatability introduced by a human driver on the open road.

Figure 1 Picture of vehicle dynamometer showing the shaft and rolls.

Figure 1 shows the structure of two twin-roll dynamometers before the operating floor level has been assembled. Two large diameter rolls with their interconnecting shaft, D.C. motor and drive shaft can be clearly seen in the picture. Testing using a rolling-road dynamometer requires the use of large diameter rolls to minimise the tyre deflection. This has the purpose of keeping the rolling resistance of the tyre, caused by hysteresis, as close as possible to the resistance on a flat road. Dynamometers can operate on two wheels of the test vehicle, which can be achieved using either a single twin roll dynamometer or individual single roll dynamometers. Rolling-Road dynamometers often undergo an arduous duty cycle comprising rapid acceleration to reach steady-state speed, followed by load changes due to vehicle testing and rapid retardation upon completion. These duty cycles invariably induce torsional oscillations in the connecting shaft, the motor drive shaft and, to a lesser extent, in the rolls.

3 VEHICLE DYNAMOMETER SIMULATION MODEL

Tighter tolerances for vehicle dynamometer specifications are continually requiring improvements in dynamometer performance with consequent changes in machine and control design. A simulation model will enable the design engineer to analyse a motor/shaft/roll system giving the capability of predicting a dynamometer's performance and thus enableing the control system to be tuned for higher performance rather than simply compensating for a less than optimal mechanical arrangement.

The vehicle dynamometer model could not sensibly be addressed using a point-wise analysis owing to the dispersed nature of the system and use of variable – geometry, wholly distributed analysis using partial differential equations throughout would not be advisable. Due to the dimensions of the arrangement it is evident that a reasonable proposition is to treat the end trunions, motor armature inertia and bearings as lumped point-wise entities. Equally the rolls with their wide face and wall thickness should be considered as distributed elements. The motor drive shaft and rolls connecting shaft should also be considered as distributed elements.

The system in Figure 2 shows a schematic arrangement of a vehicle dynamometer. Owing to the topology of the dynamometer, the hybrid modelling techniques can be employed for the modelling of the vehicle dynamometer.

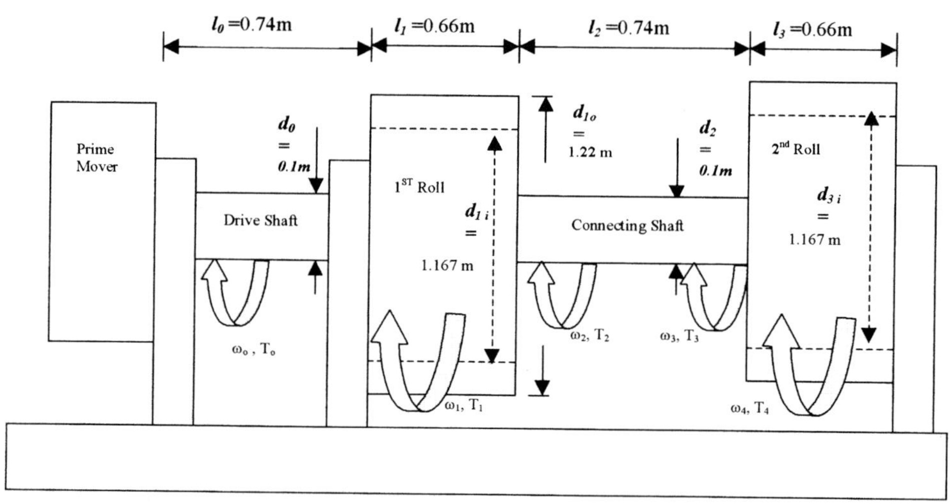

Figure 2 Schematic representation of the vehicle dynamometer.

For each shaft and roll elements the partial differential equations representing the relationship between the angular velocity and the torque can be expressed as

$$\left.\begin{aligned}\frac{\partial \omega_j}{\partial x}(x,t) &= -J_j\rho_j\frac{\partial T_j}{\partial t}(x,t)\\ \frac{\partial T_j}{\partial x}(x,t) &= -\frac{1}{G_jJ_j}\frac{\partial \omega_j}{\partial t}(x,t)\end{aligned}\right\} \quad \textbf{(1)}$$

j = i element - 1
i = 1,2,3,...

The analytical solution to equation (1) is given by Bartlett and Whally [6], and can be expressed in finite-delay impedance form as

$$\begin{bmatrix}\omega_j\\ \omega_{j+1}\end{bmatrix} = Z_j\left(z_j^{-1}\right)\begin{bmatrix}T_j\\ T_{j+1}\end{bmatrix} \quad \textbf{(2)}$$

and in admittance form as

$$\begin{bmatrix}T_j\\ T_{j+1}\end{bmatrix} = Z_j^{-1}\left(z_j^{-1}\right)\begin{bmatrix}\omega_j\\ \omega_{j+1}\end{bmatrix} \quad \textbf{(3)}$$

where

$$Z_j\left(z_j^{-1}\right) = \xi_j\left\{\begin{matrix}\dfrac{1+z_j^{-1}}{1-z_j^{-1}} & -\dfrac{2z_j^{-1/2}}{1-z_j^{-1}}\\ \dfrac{2z_j^{-1/2}}{1-z_j^{-1}} & -\dfrac{1+z_j^{-1}}{1-z_j^{-1}}\end{matrix}\right\}, Z_j^{-1}\left(z_j^{-1}\right) = \xi_j^{-1}\left\{\begin{matrix}\dfrac{1+z_j^{-1}}{1-z_j^{-1}} & -\dfrac{2z_j^{-1/2}}{1-z_j^{-1}}\\ \dfrac{2z_j^{-1/2}}{1-z_j^{-1}} & -\dfrac{1+z_j^{-1}}{1-z_j^{-1}}\end{matrix}\right\}$$

$$z_j^{-1} = e^{-2\Gamma_j l_j} \quad \textbf{(4)}$$

$$\Gamma_j = s\sqrt{\frac{\rho_j}{G_j}} \quad \textbf{(5)}$$

$$\xi_j = J_j\sqrt{\rho_j G_j} \quad \textbf{(6)}$$

Defining the 1st element as the motor drive shaft, the 2nd element as the 1st roll, the 3rd element as the rolls connecting shaft and the 4th element as the 2nd roll gives, the subscripts in equation 1, j = 0,1,2 and 3 respectively. Therefore, the distributed elements of the vehicle dynamometer can be represented and modelled using equations (2) and (3), j=0, 1, 2 ,3.
The non-distributed elements (lumped elements), such as damping associated with the bearings, can be represented as

$$\omega_4 = c_d T_4 \quad \textbf{(7)}$$

For the prime mover a D.C armature controlled motor is used with the following Input/Output relationship

$$\omega_o = \frac{K_m}{(1 + s\tau_m)}[KE_a - T_o] \quad \textbf{(8)}$$

Equations (1) to (8) yield a complete simulation model of the vehicle dynamometer, as shown in Figure 3.

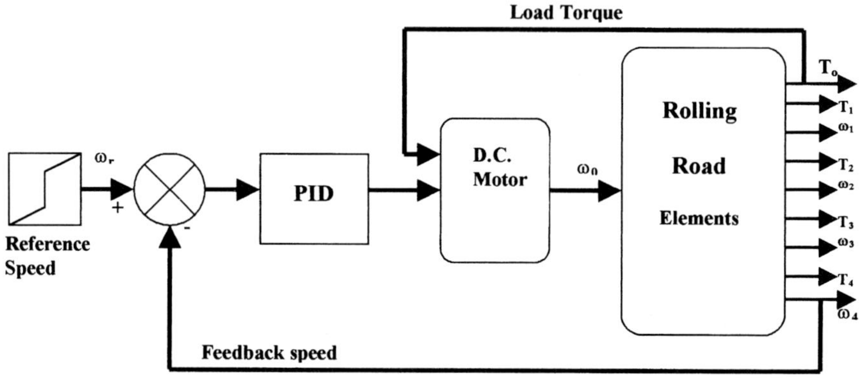

Figure 3 The complete vehicle dynamometer simulation model

4 VEHICLE DYNAMOMETER TESTING SPECIFICATIONS

There are four empirical design specifications and tolerances that specify the dynamic performance of the vehicle dynamometer before any vehicle testing is carried out. These are the steady-state error, ramp overshoot, settling time and the maximum torque fluctuations. These empirical specifications can be summarised as follow:

a. The steady-state error is the difference between the actual and measured dynamometer speed, which must be no more than 1% of the maximum operating speed of the rolls.

b. The overshoot is the difference between the maximum dynamometer speed after the input ramp has finished. This value must be no more than 2% of the maximum roll speed.

c. The settling time is the time period which it takes the dynamometer speed to reach the steady-state speed after the input ramp has stopped. This value must not exceed 1 second.

d. Torque fluctuations are measured when the control system has achieved steady-state running at a constant speed. These fluctuations are caused by the control system continually trying to correct the speed and hold it at its speed demand. This value must be less than 0.5% of the maximum torque rating of the motor.

5 THE CONTROL SYSTEM

In the vehicle dynamometer model, the steady-state error, peak overshoot and settling time were chosen to be the major characteristics to develop the control system of the dynamometer. In order to achieve this, the control system of the dynamometer was operated in closed loop control using the dynamometer 2nd roll speed as the feedback variable for comparison with the required speed demand. Consequently, a PI controller was used. The proportional band and the integral action gains K_P and K_I respectively were calculated experimentally through the calculation of the natural frequency and damped natural frequency of the system.

6 VEHICLE DYNAMOMETER MODEL DATA

The vehicle dynamometer under investigation had the parameter values listed in Table 1.

The Motor	
K_m motor gain constant	2.91×10^{-3} rad/Nms
K gain constant	342 Nm/v
τ_m motor time constant	2.1×10^{-5} s
Rolls & Shafts material specifications	
G modulus of rigidity	80×10^{9} N/m^2
ρ material density	7800 kg/m^3
Final termination	
c_d bearing damping coefficient	42 rad/Nms
Control system gains	
K_P proportional band gain	0.6
K_I integral band gain	77.6 s

Table 1. Typical vehicle dynamometer parameters values.

Simulation results were obtained and analysed for the dynamic performance of the vehicle dynamometer using the data of Table 1. The dynamometer's dynamic performance was analysed by carrying out the following tests,

1. 50 km/h roll speed with 5 m/s input ramp rate.
2. Step change in the motor input speed.
3. Vehicle brake testing (impulse load change).

7 RESULTS AND DISCUSSION

Section 3 described the specification for the vehicle dynamometer. In order to achieve these specifications it was necessary to tune the PI controller using the conventional method of Ziegler-Nichols which is based upon determining the undamped natural frequency of the system. The PI values were evaluated as 0.6 and 77.6 respectively. Using these values the ramp, step and impulse tests were carried out in order to investigate the dynamic performance of the dynamometer under closed loop control.

Figure 4 shows the dynamic performance of the dynamometer when subjected to a ramp input rate of 5m/s with a desired steady-state roll speed of 50 km/h. It can be seen that the control system achieves steady-state in less than 7 seconds following the completion of the ramp input. The dynamic response of the rolls speed, shown in Figure 5a, describes a damped oscillatory response with less than 2% maximum overshoot.

Figure 5b shows the dynamic responses of the torque in the system. It can easily be seen that during the dynamic period there is a significant difference between the torque developed across both sides of the rolls thereby giving rise to dynamic shear stresses in the shafts, as shown in Figure 5c. It can also be seen that the dynamic load torque, T_0, oscillates with an overshoot that exceeds the steady-state torque by 300%. Increasing the ramp input rate above 5m/s would increase the load torque to the extent that the motor would be unable to deliver the necessary driving torque due to current limits on the motor.

Although ramp inputs of greater than 5m/s are seldom used, occasionally a step change in the desired roll speed is required. Figure 5 shows the dynamic performance of the dynamometer after a unit step change in roll speed. The interesting feature of these results is shown in figure 5b where the load torque oscillation T_0 peak at 3500Nm per unit change in roll speed. Consequently, changes in the roll speed of more than 1 unit change will result in excess motor torque being developed. Unless this is limited, either the motor will burn out or the drive shaft will fail due to excessive shear stresses. The torque developed at T_2 and T_3 shown in Figure 5b shows a complex interaction resulting high frequency oscillations being superimposed on a damped oscillatory response. These high frequency oscillations could give rise to severe noise or vibration problems and therefore requires further investigation. Identifying these high frequency oscillations would not have been possible if a totally lumped model had been used.

Figure 6 shows the dynamic response of the dynamometer following a change in load torque, often referred to as braking torque. Figure 6a shows the speeds ω_r, ω_1 and ω_3 following a 500Nm step change in braking torque applied to each roll simultaneously. It can be seen that the change in roll speed is less than 2% of its steady-state value. It is interesting to note from Figure 6b that the high frequency oscillations in the torque responses T_2 & T_3 are clearly visible once again.

8 CONCULUSIONS

This paper has clearly shown that hybrid modelling techniques offers a viable method to modelling and simulating systems that are partially distributed and partially lumped. Using this technique on a vehicle dynamometer has highlighted the presence of a high frequency torque oscillations which could not be identified using conventional lumped parameter models.

REFERENCES

1. Bartlett, H. and Moon, S. E., " Vibration analysis and control of a vehicle dynamometer ", ASME, Structural Dynamics and Vibration, 1993, PD - Vol. 52, pp. 101 - 108.
2. Dutkiewicz, D. and Nates, R., " Design and construction of a rolling - road dynamometer for road simulation ", South African Mech. Engrs. 1985, Vol. 35, No. 3, pp. 76 - 79.
3. Whalley, R., " The response of distributed - lumped parameter systems ", Proc. Inst. Mech. Engrs. 1988, Vol. 202, No.C6, pp. 421 - 429.
4. Whalley, R., " Interconnected spatially distributed systems ", Trans. Inst. MC, 1990, Vol. 12, No. 5, pp. 262 - 270.
5. Bartlett, H. and Whalley, R., " Analogue solution to the modelling and simulation of distributed - lumped parameter systems ", Proc. Inst. Mech. Engrs. 1998, Vol. 212, part I, pp. 99 - 114.

a. Rolls speed compared with reference input speed (ω_r).

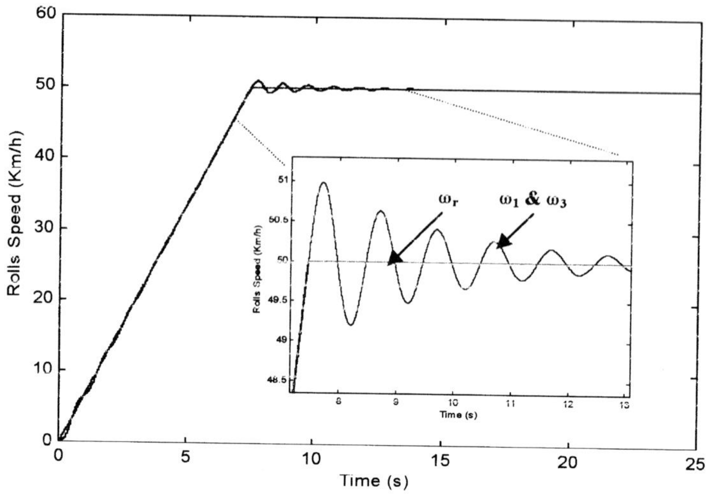

b. Dynamometer element torque variations.

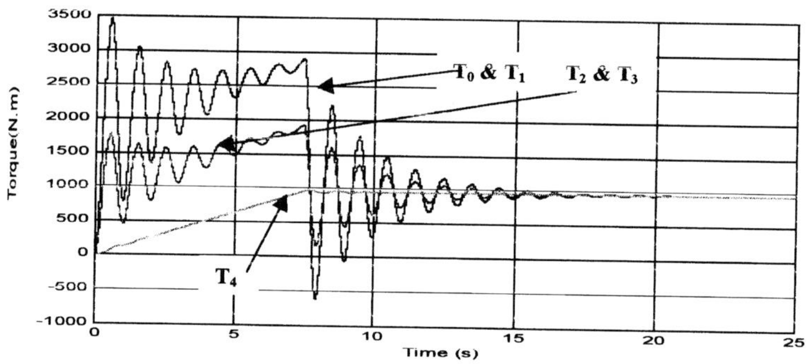

c. The shear stress of the driving and connecting shaft.

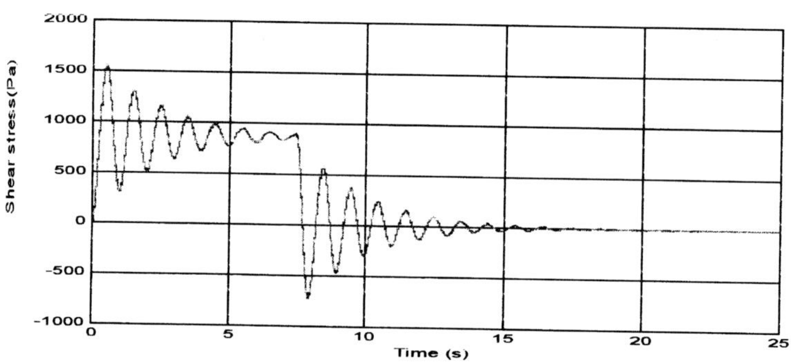

Figure 4, The vehicle dynamometer dynamic performance during a roll speed of 50 km/h with a 5 m/s input ramp rate.

a. Rolls speed changes.

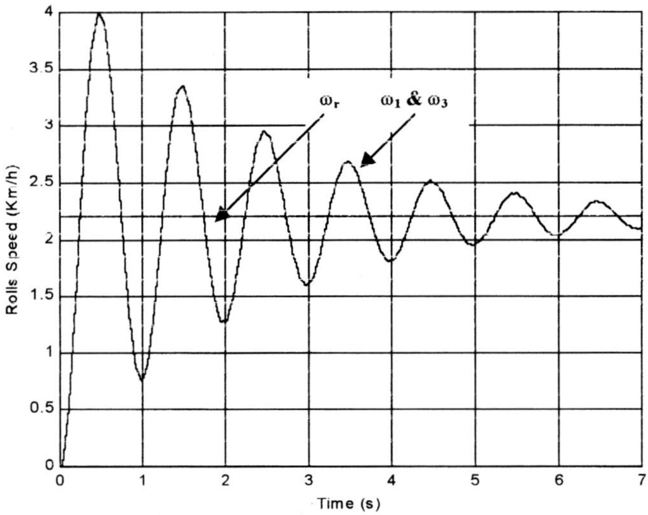

b. Dynamometer element torque variations.

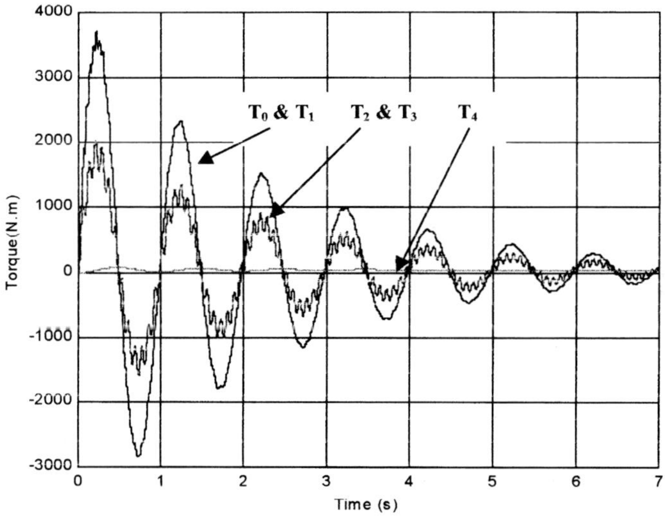

Figure 5 The vehicle dynamometer dynamic performance for a step change in the roll speed.

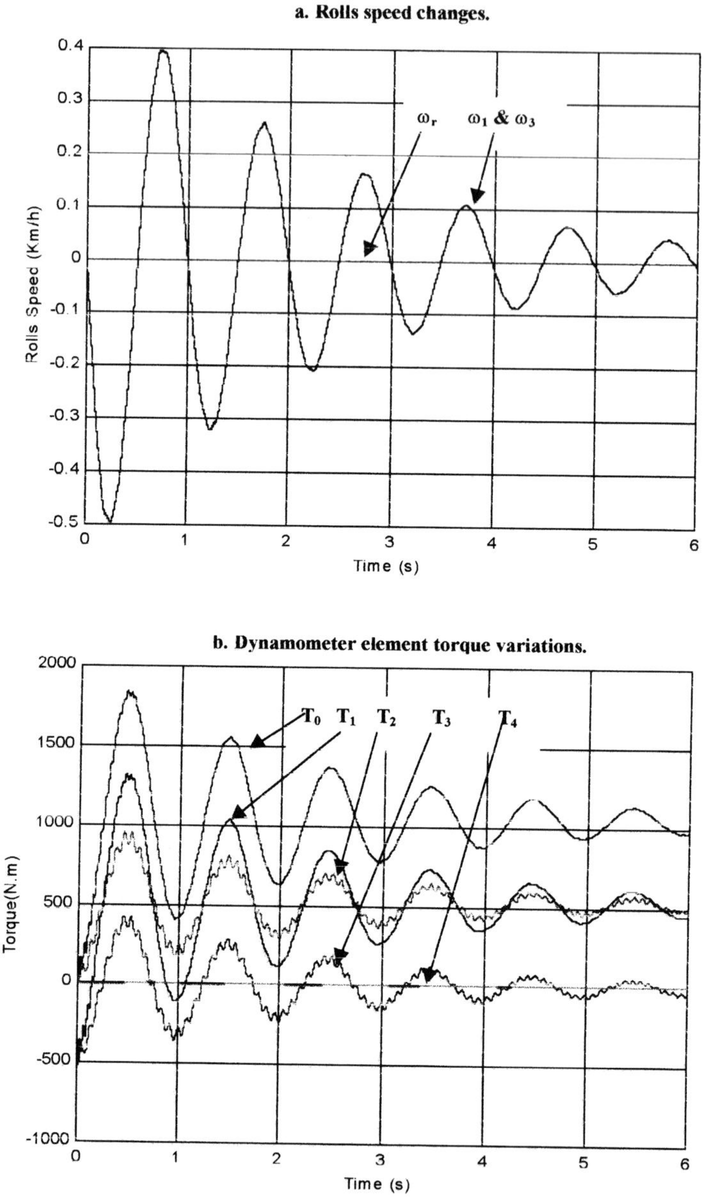

Figure 6 The vehicle dynamometer dynamic performance during an impulse disturbance (vehicle brake testing).

End milling and its effects on the spindle drive mechanism

Y HADI, M EBRAHIMI, and **H QI**
Department of Mechanical and Medical Engineering, University of Bradford, UK

SYNOPSIS

A mathematical model of the cutting process and spindle drive system in end milling is presented. The model predicts variations of the cutting forces according to a range of cutting conditions, cutter geometry's, the relationship between the cutting forces and the chip load, and geometry of cutting process, including relative positions of cutter to workpiece. In order to verify the force model, a series of experiments has been carried out. A block diagram model of the cutting process in milling operations has been developed through an analytical approach in order to predict the cutting forces. The model includes the overall deflection in the spindle drive mechanism. The model parameters were measured experimentally so that the model can be simulated.

1. INTRODUCTION

The data from the metal cutting process contain valuable information about the machine tool dynamics. This is because the cutting process is the one relationship of the whole machine tool dynamic which closes the loop between the axis feed subsystem and the spindle subsystem [1]. Therefore cutting process that affects the control of the servomechanism by generating the disturbance torque on the feed and the spindle motor.

In the milling operations the investigation of the cutting forces involved is more difficult than in other cutting operations due to the nature of the milling process. Milling is a cutting process during which the values of the cutting forces are changing periodically depending on the number of cutting edges in contact at any instant. The cutting forces produced by the milling process are dependent upon the metal contact conditions. These conditions are related to the type of milling operation and the tool employed. Many researchers, Koenigsberger and Sabberwal [2] studied the cutting force pulsation by using three component dynamometer. The most recent works by Tulsty and MacNeil [3] in end milling and by Gygax [4,5] in face milling introduced a much more detailed analysis of the cutting process in milling operations. The process can be modelled as a process with two inputs; feed rate f and spindle speed n, as shown in figure 1. The outputs of the model, the Y-direction component of the cutting force F_Y acting on the spindle and the X-axis component acts on machine table. During the cutting deflection in the spindle mechanism, causes cutter displacement; this effect is included in the model by considering a lumped stiffness of the spindle mechanism corresponding to the radial cutting force component.

The aim of this paper is to analyse the process in physical terms and hence to develop a cutting process model that relates the feed rate, spindle speed, and depth of cut to the cutting forces generated. The model is based upon the accepted methods of predicting the force in milling operations presented by Tulsty [6] and Gygax [7]. First there is a mathematical analysis of the cutting process, second the experimental methods for measuring the physical parameters are discussed, and finally a block diagram model of the process is generated that is programmed in the Simulink toolbox of the Matlab software package [16].

2.CUTTING PROCESS DYNAMICS

The following methods for predicting forces in up-milling operations were presented by Martellotti and DeVor [8,9]. The force diagram during the cut of a single tooth is given in Figure 1, where θ, is the instantaneous angular position of the cutter. Before commencing an investigation into the nature of the cutting forces and analysis of the work done during milling operation, the geometry of the chip thickness has to be studied. For the case in which the tooth path is assumed to be circular, the analytical expression is assumed, and using Martellotti principal equations. The chip thickness of the cross section of the unreformed chip is deriving. The chip thickness that lies in the front of the tooth at any instant is calculated in discrete time from the instantaneous position of the tool edge and the workpiece surface where the edge is in contact.

$$h = \frac{f_t\left[\sqrt{(2rR_d - R_d^2)} - f_t\right]}{\sqrt{\left[r^2 - 2f_t\sqrt{(2rR_d - R_d^2)} + f_t^2\right]}} \tag{1}$$

In case of slot mill (full immersion) process the width of cut or the radial depth of cut is fixed for all cutting paths, and the thickness of the chip changes with respect to the changes of spindle speed and feed rate.

For $R_d = r$, $h = f_T$

$$\text{For } R_d = 2r, h = \frac{-f_t^2}{\sqrt{r^2 + f_t^2}} \tag{2}$$

For the straight tooth cutter, chip area at any instance $= R_d h$.

The feed per tooth directly determined from measurements of the servo drive system and spindle, it is extremely important to understand the fast increase of the cutting force. For two-fluted cutter the maximum is reached at the depth of cut, which is axial depth A_d, and this depends on cutter radius r and width of cut R_d [10]. Assume the helix angle β is equal 30°, thus:

For $R_d = r$, $A_d = 0.04r$

For $R_d = 2r$, $A_d = 0.17r$

For $\alpha_{ex} \leq \theta \leq \alpha_{en}$ where α_{en} is the angle at which the cutter enters the cut, and α_{ex} is the angle at which the cutter exists the cut. The angle at exist is approximately zero, $\alpha_{ex} = 0$, and an expression for α_{en} is evaluated from the next equation.

$$\alpha_{en} = \cos^{-1}\left[1 - \frac{R_d}{r}\right] \tag{3}$$

Angular spacing between flutes on the cutter is, $\gamma = \dfrac{360}{N}$. The forces shown in figure 1, are those exerted by the cutter on the workpiece. The total force F_A that acts on the tool edge in the working plane, is the sum of tangential component F_T, and radial component F_R. F_T is usually assumed to be proportional to the instantaneous chip thickness h, and F_R is set to be proportional to F_T. The chip section that lies in front of the tooth at any instant is the product of the chip thickness and the axial depth of cut A_d, at the instant under consideration. The chip section multiplied by the specific cutting pressure K_T at that instant gives the instantaneous magnitude of the force acting at right angles to the chip area

$$F_T = K_T A_d \cdot h \tag{4}$$

$$F_R = K_R F_T \tag{5}$$

According to equation 4. The force on the cutter tooth is proportional to the frontal area of the chip being removed. Considering the relevant deflection of the spindle is due to the F_R component of the cutting force, figure 3, and this causes an error in the chip thickness h, which is equal to the deflection of spindle nose X_S.

$$X_S = \frac{F_R}{K_S} \tag{6}$$

Where K_S is the total stiffness of the spindle nose. Then equation 4 becomes:

$$F_T = K_T A_d (h - \frac{F_R}{K_S}) \tag{7}$$

Thus, the instantaneous chip thickness for any particular increment is affected by spindle deflection. The block diagram model of the cutting process can be derived from the mathematical equations, and the overall model is given in figure 2. The cutting process parameters were determined from a series of cutting tests, where the cutting forces components (F_X and F_Y) can be measured directly by using a force dynamometer.

3. DETERMINATION OF THE MODEL PARAMETERS

To measure the cutting force components a Kistler Piezo-Electric three component dynamometer was used, which allows the measurement of the cutting forces in X, Y and Z direction. The cutting force components F_X and F_Y have been used to estimate the cutting process parameters K_T and K_R, for a given work material and range of cutting conditions. In order to provide the value of parameters Gygax's [5,7] proposals have therefore been followed. The maximum force $F_{\max}$ exerted by the cutter to work-piece. Gygax [7] states that the cutting force has a maximum value when $\theta = 90^o$, giving

$$F_{\max} = K_T A_d f_t \tag{8}$$

Thus two important specific cases are given by the angular positions corresponding to the zero crossing of the cutting force components F_X and F_Y. The values for K_T and K_R were calculated by extracting the values of $F_{\max}$ from the cutting force graph (figure 4). Measurement has shown that K_T and K_R are not purely material constants and depend basically on the cutting parameters. Applying direct force to the spindle nose and measuring the nose deflection attained the static lumped stiffness of the spindle.

4. CALCULATION OF K_T & K_R

The specific cutting stiffness K_T at any instant is a coefficient depends upon the workpiece material, the geometry of the tool, and cutting conditions. It is determined experimentally as a function of the chip thickness [2]. F_R The cutting force ratio is assumed to be constant for a particular cutting conditions. K_T and K_R are the cutting coefficients varied with feed rate, axial depth of cut and radial depth of cut, estimated from the experimental data. Most of the researchers, such as Tlusty and Gygax [3,4,5] have assumed $K_T = 0.3$. Since it is depend on the cutting conditions and tool geometry, so $F_R = 0.3F_T$.

The chip section that lies in front of the tooth at any instant is the product of the chip thickness and radial depth of cut at the instant under consideration. As the result of chip section multiplied by K_T, the specific cutting pressure at that instant gives the instantaneous magnitude of the force acting at right angles to the chip area. Whereas the cutting forces have been measured, the specific cutting pressure is calculated by dividing the force by the area of the chip section.

$$K_T = \frac{\text{force at any instant}}{\text{width of cut} \times \text{chip thickness at any instant}} = \frac{\text{force at any instant}}{R_d f_t \sin(\theta)}$$

In end milling the cutting begins when $\theta = 0^o$ and the chip thickness increases to maximum at $\theta = 90^o$. The specific cutting pressure is a function of the chip thickness according to the relationship:

$$K = C h^W . \quad (9)$$

Where C and W are constants depending on the workpiece material and the milling cutting tool. The forces are deduced from the original forces by means of matrix expression:

$$d = \begin{bmatrix} -\cos(\theta) & \sin(\theta) & 0 \\ \sin(\theta) & \cos(\theta) & 0 \\ 0 & 0 & 1 \end{bmatrix}, \quad v = \begin{bmatrix} F_X \\ F_Y \\ F_Z \end{bmatrix}, \quad \begin{bmatrix} F_R \\ F_T \\ F_Z \end{bmatrix} = d^{-1} \cdot v \quad (10)$$

From equations 4, 5, and 10 for non-helical teeth, cutting components can be written as a functional of tool position [11]:

$$F_X = K_T A_d f_t (\cos(\theta) - K_R \sin(\theta)) \sin(\theta) \quad (11)$$

$$F_Y = K_T A_d f_t (\sin(\theta) + K_R \cos(\theta)) \sin(\theta) \quad (12)$$

The cutting torque T_C acting on the spindle is proportional to the tangential cutting force component:

$$T_C = R_C F_T \quad (13)$$

5. EXPERIMENTAL WORK

In order to establish the accuracy of the dynamic mechanistic model predictions of both cutting forces, a series of milling experiments were performed on CNC Bridgeport Interact 720 machine centre equipped with a dc motor table drive. The X and Y direction force components were measured with a three components piezoelectric force transducer dynamometer. These experimental were run with a 10mm, M42 High-Speed Steel (HSS) end-mills and Clarkson autolock collect, Aluminium and EN. IB 20, carbon steels workpiece. In each machining test on a CNC Bridgeport interacts 720-machine centre, a cutting period t=0.1 seconds with 1000 sampling rate. Over the steady state of each experiment during the first and last second values of the chip thickness h, radial specific cutting force coefficient K_R, and tangential specific cutting force coefficient K_T calculated according to the procedure of [9,15]. In order to compare the experimental and simulated transient force responses, it is necessary to first develop the appropriate structure of transfer functions of overall the system. The Forces dynamics can be quantified in terms of few parameters, such as the cutting speed (rev/sec) n converted to the instantaneous angular position of the cutting edge θ. In [15], an approach has been described where the parameters of the force model, K_T and K_R, may be estimated from measured X and Y forces. The parameters K_T and K_R are obtained for a data base of machining conditions and then empirical model are developed from these data to predict K_T and K_R for any set of machining conditions within the data base. The error between the experimental and test measured can be calculated and fitted. The procedure described in [2, 9, 15] has been used to solve for K_T and K_R. The presence of runout has a little effect on the solution of K_T and K_R from measured forces, thus the same estimates for the both coefficients are obtained.

6. SIMULATION AND RESULTS

Two types of material, Aluminium and Carbon steel are chosen for carrying out the end milling tests with three alternative parameters-spindle speed, feed rate, and axial depth of cut-on CNC Bridgeport machine centre. Sixteen cutting tests are executed by using 2^3 factorial design, considering the two types of material and the cutting parameters used.

The simulation model has been programmed in Simulink, based on the block diagram model of the cutting process, figure 2. The input data to the simulation program are the simulated feed rate f, the spindle speed n, and the axial depth of cut A_d. The output from the simulation program includes the cutting force components F_X, F_Y, F_T, F_R, cutting torque T_C acting on the spindle drive system and the spindle nose deflection X_S. To study the relationship between the feed rate and spindle speed a series of simulations were carried out.

The result of the simulations for one tooth-engaged cutting (two teeth cutter) is given in figures 4, which are show the X and Y-direction forces, cutting torque, and the spindle nose displacement for different feed rates and spindle speeds.

CONCLUSION

To determine the block diagram model of the cutting process the mechanics of cutting and the forces involved in milling process has been identified. This is a physical model of the process where the model parameters can be measured experimentally. The cutting force components F_X and F_Y have been used to estimate the cutting process parameters K_T and K_R for a given work material and range of cutting conditions.

Once the process parameters were known the model was simulated using a standard mathematical package. At present the model is very simple and capable of simulating multi teeth cutters. However the model can be easily expanded to accommodate any tool configuration and, it can be also extended to take account of other tool characteristics. The deflection of the spindle nose is calculated using the static lumped stiffness in the system that is measured off-line. The model can be further improved by applying the dynamic stiffness of the spindle. The model can be used to estimate the forces acting on the drive systems. Therefore it can be used in the dynamic analysis of the machine tools. Also it is proposed to use the model for chatter analysis.

NOTATION:

A_d	Axial depth of cut
F_A	Total cutting force (active force)
F_R	Radial component of F_A
F_T	Tangential component F_A
$F_X(i,n)$	X-forces acting on tooth i at angle of cutter rotation n
$F_Y(i,n)$	Y-forces acting on tooth i at angle of cutter rotation n
F_{AX}	Average x-force
F_{AY}	Average y-force
f	Feed rate (mm/sec)
f_t	Feed per tooth
h	Instantaneous chip thickness
K_T	Specific cutting stiffens
K_R	Force ratio (relates F_R & F_T)
K_S	Total stiffness of the spindle nose
n	Spindle speed (rev/sec)
N	Number of teeth on cutter
R_d	Radial depth of cut
r	Radius of the cutter
T_C	Cutting torque
X_S	Spindle nose displacement
θ	Angular displacement of cutter

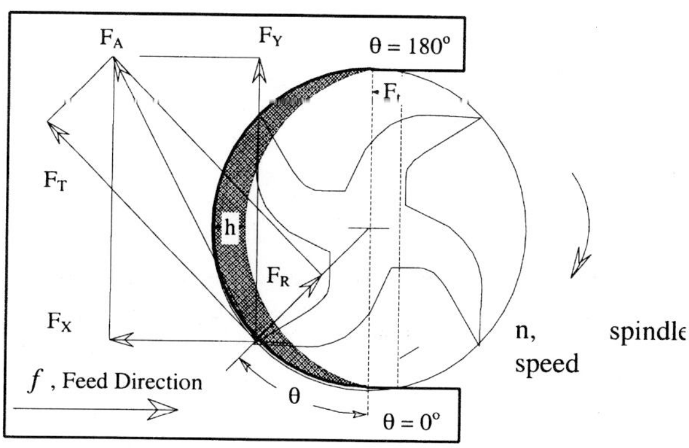

Workpiece material

Figure 1 Cutting Process Force Diagram

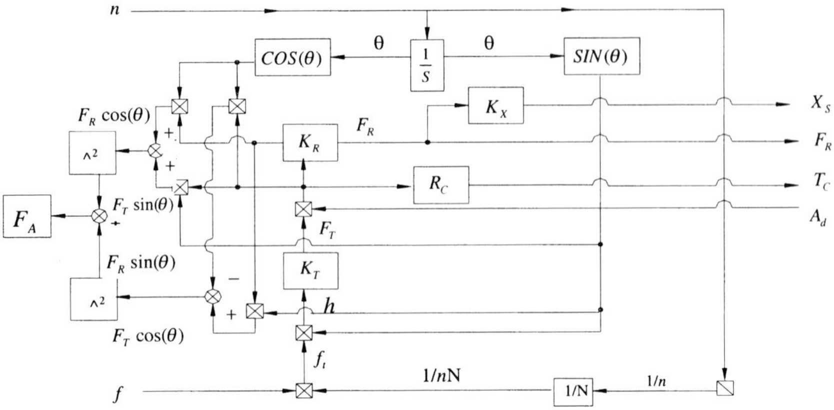

Figure 2 Block diagram model of cutting process for multi teeth cutter

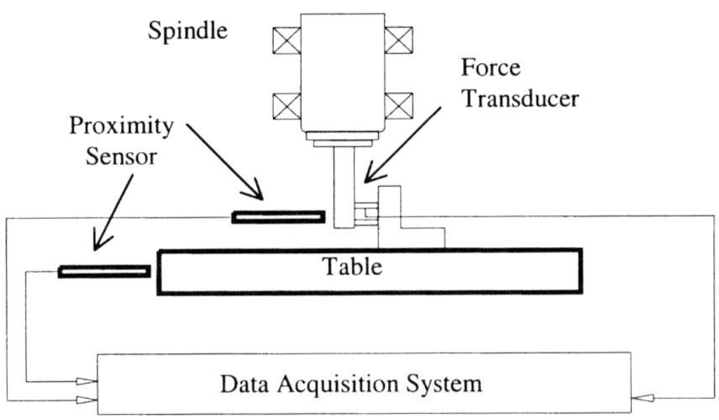

Figure 3 Deflection Measurement Test

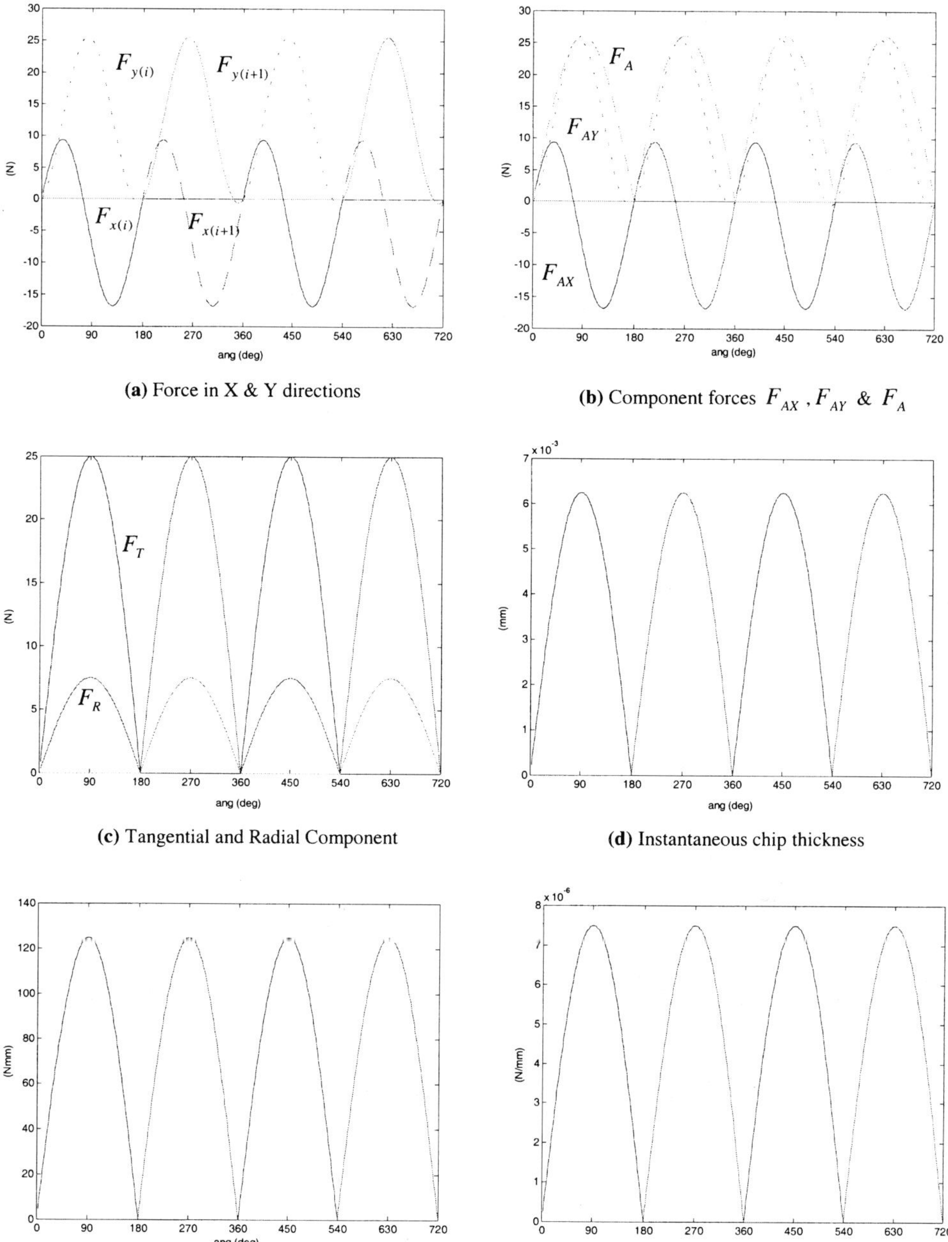

(a) Force in X & Y directions

(b) Component forces F_{AX} , F_{AY} & F_A

(c) Tangential and Radial Component

(d) Instantaneous chip thickness

(e) Torque acting on spindle

(f) Deflection of the spindle

Figure 4. Simulation of exerted cutting forces, chip thickness, cutting torque, and spindle deflection. n =2400 rev/sec, f =30 mm/sec, A_d=2 mm, R_d=10 mm, N =2 teeth.

REFERENCES:

1. Martin, K.F. and Ebrahimi, M., "Dynamic Modeling of a CNC Milling Machine", 7th Proceeding of IASTED, International Conference on Modeling & Simulation, July 1989, Lugano Switzerland.

2. Koenigsberger, F. and Sabberwal, A.J.P., "An Investigation into The Cutting Forces Pulsation During Milling operations", Int. J. of Machine Tool Design and Research, Vol. 1, 1961, pp. 15-33.

3. Tulsty, J. and MacNeil, P. "Dynamics of Cutting Forces in End Milling", CIRP Annals, 1975, pp. 20-25.

4. Gygax, P. E. "Dynamics of Single-Tooth Milling", CIRP Annals, 1979, pp. 65-70.

5. Gygax, P. E. "Cutting Dynamics and Process-Structure Interactions Applied to Milling ", Wear, Vol. 62, 1980, pp. 161-184

6. Tulsty, J., "Dynamics of High-Speed Milling" ASME Journal for Industry, Vol. 108, May 1986, PP. 59-67.

7. Gygax, P. E., "Experimental Full Cut Milling Dynamics", Annals of the CIRP, Vol. 29, 1980, pp. 61-66.

8. M. Martellotti, "*An Analysis of the Milling Process*", Trans. ASME, Vol. 63, pp. 677-701, Nov. 1941.

9. W. A. Kline and R. E. DeVor, " *The effect of runout on cutting geometry and forces in end milling*", Int. J. Machine Tool Design and Research, Vol. 23 No.2/3, pp. 123-140, 1983

10. J. Tlusty, P. MacNeil, "Dynamics of Cutting Forces in End Milling", CIRP Annals, Vol.24, 1975.

11. K. F. Martin, M. Ebrahimi, "*Modeling and Simulation of Milling Action*", Proc. Inst. Mech. Engrs, Vol.213, pp.539-553, 1999.

12. Jung-Hoon Cho and Suk-Hwan Suh, "Experimental Verification for Path *Modification Scheme Toward Net Shape Machining*", Pacific Conference on Manufacturing, P 31-36, 1996

13. Martellotti, M., "Analysis of Milling Process, part II Down Milling" Trans. of the ASME, Vol. 67, 1945, pp. 233-251.

14. Jung-Hoon Cho and Suk-Hwan Suh, "Experimental Verification for Path Modification Scheme Toward Net Shape *Machining*", Pacific Conference on Manufacturing, P 31-36, 1996

15. W. A. Kline, R. E. DeVor and J. R. Lindberg, " *The prediction of cutting forces in end milling with application to cornering cuts*", Int. J. Machine Tool Design and Research", Vol. 22, No.1, pp7-22, 1982.

16. The Math Works, Inc. Matlab User's Guide.

Authors' Index

S

T

V

W

Z

18482